I0055524

# Structural Analysis
## and
# Design
## of
# Airplanes

## Engineering Division
## McCook Field
## Dayton, Ohio

ISBN 1-929148-46-1

Wexford College Press
2005

# PREFACE
# STRUCTURAL ANALYSIS AND DESIGN

The purpose of this book is to give designers reasonable methods for the structural analysis and design of the component parts of the airplane structure. The acceptance and use of a single system of analysis should tend towards uniformity in design, and ease in checking calculations and in making comparisons between different airplanes.

In the first chapter is given a brief development of the fundamental principles of graphical and analytical structural mechanics, and strength of materials, supplemented by detailed numerical examples. Under the head of "General Considerations," Chapter II deals with matters pertaining to an airplane cellule. In a brief discussion a few important features of structural design are considered, such as the location of interplane struts and spars, wing loading, aspect and gap-chord ratios, and stagger. Suggestions as to the choice of wing and strut sections are made. The work preliminary to the main analysis, that is, the distribution of load between the wings, along the chord and on the wing tips, and the resolution of forces is explained. This is followed by concise statements of the general principles and methods of analysis, the detailed applications of which appear in the succeeding chapters.

The general procedure followed in this book is to take as examples of the different structural units of an airplane, actual standard designs chosen to illustrate as many phases of the work as possible, and to follow through each step of the analysis and structural design. All the main equations and computations are given in full. These are supplemented by explanations, drawings, and diagrams where necessary for clarifying the text.

The appendix gives equations for the moments and deflections of continuous beams with various types of loading. For the more unusual cases the derivation of the equations is given. Data necessary for design purposes, such as properties of strut sections, cable capacities, and ultimate strength of materials, are also placed in the appendix.

To entirely separate stress analysis from design is impracticable, as the two are very closely correlated. For this reason a considerable amount of space has been devoted to a discussion of design in so far as it concerns structural problems.

Three broad principles which are mentioned in the text deserve emphasis here. First, one of the two main purposes of any stress analysis is to enable a designer to so proportion structural members that there will be the correct "follow through," or that the strength of the members in one part of an airplane will bear the proper relation to the strength of the members in all its other elements. Second, wherever it is possible, and particularly in the larger types of airplanes, engineers should endeavor to design statically determinate structures even at the cost of a slight additional weight. There are always uncertainties in an indeterminate structure which occasionally cause serious failures. Third, every effort should be made to reduce secondary stresses. Sometimes these stresses become as large as the main stresses. Elimination of eccentricities and, in the case of spars, avoidance of extremely long spans in which the deflections are great are two important means of minimizing secondary stresses.

Neither an economical design for a proposed airplane, nor a determination of the strength of an existing structure closer than 10 to 20 per cent can be made on the basis of approximate computations. Complete calculations for the structural members in an airplane require considerable time and care, yet it is being generally recognized that to restrict the structural engineering of airplanes is poor economy from the standpoint both of performance and safety. Furthermore, since a careful analysis can be depended upon to give an accurate prediction of the strength of a structure, the testing to destruction of full sized airplanes, especially the larger types, becomes less necessary.

That the methods of analysis presented in this book are entirely satisfactory, or represent in any sense an ultimate development of the subject is not to be expected. They are, however, based in every case on sound engineering principles and practice, and may, therefore, be relied upon to give good results. The Engineering Division, basing its opinion on the stress analyses of the wing cellules of numerous airplanes which have been subjected to sand test, believes that the strength of a wing structure of conventional design and of a known quality of material can be predicted with an error of less than 5 per cent. The most difficult part of the work is not the calculation of the stresses so much as it is the determination of the ultimate allowable stresses in the material. Further experimental work will reduce the uncertainty in regard to some of these values.

# TABLE OF CONTENTS

CHAPTER I

## Principles of Applied Mechanics and Strength of Materials

CHAPTER II

## General Considerations

CHAPTER III

**Wing Stress Analysis**

CHAPTER VI

## Control Systems

CHAPTER VII

## Control Surfaces



CHAPTER VIII

## Fuselage

APPENDIX

## CHAPTER I

# PRINCIPLES OF APPLIED MECHANICS AND STRENGTH OF MATERIALS

## I. Applied Mechanics

1. *Principles of Statics*—The whole subject of statics, that part of mechanics dealing with bodies in equilibrium, is based on Newton's first law: "Every body continues in its state of rest or uniform motion in a straight line unless acted upon by an outside force." The corollary of this: "If any unbalanced force acts upon a body, the body is given an accelerated motion," states a condition when statical equilibrium no longer exists. Therefore, if a structure, or any part of a structure, is in equilibrium the algebraic sum of all the forces and of all the moments acting upon it must equal zero. This may be briefly expressed by the following equations, $\Sigma$ denoting "the sum of": $\Sigma X = 0$, $\Sigma Y = 0$, $\Sigma Z = 0$, and $\Sigma M = 0$, in which $X$, $Y$ and Z represent the components of the forces parallel to the X, Y and Z axes respectively, and $M$ the moment of these forces about any point. As forces frequently lie in a single vertical plane they may be resolved into horizontal and vertical components. The fundamental equations of equilibrium then become: $\Sigma H = 0$, $\Sigma V = 0$, and $\Sigma M = 0$.

2. *Law of Signs*—Care must be taken in problems to give each force or moment its proper sign. The following convention is in general use: forces acting to the right are positive, those to the left negative; forces acting upwards are positive, those acting downwards negative; clockwise moments are positive, anti-clockwise moments negative. It should also be noted that a force has three characteristics: magnitude, direction, and point of application.

The stresses in any statically determinate structure can be calculated by various applications of the equations just given. Statically indeterminate structures are much more complicated. Two of the most useful methods for solving such cases will be discussed. For determinate structures two general methods of solution may be followed, either the analytical or the graphical. Frequently a combination of the two is most convenient. The analytical solution will be taken up first.

### *THE SOLUTION OF A WARREN TRUSS*

3. *General Conceptions*—A truss is a structure, built up of members so arranged that they are subjected primarily to direct tension or compression, which transmits loads imposed upon it to the two or more points at which the truss is supported. If there are more than two supports the truss is termed continuous. This type will not be considered here. Trusses are assumed to be in a single plane which is usually that of the forces acting upon the truss. In calculating the stresses in

a truss it is assumed that the truss is "pin-jointed." By this is meant that at their points of intersection, the members are secured by a pin upon which they are free to turn. This condition seldom or never occurs in practice; the error that is made by the assumption is small. One important conception results from this assumption; namely, that any stress or load which goes into a member at a joint must act in the direction of the axis of the member. No stress can be applied to a member except at a joint. The stresses acting at each end of a member must be equal, and may each be considered as a single force. Such a member is known as a "two-force member," as shown in Fig. 1.

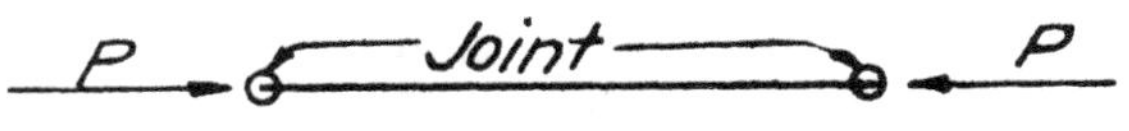


Fig. 1. Two Force Member

4. *Characteristics of Reactions*—The first step in the solution of a truss is the determination of the known external loads upon the structure. These should be shown upon a line sketch of the truss, together with their direction and point of application (see Fig. 2). When this is accomplished it is customary to solve for the supporting forces or reactions. The three equations of equilibrium, $\Sigma V = 0$, $\Sigma H = 0$, and $\Sigma M = 0$, which must be satisfied, enable us to learn three things about the reactions. As each of the two reactions has three characteristics; namely, magnitude, direction, and point of application, three facts, and only three, must be given if the structure is to be stable and statically determinate with respect to the external forces. Generally the points of application of both reactions, $l_0$ and $l_2$, and also the direction of one of

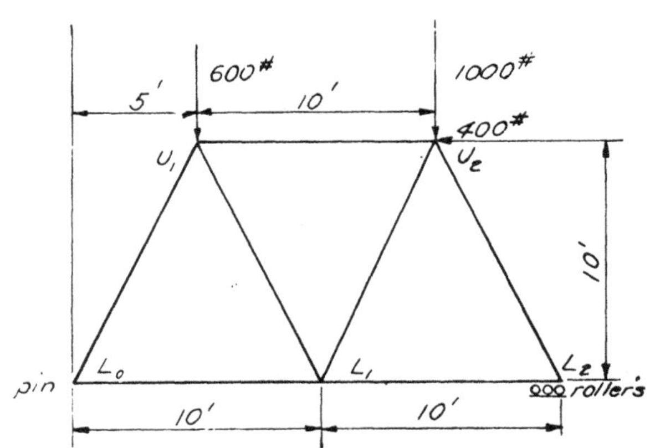


Fig. 2. Warren Truss with External Loads

them are known. Where rollers are used the reaction is always normal to the supporting surface. The conditions of equilibrium are sufficient to determine the magnitude of each reaction, and the direction of one. Should there be three reactions, or should another condition such as the direction of both reactions be unknown, the structure ceases to be statically determinate. If, on the other hand, only two unknown conditions regarding the reactions exist, the structure is unstable. It is important at the outset to be able to tell whether the supporting forces can be calculated by statics.

5. *Determination of Reactions*—The next step in computing the reactions is to take the moments of all the external forces acting upon the structure, including the reactions, about an axis passing through one of the reactions, usually the one whose direction is unknown. Since the algebraic sum of these moments equals zero, the reaction appearing in the equations may be found, as it is the only unknown. The horizontal and vertical components of the other reaction can now be obtained by resolving all the loads into their horizontal and vertical components, and by writing and solving the two equations, $\Sigma H = 0$ and $\Sigma V = 0$. A check on the calculation of the reactions is obtained by taking moments about the other reaction point. Incidentally, it is frequently much simpler to calculate the moments of the components of forces than of the forces themselves, especially if one of the components is eliminated by taking advantage of the fact that the moment of a force about an axis is the same for all points along the line of action of the force. An illustration of this is given in equation 4, in which the moments of the components $h_1$ and $v_1$ of the left hand reaction $r_1$ are taken, rather than the moment of the reaction itself, which would be difficult to compute. In this manner the horizontal component, which passes through the axis of moments, is eliminated. In the same way it is much easier to obtain the moments of the two forces at $u_2$ than the moment of their resultant.

## *CALCULATION OF REACTIONS FOR WARREN TRUSS*

(1) $$\Sigma M_{10} = 600 \cdot 5 + 1000 \cdot 15 - 400 \cdot 10 - 20 r_r = 0$$
$$r_r = +700 \text{ lbs.}$$

(2) $$\Sigma H = -400 + h_1 = 0$$
$$h_1 = +400 \text{ lbs.}$$

(3) $$\Sigma V = -600 - 1000 + 700 + v_1 = 0$$
$$v_1 = +900 \text{ lbs.}$$

$$r_1 = \sqrt{400^2 + 900^2} = +985 \text{ lbs.}$$

In Fig. 3 both $r_1$ and its components are shown, but, of course, the components take the place of the reaction itself.

Check equation.

(4) $$\Sigma M_{12} = 20 \cdot v_1 + 0 \cdot h_1 - 600 \cdot 15 - 1000 \cdot 5 - 400 \cdot 10 = 0$$
$$v_1 = +900 \text{ lbs.}$$

The solution for the stresses in the members may be either analytical or graphical.

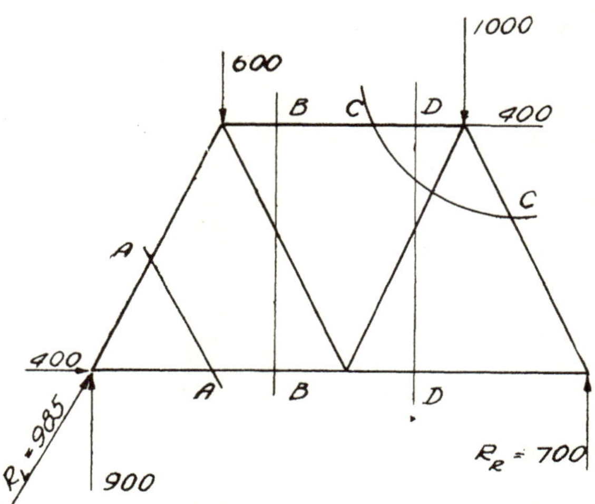


Fig. 3. Warren Truss and Reactions

## *THE ANALYTICAL SOLUTION FOR THE TRUSS STRESSES*

6. *Method of Joints*—The three methods illustrated by this problem, those of joints, of shear, and of movements, are but different applications of the three equations of equilibrium. Consider first the joint $l_0$, Fig. 3, at which there are only two unknown stresses. So far as its equilibrium is concerned, the members $l_0\ u_1$ and $l_0\ l_1$ may be cut, if they are replaced by the respective stresses acting in these members, as shown in Fig. 4. By applying $\Sigma V = 0$, in equation 5, the vertical component of the stress in $l_0\ u_1$ is obtained. The slope of this member gives the relation between $v_1$ and $h_1$, and hence the value of $h_1$. Equation 7 gives directly the stress in $l_0\ l_1$. This process is known as the method of joints. It is very simple and convenient in many cases, but for the more complicated joints it should be used in conjunction with the methods of shear and moments. No joint is capable of solution by this method at which there are more than two unknown stresses. The analytical method of joints is usually simplified by using the horizontal and vertical components of the known forces and unknown stresses.

Joint $l_0$

Solution for stress in $l_0\ u_1$ and $l_0\ l_1$ by the method of joints.

(5) $\Sigma V = 900 + v_1 = 0$

$v_1 = -900$

(6) $h_1 = v_1/2 = -450$

The stress in $l_0\ u_1 = -\sqrt{450^2 + 900^2} = -1006$ lbs. compression.

(7) $\Sigma H = 400 - 450 + h_2 = 0$

The stress in $l_0\ l_1 = h_2 = +50$ lbs. tension.

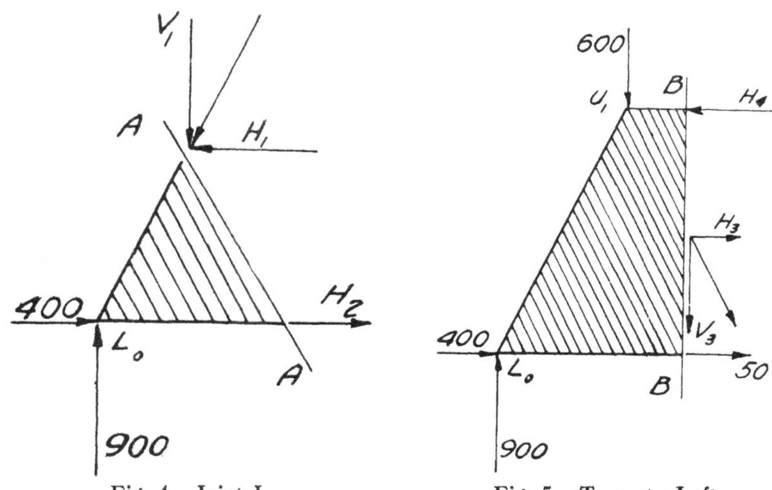


Fig. 4. Joint $L_0$

Fig. 5. Truss to Left of Section B—B

7. *Character of Stress*—In using the equations of equilibrium for calculating stresses the sign of the unknown stress or of its components should be written plus. Then the sign that is obtained upon solution of the equation determines the direction of the stress, whether toward or away from the joint, in accordance with the rules given in Art. 2. The nature of the stress in a member is thus automatically determined.

8. *Method of Shear*—A process frequently used to obtain the stress in the web members of a truss where the chords are parallel is that of shear, illustrated by the solution for the stress in $u_1\ l_1$. It consists simply in taking a section through the truss which cuts the two parallel chords and the diagonal in question, and isolating part of the truss, as in Fig. 5. The condition for equilibrium of the vertical forces is now applied and, as $v_3$ is the only unknown vertical force, its value may be obtained. As before, $h_3$ is computed from $v_3$.

*Section B—B*

Solution for stress in $u_1\ l_1$ by the method of shear.

(8) $\Sigma V = 900 - 600 + v_3 = 0$

$v_3 = -300$

$h_3 = v_3/2 = +150$

Stress in $u_1\ l_1 = +336$ lbs. tension.

9. *Method of Moments*—The condition for equilibrium of the horizontal forces $\Sigma H=0$, Fig. 5, gives directly the stress in $u_1\ u_2$. This may also be calculated by the method of moments. In this method a section, *B—B*, is taken, and the part of the truss to the left of the section is isolated. All the members cut, except the one in which the stress is to be determined, must intersect, in this case at joint $l_1$, Fig. 6. Equation 9 gives the moments of all the forces acting on the

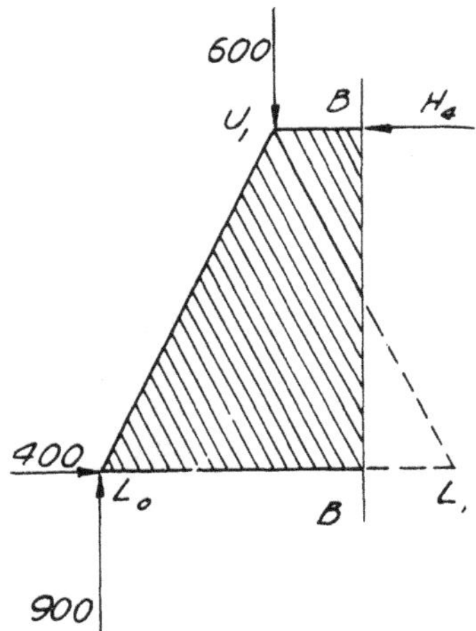


**Fig. 6. Truss to Left of Section B—B**

part of the structure under consideration, including that of the stress in $u_1 u_2$. This method is the most useful of the three explained, and should be thoroughly understood.

*Section B—B*

Solution for stress in $u_1 u_2$ by the method of moments.

(9) $\Sigma M_{l_1} = 900 \cdot 10 - 600 \cdot 5 + 10 \cdot h_4 = 0$

Stress in $u_1 u_2 = h_4 = -600$ lbs. compression.

10. *Methods of Moments and Joints*—As further examples of the method of joints and moments the solution for the stresses in $l_1 u_2$, $u_2 l_2$, $l_1 l_2$ is given, the procedure being the same as before.

*Joint $u_2$*

Solution for stress in $l_1 u_2$ and $u_2 l_2$ by method of joints.

(10 and 11) $v_6 = 2h_6$ ; $v_5 = 2h_5$

(12) $\Sigma V = v_6 + v_5 - 1000 = 0$

(13) or $2h_6 + 2h_5 - 1000 = 0$

(14) $\Sigma H = h_6 + 600 - 400 - h_5 = 0$

Solving equations 13 and 14 simultaneously.

$$\begin{cases} h_6 = +150 \text{ and } h_5 = -350 \\ v_6 = +300 \text{ and } v_5 = +700 \end{cases}$$

Stress in $l_1 u_2 = -\sqrt{150^2 + 300^2} = -336$ lbs. compression

$u_2 l_2 = -\sqrt{350^2 + 700^2} = -789$ lbs. compression

*Section D—D*

Solution for stress in $l_1 l_2$ by method of moments.

(15) $\Sigma M_{u2} = -700 \cdot 5 + 10 \cdot h_7 = 0$

Stress in $l_1 l_2 = h_7 = +350$ lbs. tension.

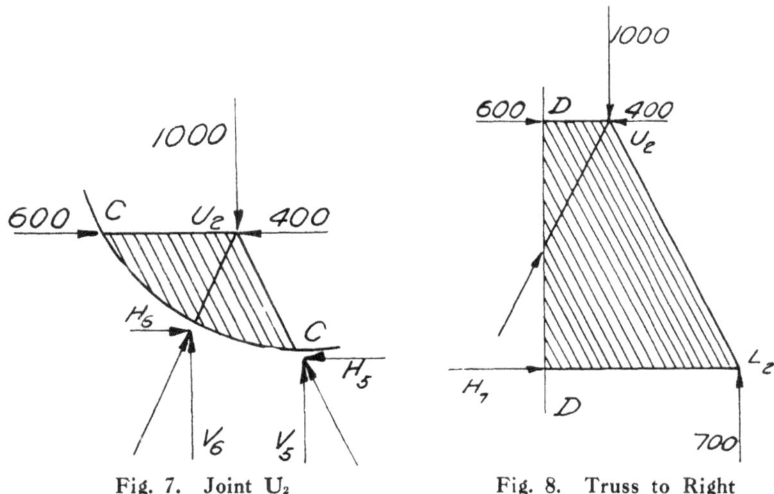


Fig. 7. Joint $U_2$

Fig. 8. Truss to Right of Section D—D

11. *Free Body Method*—It will be observed that in the solution just given the method employed in each case was the isolation of the structure, in whole or in part, with all the external loads acting upon it, and the substitution for any members cut of the stresses in those members. This is a very general and important procedure applicable to nearly all problems in statics or kinetics, and is known as the "free body method." When a body is so isolated it is comparatively easy to tell whether it is statically determinate and whether it is in a condition of stable equilibrium.

## *THE GRAPHICAL SOLUTION FOR THE TRUSS STRESSES*

12. *Scope of Graphics*—Any problem in statics that can be solved analytically is capable of a graphical solution. The graphical method is seldom used to obtain reactions. But often for stresses, particularly in a truss in which the web members have different slopes or the chords varying inclinations, it offers a ready solution to an otherwise tedious problem. This method is based on the proposition that, if the forces acting upon a structure are in equilibrium, they may be represented by the sides of a closed polygon. A force polygon is constructed by taking in order the forces acting upon a structure, and representing them, in both magnitude and direction, by the sides of the polygon. In Fig. 10, ABCDEFA forms the force polygon for the external forces applied to the truss.

13. *Lettering of Truss*—The first step in the solution is to make an accurate line drawing (Fig. 9) of the structure, upon which all the external forces are shown acting in the proper direction and at the correct

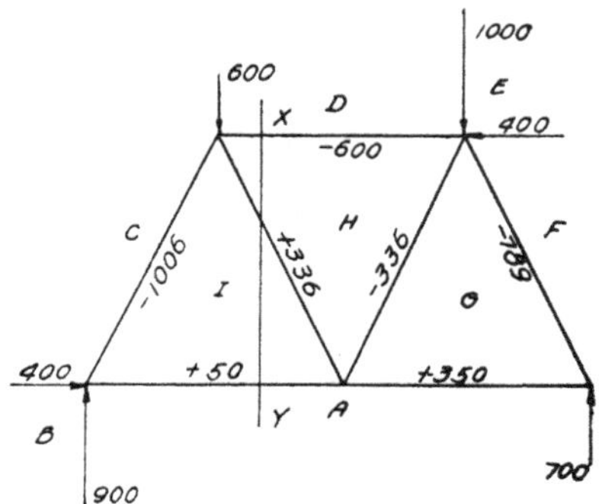


Fig. 9. Warren Truss for Graphical Solution

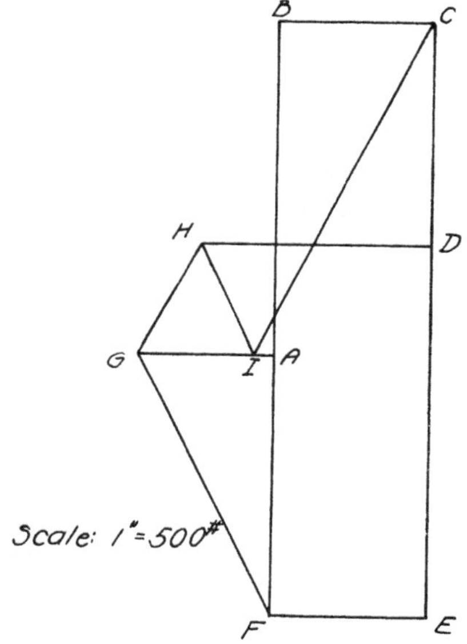


Fig 10. Stress Polygon for Warren Truss

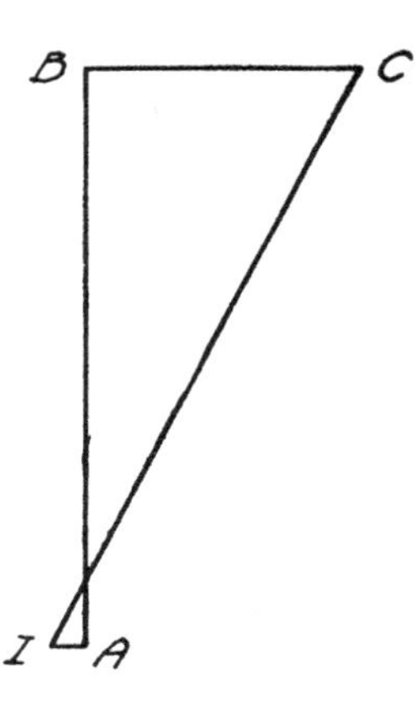


Fig. 11. Part of Stress Polygon Shown in Fig. 10

location. Next, letter the truss in such a manner that on each side of every force and member there is one letter, and only one letter.

14. *Construction of Force Polygon*—The truss in Fig. 9 illustrates the correct method. Now start with one of the external forces, *A—B* and begin the construction of the force polygon for the external forces. In case all these forces are parallel, this polygon reduces to a straight line. Next, select a joint at which there are only two unknown members.

Take, for example, joint $l_0$, Fig. 2. The members *C—I* and *I—A* may be considered cut, and replaced by the stresses in them. The joint $l_0$ must be in equilibrium under the action of the four forces, *B—C, C—I, I—A and A—B*, which may therefore be represented by the sides of a polygon. This is constructed by drawing from point *C*, Fig. 11, a line parallel to the member *C—I* and indefinite in length, and from point *A* the line *I—A*, which is prolonged until it intersects line *C—I* at *I*. In the closed polygon thus formed *C—I* represents, in magnitude and direction, the stress in member *C—I*, and *I—A* the stress in member *I—A*. In a similar manner, the stresses in the members *D—H* and *H—I* are found by considering all forces to the left of section *X—Y*. These too, being in equilibrium, may be represented by a polygon, which is *A—B—C—I—H—D—C*, of Figure 10. By the same method point *G* is located and the polygon completed.

15. *Character of Stresses*—If each side of the polygon is scaled, the lengths of its sides give the stresses in all the members of the truss. To determine whether the stress in any member is tension or compression, read the letters which designate the member in a clockwise order with regard to the joint at either end of the member. Then, with the letters in the same order, read the stress in the member on the force polygon. If the stress, read in this manner, acts toward the joint, it is compression; if away from the joint, tension. For example: take members *I—H* and *H—G*, and use the joint $l_1$ at their intersection. The stress in *I—H* is tension, because the line *I—H* in the polygon runs upward to the left away from the joint; that in *H—G* is compression, because the line *H—G* runs downward to the left toward the joint.

## II. Strength of Materials

16. *Development of Beam Theory*—This subject is the division of mechanics which deals with the internal stresses in the members of a structure, and the manner in which the members resist the external forces acting upon them. The foundation for this entire section of mechanics is the *beam theory*, which will be briefly developed. The purpose of the beam theory is to enable the stress at any point in any cross section of a rigid body subjected to bending to be calculated. The simplest case will be considered first. Let the beam in Fig. 12 be acted upon by a set of forces as shown. The forces and the beam have the following conditions imposed upon them:

I. All the forces are parallel, and lie in a single, vertical plane passing through the axis of the beam and dividing each cross section symmetrically.

II. The forces are perpendicular to the axis of the beam.

III. The material of the beam is homogeneous.

IV. The beam is of uniform cross-section throughout.

V. The axis of the beam is a straight line containing the center of gravity of every cross-section. (The term cross-section denotes a plane section perpendicular to the axis of the beam.)

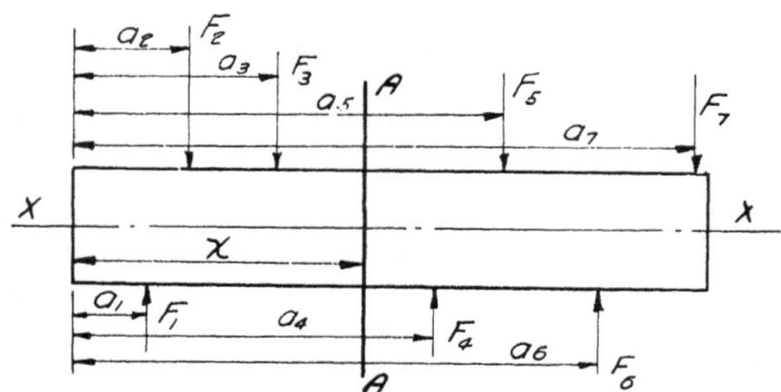


Fig. 12. Beam with External Loads

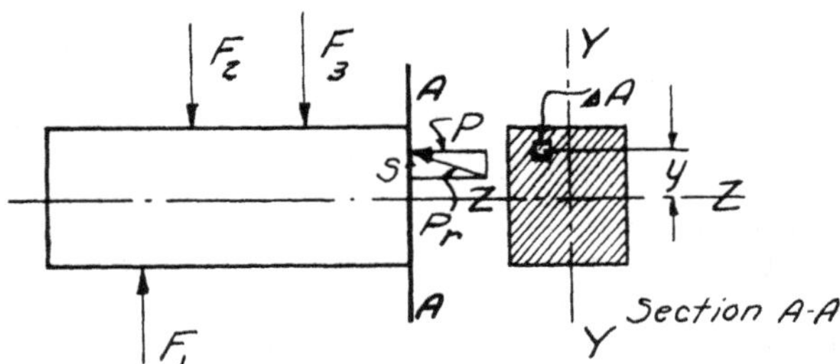


Fig 13. Beam to Left of Section A—A

In a beam conforming to these conditions, in shape, material, and loading, take a section *A—A* (Fig. 12), and isolate that portion of the beam to the left of the section (Fig. 13). Since the beam is in a condition of equilibrium the internal forces acting upon the section must be such that the equations of equilibrium, $\Sigma H = 0$, $\Sigma V = 0$, and $\Sigma M = 0$, are satisfied. Since there are no external forces acting upon the beam in the direction of the Z—Z axis there can be no component of the internal stress along this axis. That $\Sigma V$ may equal zero a vertical component of the internal stress must exist so that $f_1 - f_2 - f_3 - S = 0$ where $S$ is the internal shear at the section. In the same manner the internal resisting moment of the components of the internal stress normal to the section must be such that $f_1(x - a_1) - f_2(x - a_2) - f_3(x - a_3) + M = 0$, where $M$ is the internal resisting moment at the section. Furthermore, as none of the external forces have any horizontal component along the beam axis, the algebraic sum of these normal components of the internal stress must equal zero. They therefore form a couple.

Consider any small element of area in the cross-section, $\Delta A$, and let $p_r$ be the resultant internal stress acting upon it. $p_r$ may be resolved into two components, $s$, a shear stress, and $p$, a normal stress.

Then $\Sigma \Delta A \cdot s = S = \Sigma F$, and $\Sigma \Delta A \cdot p \cdot y = M = \Sigma F\ (x - a)$, and $\Sigma \Delta A \cdot p = 0$, $\Sigma$ denoting "the sum of."

There are three fundamental assumptions in the common theory of beams.

I. A cross-section which is a plane before bending remains a plane after bending.

II. "Hooke's Law" holds; that is, the stress is proportional to the strain throughout the beam.

III. The ratio between the normal component of the stress at any point in a given cross-section to the strain in the direction parallel to the axis of the beam at that point is the same for both tension and compression, and has the same value that would be obtained from a bar of like material subjected to a uniform tension or compression. In other words the moduli of elasticity in tension and compression are equal.

To make these assumptions perfectly clear the words stress and strain, as used in this connection, will be defined. Stress is the force of tension or compression acting upon a unit of area, and is expressed as pounds per square inch. Strain is the elongation or deformation per unit of length, or the total deformation divided by the length within which that deformation occurs; it is expressed in inches per inch. "Hooke's Law" merely states that until the unit stress reaches a certain value, known as the elastic limit of the material, the ratio between the stress and the strain is a constant quantity, called the modulus of elasticity, which is expressed in pounds per square inch.

Consider now a portion of a bent beam (Fig. 14), and take two plane cross-sections $A$—$B$ and $C$—$D$ which before bending were a distance $L$ apart. If assumption $I$ holds, some longitudinal layer $mn$ will, after bending, have a length $L_1$ and the length of all other layers

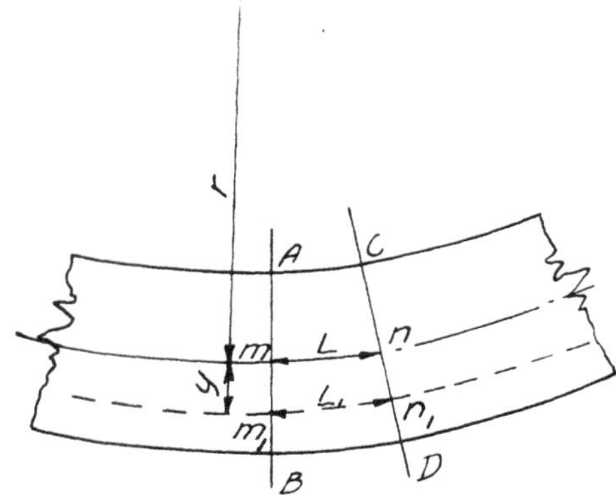


Fig. 14 Illustration of Beam Curvature

parallel to $mn$ will be proportional to their distance from a fixed center. If the distance $mn$ is taken short enough, the intersection of the neutral layer $m$ with the plane of the forces can be considered as an arc of a circle of radius $r$. The length $L_1$ can be found from the proportion $\frac{L_1}{L} = \frac{r + y}{r}$. Let $\delta$ equal the strain in the layer $m_1n_1$, and then $L_1 = L + \delta \cdot L$. From this $\frac{L + \delta \cdot L}{L} = \frac{r + y}{r}$, or $\delta = \frac{y}{r}$. Hence, the strains in the various layers of the beam are proportional to their distances from the neutral layer. The intersection of the neutral layer with the plane of the forces is known as the neutral axis of the beam.

It follows from assumptions II and III that the intensity of the normal component of the stress on any cross-section is zero at the neutral axis, and at any other point in the section is proportional to the distance of that point from the neutral axis. That is, the stress varies uniformly, and since the resultant is a couple, the neutral axis must pass through the center of gravity of the section. This variation in normal stress over the cross-section may be represented graphically, as in Fig. 19.

If $a$ equals the intensity of the normal stress at a unit's distance from the neutral axis, then the intensity of stress at a distance $y$ equals $p = ay$. Substituting this value for $p$ in the expression for $M$, $\Sigma \Delta A \cdot p \cdot y$, previously given, we have $M = \Sigma \Delta A \cdot a \cdot y^2 = a \cdot \Sigma \Delta Ay^2$. Since $I$, the moment of inertia of the section, equals $\Sigma \Delta Ay^2$, $M = aI$, $= p \cdot I/y$. From this $p = My/I$, where p is the normal stress at any point at a distance $y$ from the neutral axis, and $M$ is the external bending moment at the section. $p$ is called the "fiber stress." When $y$ equals the distance from the neutral axis to the outside layer of the beam at the section of greatest external bending moment, $p$ is a maximum, and is called the "*maximum fibre stress.*"

17. *Calculation of Moments and Shears*—The problem which follows illustrates nearly all the points connected with the calculation of moments and shears, and the plotting of moment and shear curves. As the work is given in detail no further explanation is necessary. The usual graphical representation of bending moment and shear given in Fig. 18 will be found of help in determining the sign.

The convention used in statics of calling clockwise moments plus and the reverse minus does not always hold. The standard, almost universally adopted, is that moments which produce compression in the top of a beam are positive and moments which produce tension in the top of a beam are negative. Shear is positive when the left half of the beam tends to slide up past the right; negative when the right tends to slide up past the left. These conventions of course apply not only to the total moment or shear at any point, but to the moment or shear caused by any individual load.

The complete computations for the reactions, shears, and bending moments for a beam loaded as shown in Fig. 15 are as follows:

*CALCULATION OF REACTIONS*

$$\Sigma M_f = -\frac{125^2 \times 1}{2} - 5\times70\times90-100\times125-50\times80-90\times55-200\times15+r_1\times100=0$$

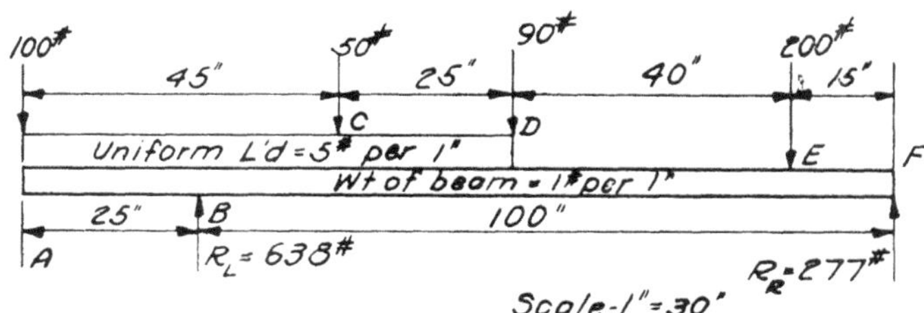


Fig. 15. Beam with External Loads

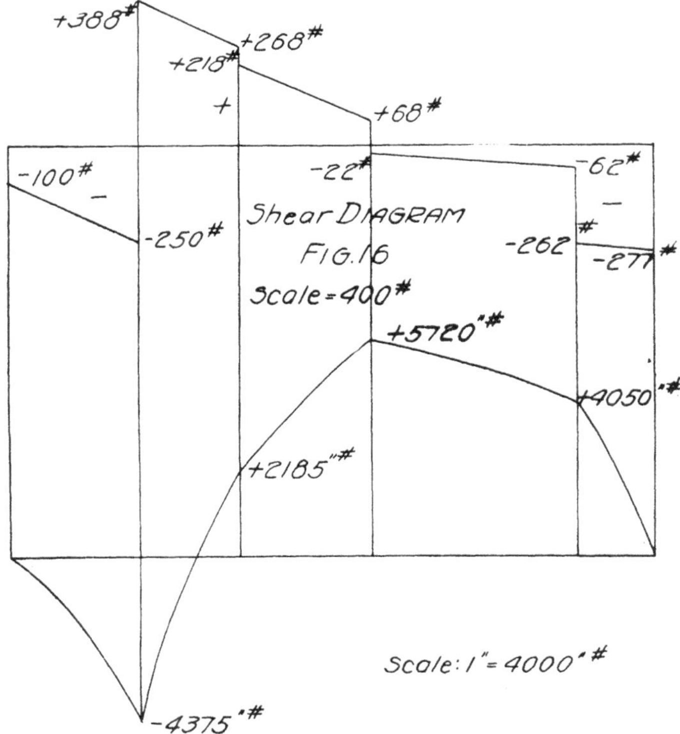


Fig. 17. Moment Diagram

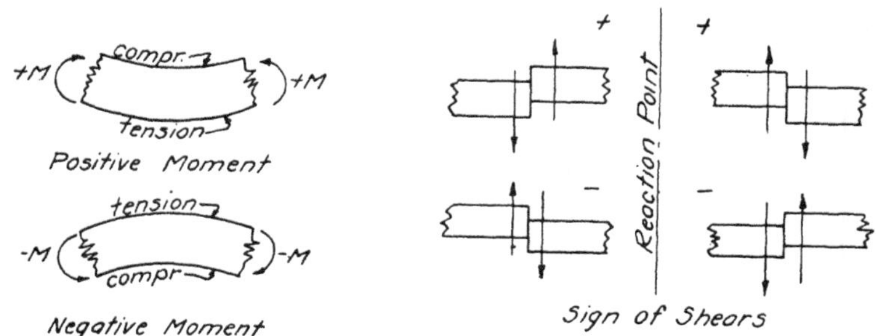


Fig. 18. Graphical Representation of Sign for Moments and Shears

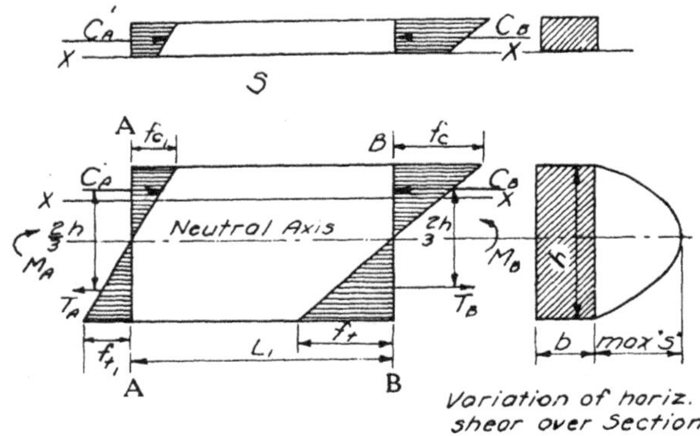


Fig. 19. Portion of Beam Showing Internal Bending and Shear Stresses

$r_l = +638$ lbs. = left reaction.

$\Sigma V = +638 - 100 - 50 - 90 - 200 - 5 \times 70 - 1 \times 125 r_r = 0$

$r_r = +277$ lbs. = right reaction.

*Check*—

$$\Sigma M_b = -100 \times 25 + 5 \times 70 \times 10 + 50 \times 20 + 90 \times 45 + 200 \times 85 + 125 \times 1 \times 27.5 - r_r \times 100 = 0$$

$r_r = +277$ lbs. = right reaction.

## *CALCULATION OF SHEARS*

Shear at *—A = 0

Shear at *+A = —100

Shear at —B = —100 — (6×25) = —250

Shear at +B = —250 + 638 = +388

Shear at —C = +388 — (20×6) = +268
Shear at +C = +268 — 50 = +218
Shear at —D = +218 — (6×25) = +68
Shear at +D = +68 —90 = —22
Shear at —E = —22 — (40×1) = —62
Shear at +E = —62 —200 = —262
Shear at —F = —262 — (15×1) = —277
Shear at +F = —277 + 277 = 0
*+ designates the right side of section, — the left side.

*CALCULATION OF BENDING MOMENTS*

Moment at A = 0

$$\text{Moment at B} = -100\times 25 - \frac{6\times 25^2}{2} = -4375 \text{ in. lbs.}$$

$$\text{Moment at C} = -100\times 45 - \frac{6\times 45^2}{2} + (638\times 20) = +2185 \text{ in. lbs.}$$

$$\text{Moment at D} = -200\times 40 + (277\times 55) - \frac{55^2\times 1}{2} = +5720 \text{ in. lbs.}$$

$$\text{Moment at E} = +277\times 15 - \frac{15^2\times 1}{2} = +4050 \text{ in. lbs.}$$

Moment at F = 0

18. *Plotting of Moments and Shears*—In constructing moment and shear diagrams it is convenient to remember that with concentrated loads only the shear is constant between loads and the moment varies as a straight line; with uniform distribution of loads the shear varies as a straight line and the moment as a parabola; with uniformly varying distributed loads the shear curve is a parabola, while the moment curve is a cubic parabola. In all cases for downward loads the moment curve is convex upward for both a negative cantilever moment and the moment within the span. One very common mistake made both in plotting bending moments, and in using them in beam design, is the failure to convert moments expressed in foot-pounds into inch-pounds when the other units in the problem require this conversion.

19. *Relations Between Moments and Shears*—Several extremely important relations exist between the bending moment and shear in a beam. The bending moment at any point, or the ordinate to the moment diagram, is equal to the algebraic sum of the areas under the shear curve to that point multiplied by the scales to which the shear curve is constructed. For example: referring to Figs. 15 and 17, the cantilever

$$\text{moment at B} = m_b = -\frac{(.25 \text{ in.}+.625 \text{ in.})}{2} \times .833 \text{ in.} \times 30 \times 400 =$$

—.364 sq. in. × 12,000 = —4375 in. lbs.; or the moment at C = $m_c$ =

$$\left[-.364 + \frac{(.97 + .67)}{2} \times .667\right] \times 12000 = +2185 \text{ in. lbs.}$$ As a corollary of this relation we have the fact that the maximum bending moment occurs at the point where the shear curve changes sign; that is, where the external shear is zero. (See Figs. 16 and 17.) Very often the only moment desired is the maximum one, and since the shear curve shows at once the one or two points where this moment can occur, the moment at these points can be computed without the necessity of drawing the moment curve. One more relation exists between the moment and shear that is much used. By it, it is possible to compute the moment at any section in a beam, knowing the moment at any other section, the shear at that section, and the loads between the two sections. The equation is: $m_x = m_b + s_b \cdot x + \Sigma F \cdot a$, $\Sigma F \cdot a$ being the moment about section $X$ of all the loads between the sections. Whether the plus or minus sign is used with the second and third terms depends on the direction of the shear and the loads. If these are in such a direction as to cause a positive bending moment, or one that would produce compression in the top of the beam, their sign is positive, otherwise it is negative. In this connection a positive shear can always be considered as an upward load, and a negative shear as a downward load. Example:

$$m_d = -4375 \text{ in. lbs.} + 388 \text{ lbs.} \times 45 \text{ in.} - \frac{6 \times 45^2}{2} - 50 \text{ lbs.} \times 25 \text{ in.} =$$

$+5720$ in. lbs. (See Figs. 15-17.)

20. *Horizontal Shear*—In the explanation of the theory of beam action the question of stresses due to shear was not touched upon. It was shown that the external shear at any section equals the algebraic sum of the forces on either side of the section. According to the last principle presented in the preceding paragraph there can be no increase in an external bending moment unless an external shear is present at the section over which this increase in moment occurs. Furthermore, in order that the internal resisting moment may increase, as the external moment becomes greater certain stresses are set up between the horizontal layers of a beam which are called "horizontal shear" stresses. Referring to Fig. 19, it is evident that if the bending moment at section *B—B* is larger than that at section *A—A* there will be a greater intensity of normal stress at all points on section *B—B*. If a portion of the beam above any horizontal plane, *X—X*, be isolated it will be seen that the condition of equilibrium for horizontal forces requires a horizontal shear, *S*, acting on the underside of the section equal to the difference between the normal forces on the ends. Letting $b$ denote the breadth of the beam, and $L$, the distance between sections *A—A* and *B—B*, the intensity of horizontal shear on the *X—X* plane $= s = \frac{S}{b \cdot L_1}$. As shown in Fig. 19, the intensity of this shear stress has a parabolic variation from

zero at the outside layer to a maximum at the neutral axis of the beam. Its value is given by the formula, $s = \frac{S \cdot Q}{b \cdot I}$, in which $S$ is the external shear at the section, $Q$ the statical moment about the neutral axis of the portion of the beam above the horizontal plane (usually the one through the neutral axis) on which $s$ is desired, $b$ the breadth of the beam, and $I$ the moment of inertia of the entire cross-section.

Fig. 20 is a graphical representation of what takes place when in-

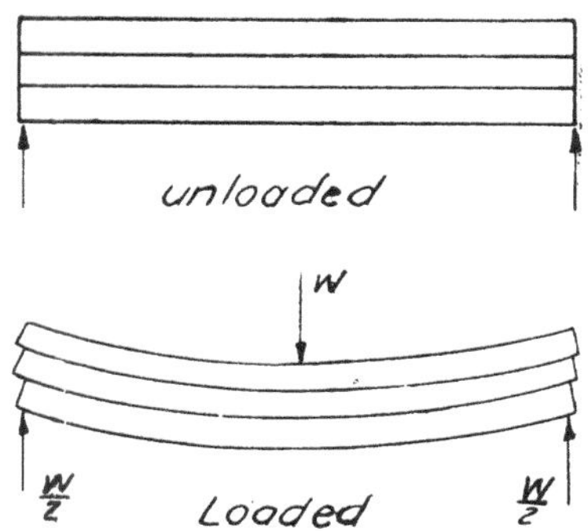


**Fig. 20. Illustration of Effect of Horizontal Shear**

stead of a single beam three separate planks are used. Because no resistance is offered to horizontal shear forces acting between the planks, they can not act as a unit, and the combination has but one-third the strength of a single beam of the same total depth and width.

Beams must always be investigated for their strength in horizontal shear, for failure will occur just as readily from this cause as from excess of normal stress, if the beams are improperly designed. The example which follows makes the application of the shear formula clear.

## *CALCULATION OF HORIZONTAL SHEAR STRESSES IN A BEAM*

Assume a beam of the section shown in Fig. 21.

Moment of inertia $= I = \frac{2 \times (3.6)^3}{12} - \frac{(2.0 - .6) \times (2.4)^3}{12} = 6.17$

Maximum statical moment $= Q = 2 \times .6 \times 1.5 + .6 \times 1.2 \times .6 = 2.24$

External shear on the section $= 1000$ pounds.

Maximum intensity of horizontal shear $= s$.

$$s = \frac{1000 \times 2.24}{.6 \times 6.17} = 605 \text{ lbs. per sq. in.}$$

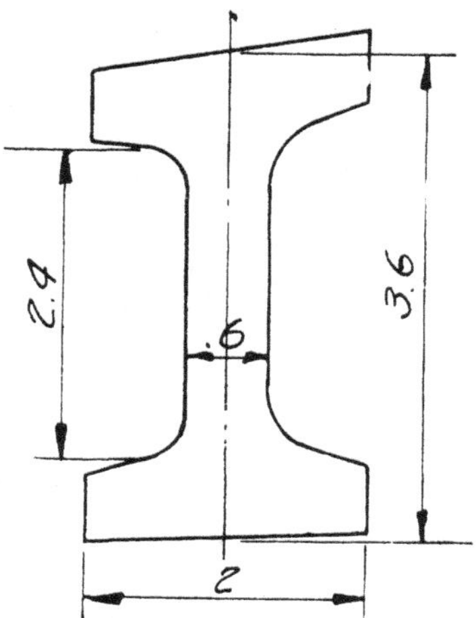


Fig. 21. Section of a Beam

In addition to horizontal shear stresses, stresses due to vertical shear are present. These stresses act in planes at right angles to the horizontal shear stresses, and at any point are equal to the latter, in accordance with the principle that shear stresses on any two planes at right angles to each other are equal in intensity. As a rule vertical shear stresses are not a limiting factor in design work.

## *THE THEOREM OF THREE MOMENTS*

21. *General Equation*—The derivation of the three moment equation from the theory of the elastic curve will not be taken up here. In the form in which it is used for uniformly distributed downward loads on a beam, with its supports either all on the same level or lying in the same straight line, the equation is as follows

$$m_1 l_1 + 2m_2 (l_1 + l_2) + m_3 l_2 = + \frac{w_1 l_1^3}{4} + \frac{w_2 l_2^3}{4}$$

22. *Determination of Signs*—The determination of the sign of the shears in the solution of the three moment equation is apt to cause more confusion than any other part of the problem. A definite rule can be given, however, that will always hold whatever the type of loading. If the portion of the beam between a reaction point and a section at which the shear is to be determined tends to slide up by the part of the beam on the other side of the section, the shear is positive. If the tendency to slide is the reverse, the shear is negative. Fig. 18 is a graphical representation of this law. When first calculating the shears from the bending moments at the reaction points by the equation, $m_2 = m_1 + s_{+1} \cdot X + \Sigma F \cdot a$, the sign of the shear term should always be written plus. If the

solution of the equation gives the shear a positive sign, the shear is positive. But if the solution results in a negative sign, the shear is negative, and must, therefore, be used with a minus sign in any further computations. It will be observed that a positive shear always tends to produce a positive moment, or one that causes compression in the top of a beam, while a negative shear tends to produce a negative moment. Attention should again be called to the fact that downward loads are negative, upward loads positive. In the right member of the general equation given above, both terms have a positive sign, but in the computations which follow, these terms become negative because $w_1$, and $w_2$ are negative.

The solution for a typical case (See Fig. 22) is given below:

*MOMENTS*

(1) $m_1 = -\frac{10 \times 25^2}{2} = -3125$ in. lbs. and $m_4 = -\frac{10 \times 20^2}{2} =$ $-2000$ in. lbs.

(2) $-3125 \times 100 + 2m_2(100+50) + m_3 \times 50 = -\frac{10}{4}(100^3+50^3)$

(3) $m_2 \times 300 + m_3 \times 50 = -2{,}500{,}000$

(4) $m_2 \times 50 + 2m_3\ (50+110) - 2{,}000 \times 110 = -\frac{10}{4}\ (50^3+110^3)$

(5) $m_2 \times 50 + m_3 \times 320 = -3{,}480{,}000$

Solve equations (3) and (5) simultaneously—

$m_2 = -6{,}695$ in. lbs; $m_3 = -9{,}835$ in. lbs.

*SHEARS AND REACTIONS*

(6) $\Sigma V = -25 \times 10 + s_{-1} = 0;\ s_{-1} = +250$ lbs.

(7) $-6695 = -3{,}125 + 100 \times s_{+1} - \frac{100^2 \times 10}{2}$

(8) $s_{+1} = \frac{-6695+3125}{100} + \frac{100^2 \times 10}{2 \times 100} = +464.3$ lbs.

(9) $r_1 = s_{-1} + s_{+1} = +714.3$ lbs.

(10) $\Sigma V = -10 \times 100 + 464.3 + s_{-2} = 0; s_{-2} = +535.7$ lbs.

(11) $-9835 = -6695 + 50s_{+2} - \frac{50^2 \times 10}{2}$

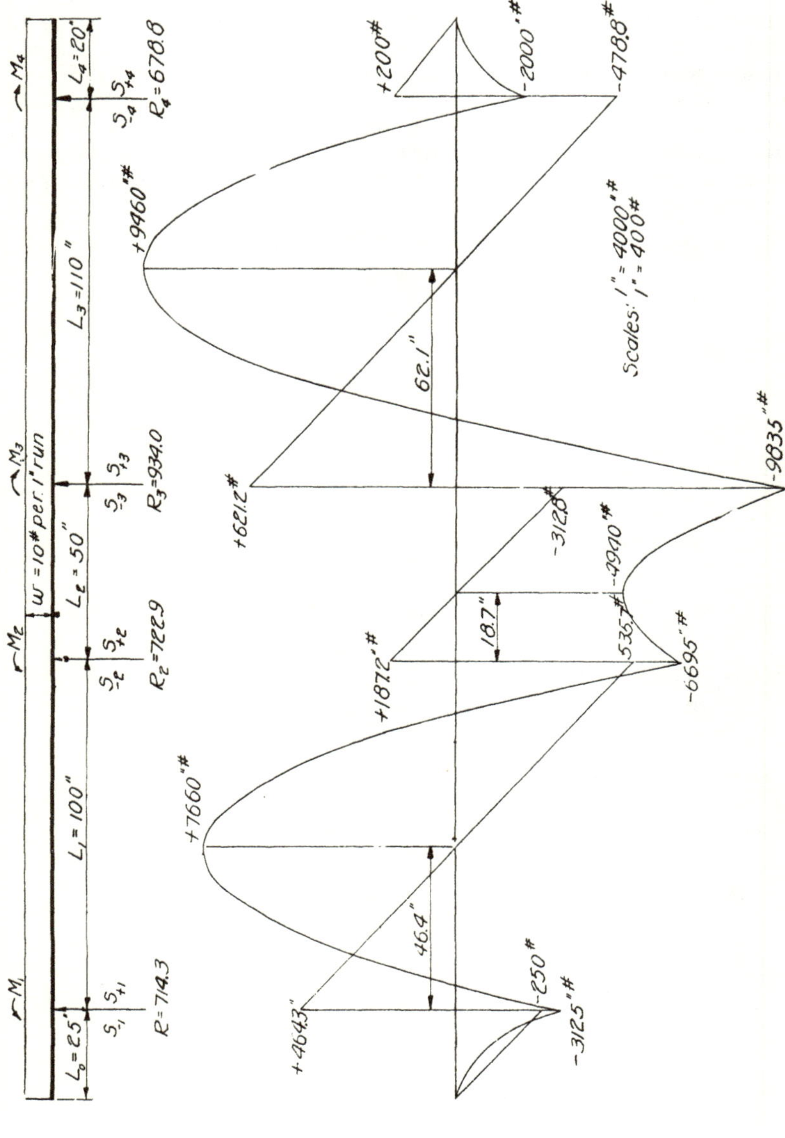


Fig. 22 (Upper). Continuous Beams with Loads and Reactions
Fig. 23 (Lower). Bending Moment and Shear Diagram for Continuous Beams

(12) $s_{+2} = \dfrac{(-9835+6695)}{50} + \dfrac{50^2\times 10}{2\times 50} = +187.2$ lbs.

(13) $r_2 = s_{-2}+s_{+3} = +732.9$ lbs.

(14) $\Sigma V = -50\times 10 + 187.2 + s_{-3} = 0;\ s_{-3} = +312.8$ lbs.

(15) $-2000 = -9835 + 110\,s_{+3} - \dfrac{110^2\times 10}{2}$

(16) $s_{+3} = \dfrac{(-2000+9835)}{110} + \dfrac{110^2\times 10}{2\times 110} = +621.2$ lbs.

(17) $r_3 = s_{-3}+s_{+3} = +934.0$ lbs.

(18) $\Sigma V = -110\times 10 + 621.2 + s_{-4} = 0;\ s_{-4} = +478.8$ lbs.

(19) $s_{+4} = +20\times 10 = +200$ lbs.

(20) $r_4 = s_{-4} + s_{+4} = +678.8$ lbs.

*CHECK*

(21) Total load $= 10(25+100+50+110+20) = r_1+r_2+r_3+r_4 =$ 3050 lbs.

*POINTS OF ZERO SHEAR*

Span 1-2 $X = \dfrac{s_{-1}}{w} = \dfrac{464.3}{10} = 46.4$ in. from $r_1$

Span 2-3 $X = \dfrac{187.2}{10} = 18.7$ in. from $r_2$

Span 3-4 $X = \dfrac{621.2}{10} = 62.1$ in. from $r_3$

*CENTER MOMENTS*

(22) $m_{1-2} = -3125 - \dfrac{46.4^2\times 10}{2} + 46.4 \times 464.3 = +7660$ in. lbs.

(23) $m_{2-3} = -6695 - \dfrac{18.7^2\times 10}{2} + 187.2\times 18.7 = -4940$ in. lbs.

(24) $m_{3-4} = -9835 - \dfrac{62.1^2\times 10}{2} + 621.2\times 62.1 = +9460$ in. lbs.

The methods of solution used above are but applications of the principles previously discussed, especially the relationship between the moments at two sections of a beam, the shear at one of the sections, and the loads between the sections. In determining the maximum moments in a span, the point of zero shear at which, as has been explained, this moment occurs, was first calculated. As in the present example, there may be two or more unknown moments over the supports, making it necessary to write the three moment formula as many times as there are unknown moments, and to solve the resulting equations simultaneously. It is first applied to the first and second interior spans at the left end of the beam, then to the second and third, and so on until the last interior span at the right end of the beam is reached. In the general equation $m_1$ is always the moment at the left end of the left of the two spans being considered, $m_2$ that between the spans, and $m_3$ that at the right end of the right span.

For purposes of plotting, the law of signs given above must be slightly modified. If that portion of the beam to the left of a section tends to slide up by the portion to the right of the section, the shear is positive. When the tendency to slide is the reverse, the shear is negative. This rule is illustrated by the part of Fig. 18 to the right of the reaction point.

23. *Combined Bending and Compression*—When the forces acting on a beam do not conform in direction and point of application to the conditions stated at the beginning of the discussions of the beam theory, the formula for simple beam action, $f = M \cdot y/I$, must be supplemented. In Fig. 24 none of the loads are both perpendicular to the axis of the

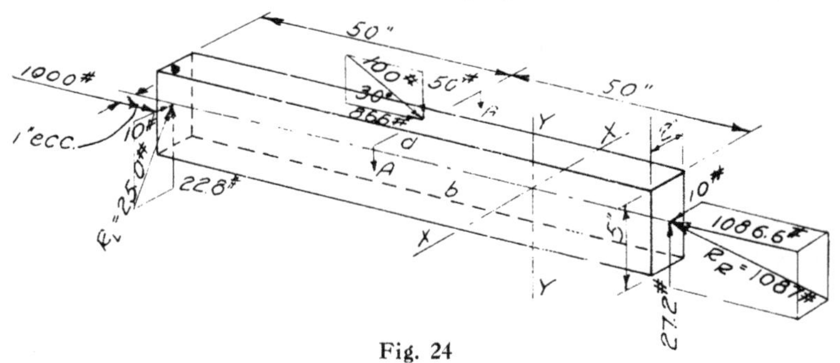


Fig. 24

beam and in a vertical plane passing through the neutral axis. The first step is to resolve all forces which are neither parallel nor perpendicular to this axis into two, or, if necessary, three components which are either parallel or perpendicular to it. Those components or forces which are at right angles to the neutral axis of the beam produce bending about that axis of the cross-section to which they are perpendicular; those components or forces parallel to the neutral axis cause axial compression in the beam, and if they are not applied at the center of gravity of the section they also produce bending about one or both axes of the cross-section, depending on the nature of their eccentricities. The calculation of the moments and reactions for the beam and loads in Fig. 24 will not be given. It should be noted that the right reaction has three components and the left, two components. The eccentricity of the 1000 lb. load causes a moment about the *Y—Y* axis. The stresses on section *A—A* are given below.

$$m_{x-x} = 1358 \text{ in. lbs.}$$

$$m_{y-y} = 500 \text{ in. lbs.}$$

$$P = 1087 \text{ lbs.}$$

$$f_{x-x} = \frac{1358 \times 2.5}{20.8} = 163.0$$

$$f_{y-y} = \frac{500 \times 1.0}{3.33} = 150.0$$

$$P/A = \frac{1087}{10} = 108.7$$

Maximum Compressive Stress = 421.7 lbs. per sq. in.
Maximum Tensile Stress = 313.0—108.7 = 204.3 lbs. per sq. in.

At point *a* the maximum compressive stress occurs, and at point *b* the maximum tensile stress.

The above procedure is satisfactory when the beam is relatively short so that the deflection of the beam is small. In the case of long, slender beams carrying a large axial load, additional moments are caused by the deflection of the beam, which may be considered as a laterally loaded column. This case will be treated in a later article on wing spars.

24. *Column Formulas*—When a column is straight, and the only load imposed upon it is axial and is applied at the center of gravity of the column, then the stress equals *P/A*, where *P* is the load, and *A* the cross-sectional area. For very short columns the maximum value of *P/A* closely equals the stress at the yield point of the material, but for long columns the *P/A* at which failure will occur is much less than this value. The purpose of column formulas is to determine the maximum allowable *P/A*. In practically all formulas the reduction in the ultimate stress for "column action" is a function of the ratio of the length of the column to its least radius of gyration.

For columns having a high value of $L/\rho$, the maximum *P/A* is given very closely by Euler's formula.

$$P/A = \frac{c\pi^2 E}{(L/\rho)^2} \quad \text{or} \quad P = \frac{c\pi^2 EI}{L^2}$$

E = Modulus of elasticity of material.
L = Length of column.
I = Least moment of inertia of column.
$\rho$ = Least radius of gyration of column.

c = A constant depending on the degree of fixity of the ends of the column.
c = 1 for round ends.
= 1/4 for one end free and one end fixed.
= 2.05 for one end round and one end fixed.
= 4 for fixed ends.

The lower value of $L/\rho$ to which this formula applies is given in the discussion below.

The column formula that agrees best with test data on columns, whose $L/\rho$ is less than the lower limit to which the Euler formula is applicable, is J. B. Johnson's parabolic formula,

$$P/A = f - \frac{f^2}{4\,c\pi^2\,E}\,(L/\rho)^2$$

f = Yield point of material
c = A constant with same values as in Euler's formula.

For a value of $L/\rho = 0$, $P/A$ equals the stress at the yield point, which for columns, coincides with the ultimate strength. As shown in Fig. 25, which is the curve for mild steel, the parabola is tangent to the Euler curve at approximately the point where columns cease to follow the Euler law. The abscissa of this point of tangency is given by the expression $L/\rho = \sqrt{\dfrac{2\,c\pi^2\,E}{f}}$; the ordinate, for all degrees of fixity, equals $f/2$.

For mild steel, the assumption that the ultimate strength of the column equals the yield point is perhaps somewhat severe. The use of 40,000 instead of 36,000 for the formula in Fig. 25 would give less conservative and yet very reasonable values. One of the important features of this formula is that the correction for column action is a function of both the yield point and the modulus of elasticity. This is especially valuable where experiment has not determined the column strength of the material, as in the case of high strength, alloy steels and duraluminum.

25. *Wood and Steel in Combination*—There are occasional cases of wood and steel in combination when it is desirable to determine the proportion of the stress which each material carries.

$E_w$ = modulus of elasticity of wood
$E_s$ = modulus of elasticity of steel
$A_w$ = area of wood
$A_s$ = area of steel
$I_w$ = moment of inertia of wood
$I_s$ = moment of inertia of steel

Case I. Beams subjected to bending.
W = total load = $W_w + W_s$
$W_w$ = load carried by wood
$W_s$ = load carried by steel

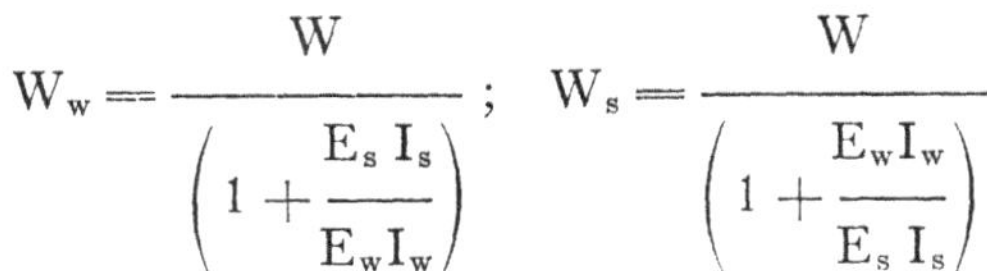

$$W_w = \frac{W}{\left(1 + \frac{E_s I_s}{E_w I_w}\right)}; \quad W_s = \frac{W}{\left(1 + \frac{E_w I_w}{E_s I_s}\right)}$$

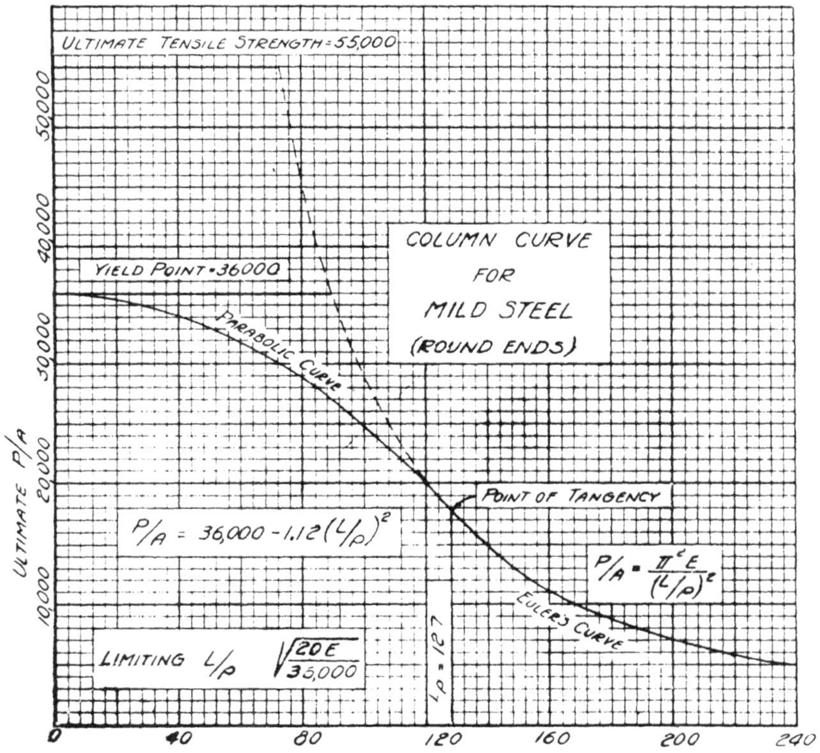

L/ρ

Fig. 25.

Case II. Short columns

$P$ = total axial load

$P_w$ = load carried by wood

$P_s$ = load carried by steel

$$P_w = \frac{P}{1 + \frac{E_s A_s}{E_w A_w}}; \quad P_s = \frac{P}{\left(1 + \frac{E_w A_w}{E_s A_s}\right)}$$

Case III. Euler columns

$$\rho \text{ of composite column} = \sqrt{\frac{E_w I_w + E_s I_s}{E_w A_w + E_s A_s}}$$

$$\text{Total EI} = (E_w I_w + E_s I_s)$$

26. *Torsion, and Combined Torsion and Bending*—In control systems and surfaces tubes are frequently subjected to torsion. The stresses due to torsion, and the torsional angle, or angle of twist for round tubes are given by the following formulas:

$f_s$ = torsional unit stress
J = polar moment of inertia of section
M = twisting moment
r = radius of the tube
$\phi$ = torsional angle in radians (this angle in degrees = $57.3 \times \phi$).
L = length of tube for which torsional angle is desired.
$E_s$ = shearing modulus of elasticity

$$f_s = \frac{M \cdot r}{J} \quad \text{and} \quad \phi = \frac{M \cdot L}{E_s \cdot J}$$

The polar moment of inertia equals the sum of the moments of inertia of the section about any two axes at right angles to each other, or, for circular sections, twice the ordinary moment of inertia.

These formulas are for solid or hollow circular sections only. For elliptical and rectangular sections the distribution of stress is different. With both types of sections the maximum stress occurs on the broad side of the section. In a rectangle the stress is a maximum at the center of each face and decreases to zero at the corners. The following formulas give the maximum stresses and angles of twist.

b = short side of rectangle, or minor axis of ellipse.
h = long side of rectangle, or major axis of ellipse.
$f_s$ = maximum stress on section.

| Rectangle | Square | Ellipse |
|---|---|---|
| $f_s = \frac{4 \cdot 5\,M}{b^2\,h}$ | $f_s = \frac{4 \cdot 5\,M}{b^3}$ | $f_s = \frac{16\,M}{\pi\,b^2\,h}$ |

The stress at the center of the short side can be computed from $f_s$ by multiplying $f_s$ by the ratio $b/h$. The angle of twist for a square section equals $\phi = 7.11\,ML/E_s b^4$.

Frequently a member is subjected to both bending and torsional stresses. These two kinds of stresses may be combined as follows:

$f_b$ = the bending stress computed from $\frac{M \cdot y}{I}$

$f_s$ = the torsional stress computed from $\frac{M \cdot r}{J}$

$$F_c = \frac{f_b}{2} + \sqrt{f_s^2 + f_b^2/4} = \text{maximum tensile or compressive stress.}$$

$$F_s = \sqrt{f_s^2 + f_b^2/4} = \text{maximum shearing stress.}$$

In using all the formulas given in this chapter the greatest care must be taken to have the units consistent. It is best to use the same units in all computations. A large part of the trouble with units is caused by the use of feet and foot units in places of inches and inch units.

References:

"Applied Mechanics," Vol. I and II, by Fuller & Johnson.
"Theory of Structures," by C. M. Spofford.
"Modern Framed Structures," by Johnson, Bryan & Turneaure.
"Mechanics of Materials," by M. Merriman.

## CHAPTER II

# GENERAL CONSIDERATIONS

27. *Visibility*—Visibility is a quality which varies in importance with the type of airplane and the service which it is intended to perform. The pilot must always have a field of view which will permit him to fly his airplane and land it with ease and safety. When the crew consists of three or more persons, a judicious arrangement of the personnel will relieve the pilot of all duties except that of flying the airplane, and a careful arrangement of the remaining members of the crew should leave no blind spots which would permit an enemy aircraft to approach unseen. As the number of the crew is reduced, the importance of a wide field of vision for each individual member is increased, and in a single-seater pursuit, where the pilot must also assume the duties of observer, visibility attains its maximum importance and can be secured only by giving the greatest care and attention to the design of the structure as a whole.

If the pilot has considerable freedom of movement, he may, by changing his position, look around almost any obstruction and reduce the blind spots to a minimum. Ordinarily, however, he is strapped in his seat with a belt so that his movements are restricted to bending at the waist and neck, and turning the head. The cockpit should be designed to enable the pilot to take the maximum advantage of such movements as are possible to improve his field of view. The cockpit opening must give ample room for movements of the head and shoulders. It should be cut out flush with the sides of a fuselage of rectangular cross section, and where the section is round or oval, the opening should be large enough to permit the pilot to look down vertically over the side. With monocoque construction it may prove impractical to cut out a large enough opening without seriously weakening the structure, but with a stick and wire or plywood truss fuselage, the cowling can be cut away as far as the top longeron without any difficulty. From a properly designed cockpit, the pilot should be able to look over the side and see directly beneath the fuselage.

A narrow fuselage, especially at the engine section of a tractor airplane, is favorable to a good field of view. It is difficult to secure good vision about the wide blunt nose of a rotary motor housing, while the narrow section possible where the cylinders of the engine are vertical and in line creates a very good field of view forward. A large nose radiator increases the blind area in front, the most favorable condition being reached when the engine cowling can be run down to a rough conical shape with the point at the hub of the propeller. While it is desirable to have an engine housing which will protect the engine from the weather as much as possible, the visibility can be greatly improved by reducing the volume enclosed by the cowling to a minimum. The cylinder heads may be left exposed, and along the sides of a vertical motor the cowling may be "dished in" to improve the field of view.

To secure good visibility about a wing, it is essential that the pilot be so placed that his eyes are nearly on the line of the chord. If the trailing edge of the center section be cut away to enable the pilot to be properly placed, he can uncover all the blind area obstructed by the wing unless it is of extremely thick section. Perfect visibility can be secured with a parasol monoplane, or about the upper and middle wings of a biplane and triplane respectively. The narrow chord of the wings of a triplane tends to reduce the blind area obstructed by them to a minimum. Good visibility around the lower wing of a biplane is more difficult to secure. It is impossible to eliminate this blind spot entirely. but it may be cut to a minimum by careful design. The use of a narrower chord on the lower wing than on the upper, a wide gap which will place the pilot high above the lower wing, a smaller span to the lower than to the upper wing, and the cutting away of the trailing edge near the fuselage are devices at the disposal of the designer which will cut the blind area to a minimum. The use of a tapering wing with wide chord at the fuselage and the use of large dihedral in the lower wing are both factors which tend to increase the blind area under the wing.

Stagger is an arrangement which greatly assists the designer in securing visibility by enabling him to locate the wings in the most advantageous position with relation to the pilot. A slight positive stagger is generally necessary if the pilot of a single-seater biplane is to be placed with his eyes in line with the chord of the upper wing. By use of stagger the lower wing may be placed so that the area obstructed by it is in the least important part of the field of view. A maximum angle of vision over the leading edge of the lower wing is, in general, more important than a large angle over the trailing edge, and an adjustment of the stagger may be used to place the lower wing in the position which gives the best results.

External obstructions such as struts, gun installations, exhaust manifolds, radiators, etc., interfere with the pilot's view to a greater or less extent. Where the obstructing surface is small and detached from larger obstructing surfaces, as is the case with narrow struts, a very slight movement of the pilot's head will enable him to see all the area behind them so that their effect on the field of view is unimportant. But in the case of radiators, gun installations and exhaust manifolds which are directly attached to a large obstructing surface, the field of view may be materially affected. The use of short exhaust stacks and the housing of the guns inside the fuselage are desirable, while the building of the radiator into the wing so that it conforms to the wing curve undoubtedly gives the best installation from the visibility standpoint.

28. *Choice of Wing Section*—There are three fundamental considerations in the aerodynamical design of an airplane:

I. The determination of the wing section and area, and the disposition of area which will give the desired performance in steady motion.

II. The disposition of auxiliary surface area which will give proper stability and controlability.

III. The reduction of parasite resistance to a minimum.

The first problem is the chief concern of structural analysis. The selection of the wing area and section is necessary for a determination of the span, chord, strut locations and spar sizes.

For a preliminary determination of the wing section, it is sufficient to consider various sections applied in turn to a single airplane of constant weight, area, power and parasite drag. Certain criteria may thus be developed which permit evaluation of the properties of the sections for high speed, climb, ceiling, etc.

The assumption of constant area is a first approximation. For final choice the area must be considered as variable, and selection made by means of a series of performance estimates. However, the analytical tests used are adequate to determine those sections best adapted to each class of airplane.

For level flight at high speed

$$P = \delta(K_x + c)AV^3$$

where $P$ is the power of the motor.
$K_x$ is the drag coefficient of the wings.
$K_y$ is the lift coefficient for the wings.
$K_{ym}$ is the maximum value of $K_y$.
$c$ is the parasite coefficient per unit wing area.
$A$ is the wing area.
$\delta$ is the air density.
$V$ is the speed—in this case high speed.

or
$$P = \delta K_y \frac{(K_x + c)}{K_y} A V^3$$

and
$$V = \left\{\frac{P}{K_y \delta A}\right\}^{1/3} \left\{\frac{K_y}{K_x + c}\right\}^{1/3}$$

It is evident from this equation that if these airplanes, differing only in aerofoil section, be flying at the same value of $K_y$, the airplane with aerofoil of higher $K_y/K_x$ at that condition will not need all its power to maintain the speed, or it may be driven at a higher speed for the same power. A preliminary valuation of aerofoils for high speed is the value of $K_y/K_x$ at the $K_y$ reached in flight.

The landing speeds of two airplanes differing only in aerofoil section are inversely proportional to the square roots of the maximum values of the lift coefficients of the sections.

$$V = \left\{\frac{W}{\delta K_{ym} A}\right\}^{1/2}$$

$$= \left\{\frac{W}{\delta A}\right\}^{1/2} \left\{\frac{1}{K_{ym}}\right\}^{1/2}$$

The maximum unit lift has been listed, therefore, in Table I.

Speed range is defined as the ratio of high speed to landing speed. Since landing speed is inversely proportional to the square root of the lift coefficient, and high speed to the cube root of the $L/D$ at the value of $K_y$ usual to high speed, the speed range is proportional to the product of these figures:

$$V_m \propto \left[\frac{1}{K_{ym}}\right]^{1/2}$$

$$V \propto (L/D)^{1/3}$$

$$V/V_m \propto (K_{ym})^{1/2} \times (L/D)^{1/3}$$

The values of this quantity are listed in Table I, for $K_y = .0008$ and .0004. It will be observed that the aerofoils are not of the same value in speed range for low speed and high speed airplanes.

Due to the fact that this speed range factor is a first approximation, one aerofoil should not be chosen above another by reason of a small favorable difference, for finer analysis may reverse the result. However, large differences shown by this method may reasonably be taken as indicative of comparative merits.

The maximum altitude or minimum density which an airplane can reach is a complex function of its engine power and drag characteristics. In his paper on "Supercharging," Dr. de Bothezat shows that the ceiling density is expressed

$$\delta_c = \left[\frac{W\,\delta_0^{2/3}}{P_0^{2/3}\,A^{1/3}\,n^{2/3}}\right] \times \left[\frac{K_x + c}{K_y^{3/2}}\right]^{2/3}$$

where $n$ is the average propeller efficiency.

In the comparison being made, the airplane with the aerofoil possessing highest value of $\frac{K_y^{3/2}}{K_x}$ will give the highest ceiling. This quantity is given in Table I.

The rate of climb is also a complex function of the power and drag characteristics.

$$R = \frac{n\,P_0}{W} - \left[\left(\frac{P}{\delta_0 A}\right)^{1/2} \times \left(\frac{K_x + c}{K_y^{3/2}}\right)\right]$$

As long as the power and wing area are considered constant the rate of climb may be taken as dependent solely on the quantity, $K_x/K_y^{3/2}$ and in the first approximation is maximum when $K_y^{3/2}/K_x$ is a maximum. This quantity is, therefore, also a criterion of climb.

In an analysis of the cruising problem, Lt. Col. Durand shows that the maximum radius is obtained at an angle of attack of $\frac{K_x+c}{K_y}$ minimum. So long as $c$ is actually constant the aerofoil which has the highest maximum $K_y/K_x$ will give best cruising radius. $L/D$ maximum is, therefore, listed as a criterion for cruising radius.

A minimum *c.p.* movement on the aerofoil is desired in order that the load may be at all times well distributed between the spars. A large movement may be a sufficient cause for the rejection of a section even if its performance characteristics are satisfactory. The *c.p.* travel from the most forward position to $K_y$, .0008, has been computed for the aerofoils listed. This is not the range applicable to all types of airplanes, but gives a good comparative figure.

Spar depth is a very important consideration. It is possible to secure very desirable aerodynamical characteristics with very thin or very highly cambered sections which would be entirely inapplicable to any but the most lightly loaded airplanes. For most conventional two-spar designs the usual spar locations are at about 10 per cent to 15 per cent and 70 per cent of the chord. The spar depths of the sections have therefore been listed in Table I. Other things being nearly equal, the section with the greatest spar depth is to be recommended.

It is rarely the case that an aerofoil is selected for a single performance characteristic. Few airplanes require predominantly high speed, climb, or ceiling, but rather varying combinations of these qualities.

In Table II the fifteen Air Service Specification types have been divided into six general classes with the average performance requirements shown.

It is obviously impossible for one rigid aerofoil to be best "all around." Only by variable area and camber arrangement could such a condition be more nearly fulfilled. The choice must be made rather as a compromise of performance characteristics, with requisite spar depth and reasonable *c.p.* movement.

On the basis of the above figures the sections given in Table II are recommended as suitable for the six classes. It will be observed that in some cases two sections are listed, which indicates that their characteristics are very similar and little choice is possible. The complete characteristics and ordinates for these sections are given in Table III.

In a pursuit airplane speed must be combined with fast climb and high ceiling. The wing section must, therefore, be somewhat of a compromise between an extremely high speed and high climb wing. There are, however, different types of pursuit airplanes and the relative importance of speed, climb and maneuverability varies with the special service for which the airplane is intended. Where ability to maneuver is to be strongly emphasized, a section possessing high lift characteristics is employed. The U.S.A.-27 is well adapted to this use, and the depth afforded for spars makes it suitable for internally braced construction. This section is also very efficient except for extreme high speed. It is well suited to a night pursuit airplane in which high speed is not so important as climb, low landing speed and maneuverability. For a pursuit airplane that would be required to land in restricted space, as in mountainous regions or roads, it is essential to use as high a lift section as possible in order to reduce the area and hence the span. For this purpose, also, the U.S.A.-27 is the best wing. The U.S.A.-16 has better climb and ceiling characteristics than the R.A.F.-15, as is indicated by a higher value of $K_y^{3/2}/K_x$, and is therefore to be chosen if more importance is to be attached to climb and ceiling than to high speed. The U.S.A.T.S.-5 is too thick for a pursuit airplane, but is mentioned to indicate the best that can be done in adapting thick sections to speed work. The same sections are adapted to class 2, because the same requirements apply except to a lesser degree.

For both classes 3 and 4 the U.S.A.-5 was selected as the best thin section. It combines low speed with good climb and high efficiency at cruising speed, characteristics which are of more importance than high speed. Since a low landing speed is very desirable with airplanes of these types, and since also it is important to keep the span and hence the area as small as possible, the large $Ky$ of the U.S.A.-5 section is one of the most important factors to be considered. However, for airplanes in which high speed is of more importance than the considerations discussed above, the U.S.A.-15 section is to be recommended. Its structural characteristics are superior to those of the U.S.A.-5. In these types of airplanes it is very desirable to use a single instead of a double bay cellule, and the greater spar depth of the U.S.A.-15 section may make this possible in cases where, with the U.S.A.-5 section, it would be impracticable.

It will be observed that the only intermediate and thick sections chosen for any class were the U.S.A.-27 and the U.S.A.T.S.-5. An examination of Table II will show that these are far superior to the other sections in the groups of intermediate and thick wings, for all except very special work. For small or moderate sized airplanes these wings will be too deep to be economical unless at least partially internal braced construction is used.

In the larger designs of classes 5 and 6 advantage could be taken of the greater depths of these sections to increase the length of the bays, thereby reducing the inter-plane bracing. In all types of airplanes, but particularly in larger bombers and express or passenger airplanes where the value of the airplane is almost directly proportional to the useful load carried, the advisability of selecting a high lift section becomes apparent. For instance, the U.S.A.-27 has a $K_{ym}$ 42 per cent greater than the U.S.A.-16, and therefore the same wing area can support a proportionately larger load. But, as the total useful load for an airplane is rarely as much as 40 per cent of the gross weight of the airplane, the use of a high lift wing would about double its net carrying capacity. This is, of course, partly offset by the fact that the ceiling and the climb, which are closely proportional to the power loading, would be considerably affected by increasing the gross weight of the airplane. In commercial airplanes this is not as important as in military types. The change in high speed, moreover, would not be great. In commercial airplanes and in long range bombers where economy in operation and large radius of

## TABLE I—CHARACTERISTICS OF AEROFOILS

| Aerofoil | Landing Speed ($K_{ym}$) | Cruising (Max. L/D) | High Speed Pursuit L/D at $K_y = .0004$ | Speed Range $(K_{ym})^{1/2} \times (L/D)^{1/3}$ $K_y = .0004$ | High Speed (large airplanes) L/D at $K_y = .0008$ | Speed Range $(K_{ym})^{1/2} \times (L/D)^{1/3}$ $K_y = .0008$ | Ceiling, Climb $K_y^{3/2}/K_x$ (max.) | Per Cent of Chord: C.P. Movement $K_y = .0008$ to max. forward position | Per Cent of Chord: Wing Depth at 10% chord | Per Cent of Chord: Wing Depth at 15% chord | Per Cent of Chord: Wing Depth at 70% chord | Reference: Serial No. of McCook Field Report |
|---|---|---|---|---|---|---|---|---|---|---|---|---|
| 1 | 2 | 3 | 4 | 5 | 6 | 7 | 8 | 9 | 10 | 11 | 12 | 13 |
| U.S.A. 1 | .00292 | 16.7 | 6.5 | .1000 | 12.2 | .146 | .660 | 15 | 5.72 | 5.90 | 4.38 | D-52.33/82 |
| 2 | .00339 | 15.7 | 4.8 | .0985 | 8.6 | .120 | .701 | 23 | 4.87 | 5.60 | 4.35 | D-52.33/82 |
| 3 | .00324 | 16.4 | 3.4 | .0855 | 9.7 | .122 | .712 | 24 | 4.83 | 5.70 | 4.44 | D-52.33/82 |
| 4 | .00364 | 15.9 | 3.4 | .0905 | 7.4 | .118 | .681 | 24 | 5.01 | 5.60 | 4.38 | D-52.33/82 |
| 5 | .00328 | 16.2 | 4.3 | .0930 | 11.2 | .129 | .671 | 22 | 5.63 | 6.20 | 4.47 | D-52.33/82 |
| 6 | .00297 | 17.2 | 5.6 | .0965 | 12.4 | .126 | .726 | 20 | 4.50 | 5.00 | 3.59 | D-52.33/82 |
| 8 | .00317 | 13.6 | 3.2 | .0830 | 7.5 | .110 | .625 | | 12.30 | 13.70 | 10.65 | D-52.33/82 |
| 9 | .00339 | 15.5 | 5.4 | .1020 | 9.9 | .125 | .665 | | 8.22 | 9.10 | 7.12 | D-52.33/82 |
| 10 | .00287 | 16.4 | 5.0 | .0920 | 13.7 | .129 | .622 | | 6.32 | 7.00 | 5.38 | D-52.33/82 |
| 11 | .00285 | 16.6 | 5.4 | .0912 | 11.8 | .122 | .643 | | 5.85 | 6.80 | 5.00 | D-52.33/82 |
| 12 | .00326 | 13.6 | 3.0 | .0825 | 7.4 | .112 | .625 | | 12.27 | 13.90 | 10.63 | D-52.33/82 |
| 14 | .00286 | 16.4 | 6.8 | .1010 | 12.3 | .123 | .652 | 13 | 5.42 | 5.93 | 5.02 | D-52.33/82 |
| 15 | .00300 | 16.5 | 6.6 | .1080 | 13.4 | .130 | .658 | 15 | 6.11 | 6.76 | 5.32 | D-52.33/82 |
| 16 | .00252 | 18.8 | 8.8 | .1030 | 18.6 | .129 | .629 | 6 | 5.78 | 6.18 | 4.77 | D-52.33/82 |
| 17 | .00275 | 16.0 | 7.8 | .1040 | 14.7 | .128 | .630 | 8 | 5.73 | 6.38 | 4.90 | D-52.33/82 |
| 18 | .00276 | 15.4 | 6.8 | .1000 | 13.1 | .124 | .611 | 13 | 6.38 | 7.48 | 5.62 | 665 |
| 19 | .00259 | 16.1 | 7.8 | .1010 | 14.0 | .122 | .642 | 10 | 4.52 | 4.80 | 3.70 | 915 |
| 20 | .00219 | 15.7 | 10.2 | .1010 | 15.0 | .116 | .588 | 2 | 5.86 | 6.21 | 4.83 | 916 |
| 21 | .00231 | 15.1 | 8.2 | .0980 | 13.4 | .114 | .613 | 8 | 6.37 | 7.60 | 5.60 | 943 |
| 22 | .00297 | 16.1 | 7.6 | .1070 | 13.3 | .129 | .642 | 16 | 7.80 | 8.75 | 4.50 | 992 |
| 23 | .00259 | 17.1 | 8.2 | .1040 | 15.2 | .128 | .675 | 17 | 5.00 | 5.68 | 4.17 | 1036 |

| | | | | | | | | | | | | | |
|---|---|---|---|---|---|---|---|---|---|---|---|---|---|
| | 24 | .00270 | 16.6 | 8.0 | .1040 | 12.9 | .122 | .678 | 20 | 5.65 | 6.56 | 4.78 | 1037 |
| | 25 | .00316 | 15.4 | 5.2 | .0970 | 10.5 | .123 | .641 | 21 | 7.58 | 8.10 | 6.74 | 1205 |
| | 26 | .00310 | 17.0 | 6.2 | .1020 | 13.0 | .125 | .662 | 21 | 8.17 | 9.25 | 6.90 | 1205 |
| | 27 | .00359 | 16.2 | 5.0 | .1020 | 10.8 | .132 | .677 | 15 | 9.17 | 10.40 | 7.90 | 1205 |
| | 28 | .00275 | 15.8 | 6.0 | .0960 | 11.2 | .117 | .678 | 26 | 9.99 | 11.66 | 7.94 | 1205 |
| | 29 | .00331 | 14.4 | 5.4 | .1010 | 10.0 | .124 | .643 | 30 | 10.01 | 11.79 | 8.88 | 1205 |
| U.S.A.T.S. | 1 | .00296 | 10.5 | 4.8 | .0920 | 7.5 | .107 | .519 | | 16.60 | 20.10 | 16.30 | 321 |
| | 2 | .00339 | 12.2 | 5.3 | .1010 | 8.8 | .120 | .576 | 21 | 12.40 | 14.80 | 13.60 | 724 |
| | 3 | .00431 | 12.6 | 2.2 | .0855 | 4.8 | .111 | .630 | | 8.70 | 10.00 | 8.50 | 321 |
| | 4 | .00435 | 12.0 | 2.2 | .0860 | 5.2 | .114 | .612 | | 8.20 | 9.30 | 7.80 | 321 |
| | 5 | .00386 | 12.8 | 4.6 | .1030 | 8.9 | .128 | .595 | 34 | 13.20 | 15.40 | 10.60 | 724 |
| | 6 | .00413 | 13.0 | 2.4 | .0865 | 6.0 | .117 | .613 | | 11.40 | 13.70 | 9.00 | 321 |
| | 7 | .00392 | 12.8 | 2.4 | .0840 | 5.8 | .112 | .691 | | 9.60 | 11.10 | 11.20 | 321 |
| | 8 | .00405 | 12.7 | 3.0 | .0915 | 8.2 | .128 | .621 | | 11.90 | 13.80 | 9.50 | 321 |
| | 9 | .00402 | 12.2 | 2.0 | .0800 | 5.0 | .109 | .592 | | 12.30 | 14.40 | 11.40 | 321 |
| | 10 | .00442 | 12.9 | 2.4 | .0895 | 5.6 | .118 | .640 | | 11.50 | 13.50 | 10.80 | 321 |
| | 11 | .00328 | 14.0 | 2.3 | .0760 | 6.6 | .108 | .640 | | 8.40 | 10.20 | 9.20 | 724 |
| | 13 | .00330 | 13.3 | 4.6 | .0950 | 8.2 | .117 | .600 | 20 | 11.10 | 12.60 | 9.00 | 321 |
| Durand | 13 | .00408 | 13.3 | 2.8 | .0900 | 6.5 | .119 | .627 | | 10.20 | 12.00 | 10.40 | 724 |
| R.A.F. | 3 | .00348 | 15.6 | 5.3 | .1010 | 8.8 | .121 | .698 | 42 | 4.78 | 5.40 | 3.50 | D-52.33/93 |
| | 6 | .00306 | 16.7 | 5.2 | .0960 | 11.9 | .126 | .663 | 17 | 5.60 | 6.30 | 5.35 | D-52.33/70 |
| | 14 | .00326 | 15.4 | 4.6 | .0950 | 11.0 | .127 | .637 | 15 | 4.70 | 5.30 | 4.80 | D-52.33/24 |
| | 15 | .00260 | 16.8 | 9.6 | .1080 | 16.5 | .128 | .615 | 19 | 6.05 | 6.31 | 4.80 | 834 |
| | 17 | .00280 | 14.2 | 5.7 | .0950 | 11.9 | .121 | .592 | 8 | 6.31 | 6.77 | 5.38 | D-52.33/24 |
| A.E.G. | 1 | .00302 | 15.4 | 5.3 | .0960 | 11.3 | .124 | .643 | 5 | 5.60 | 6.20 | 2.20 | 669 |
| | 2 | .00330 | 15.5 | 5.8 | .1030 | 11.0 | .228 | .644 | 7 | 6.10 | 7.00 | 2.90 | 669 |

All the data listed in this table are based on wind tunnel tests made at the Massachusetts Institute of Technology. The same air speed of 30 M.P.H. was maintained in all tests. All models were 18 x 3 inches in size. Therefore, these data are comparable. Results from British and French tests are not included because they are not comparable to the same degree.

The data on all the U.S.A. thick sections are more optimistic than the data on the U.S.A.T.S. 2, 5 and 13 by 3 to 4 per cent. When these tests were run (Sept. 14, 1918) the tunnel was not properly calibrated. Later (March 12, 1919) the tests on the U.S.A.T.S. 2, 5 and 13 were re-run and the results were 3 to 4 per cent lower than before.

The data given on the R.A.F. 15 wing are for what is known as the M.I.T. section rather than the true British or N.P.L. section. The lift on the latter is larger by 3.5 per cent than the lift on the section with the M.I.T. ordinates, but L/D forthe M.I.T. section is considerably higher, especially at low angles.

*TABLE II*

*CLASSIFICATION OF AIRPLANES FOR WING SELECTION*

| Class | Military Type | Landing Speed | Climb (Ground) | Ceiling | High Speed (Ground) | Sections | | | Commercial Type |
|---|---|---|---|---|---|---|---|---|---|
| | | | | | | Thin | Intermediate | Thick | |
| I | Pursuit 1, 2, 3, 5 | 60 | 1700 | 25,000 | 150 | R.A.F. 15<br>U.S.A. 16 | U.S.A. 27 | U.S.A.T.S. 5 | |
| II | Observation Fighters,<br>Small Bombers, 8, 9, 10, 11 | 55 | 1000 | 20,000 | 120 | R.A.F. 15<br>U.S.A. 15<br>U.S.A. 16 | U.S.A. 27 | U.S.A.T.S. 5 | Mail, Passenger |
| III | Messenger | 40 | —— | —— | 100 | U.S.A. 5 | U.S.A. 27 | —— | Single-seater sport |
| IV | Training 14, 15 | 40 | 900 | 18,000 | 105 | U.S.A. 5<br>U.S.A. 15 | U.S.A. 27 | U.S.A.T.S. 5 | Training, sport, passenger, & mail |
| V | Ground Attack 4, 6, 7 | 50 | —— | —— | 100 | U.S.A. 15 | U.S.A. 27 | U.S.A.T.S. 5 | |
| VI | Large Bombers 12, 13 | 50 | 650 | 13,000 | 100 | U.S.A. 5<br>U.S.A. 15 | U.S.A. 27 | U.S.A.T.S. 5 | Express, passenger & mail. |

action are more important than high speed, the maximum $L/D$ is the chief criterion, provided the *c.p.* travel and the spar depth are reasonable.

For bombers, both U.S.A.-5 and 15 are recommended. The greater maximum lift coefficient of the U.S.A.-5 permits the wing area to be 9.5 per cent less than with the U.S.A.-15 section, with the same landing speed and nearly the same ceiling. In the Bothezat formula for ceiling density it should be noted that the ceiling is closely proportional to the cube root of the area and to the two-thirds power of $K_y^{3/2}/K_x$. Therefore, within comparatively narrow limits this formula may be used to compare the ceilings given by different wings when the wing areas are adjusted so as to give the same landing speed in each case. Since the efficiency at cruising speed, and the speed range of these two wings are nearly the same, the U.S.A.-5 section is recommended as a good alternate to the U.S.A.-15, providing the structural proportions of the airplane are such that the smaller spar depth afforded by the U.S.A.-5 is sufficient.

There are also types of airplanes which are designed for special purposes and a general purpose wing is not required. In a racing airplane the value of $L/D$ at $K_y$, .00035, is almost the sole measure of the suitability of the wing. In an airplane intended for very high ceiling, maximum value of $K_y^{3/2}/K_x$ is the important factor. For an "all purpose" airplane of moderate size the U.S.A.-15 is recommended as the best "thin" section from both an aerodynamical and structural point of view.

The disposition of wing area (whether a monoplane, biplane or triplane is used, and the aspect ratio, gap, stagger, etc.) has a definite effect in modifying the $L/D$ given by model tests. Quantitative studies of these combinations lead to the following conclusions:

I. The aspect ratio of the wing should be as great as is consistant with structural strength, visibility and other requirements. With increasing aspect ratio:
   (*a*) The lift improves steadily for all of the flying angles.
   (*b*) Maximum lift occurs at a smaller angle.
   (*c*) L/D improves steadily.

II. The gap-chord ratio should be as large as possible. With increasing gap-chord:
   (*a*) Lift improves steadily.
   (*b*) L/D improves, especially at low angles.

III. As large a positive stagger as is consistent with structural and balance requirements should be used. For by increase of stagger:
   (*a*) Lift improves.
   (*b*) L/D improves at low angles.
   (*c*) c.p. movement is decreased and made more nearly stable.

## *TABLE III*

### *U.S.A. 5*

M.I.T., March 17, 1917
Velocity—30 M.P.H. in all tests
All Models—18″ x 3″

| Characteristics | | | | Ordinates | | |
|---|---|---|---|---|---|---|
| i | $K_y$ | L/D | C.P. | Distance from L.E. | Lower Ordinate | Upper Ordinate |
| —4 | —.000326 | —1.58 | —— | 0.00 | .62 | .62 |
| —2 | .000346 | 3.64 | .753 | 1.25 | .13 | 2.10 |
| 0 | .000910 | 12.28 | .498 | 2.50 | .03 | 3.03 |
| 2 | .001355 | 15.72 | .415 | 3.75 | .00 | |
| 4 | .001740 | 15.98 | .348 | 5.0 | .03 | 4.40 |
| 6 | .002120 | 14.80 | .327 | 7.5 | .25 | 5.40 |
| 8 | .002470 | 13.52 | .315 | 10. | .57 | 6.20 |
| 10 | .002870 | 12.08 | .303 | 20. | 1.55 | 7.92 |
| 12 | .003130 | 10.84 | .300 | 30. | 2.02 | 8.30 |
| 14 | .003285 | 9.25 | .288 | 40. | 2.17 | 8.14 |
| 16 | .003205 | 7.63 | .298 | 50. | 1.96 | 7.55 |
| 18 | .003150 | 4.57 | .330 | 60. | 1.56 | 6.75 |
| | | | | 70. | 1.16 | 5.63 |
| | | | | 80. | .76 | 4.24 |
| | | | | 90. | .55 | 2.52 |
| | | | | 100. | .25 | .25 |

Radius L.E. = .36
Radius T.E. = .25

### *U. S. A. 15*

M.I.T., November, 1917

| i | $K_y$ | L/D | C.P. | Distance from L.E. | Lower Ordinate | Upper Ordinate |
|---|---|---|---|---|---|---|
| —4 | —.00035 | —2.8 | | 0.00 | 1.06 | 1.06 |
| —2 | .00008 | 1.4 | | 1.25 | .46 | 2.56 |
| 0 | .00058 | 9.97 | .480 | 2.50 | .33 | 3.44 |
| 2 | .00108 | 15.50 | .407 | 5.0 | .15 | 4.80 |
| 4 | .00147 | 16.50 | .353 | 7.5 | .03 | 5.58 |
| 6 | .00183 | 15.25 | .330 | 10. | .00 | 6.11 |
| 8 | .00220 | 13.95 | .320 | 15. | .12 | 6.88 |
| 10 | .00257 | 12.35 | .310 | 20. | .33 | 7.28 |
| 12 | .00290 | 11.30 | .303 | 30. | .86 | 7.61 |
| 14 | .00301 | 9.43 | .297 | 40. | 1.00 | 7.55 |
| 16 | .00286 | 4.98 | .340 | 50. | .72 | 7.11 |
| | | | | 60. | .28 | 6.36 |
| | | | | 70. | .00 | 5.32 |
| | | | | 80. | .11 | 3.90 |
| | | | | 90. | .23 | 2.50 |
| | | | | 100. | .50 | .83 |

Radius L.E. = .51
Radius T.E. = .20

## *TABLE III (Continued)*

### *U.S.A. 16*

M.I.T., October 15, 1917

| Characteristics | | | | Ordinates | | |
|---|---|---|---|---|---|---|
| i | $K_y$ | L/D | C.P. | Distance from L.E. | Lower Ordi-nate | Upper Ordi-nate |
| —2 | —.00003 | —0.5 | .61 | 0.0 | .29 | .29 |
| 0 | .00035 | 7.5 | .39 | 2.5 | —.64 | 2.99 |
| 2 | .00092 | 18.9 | .335 | 5.0 | —.76 | 3.88 |
| 4 | .00135 | 16.7 | .315 | 7.5 | —.83 | 4.52 |
| 6 | .00170 | 15.0 | .300 | 10. | —.83 | 4.95 |
| 8 | .00209 | 13.7 | .290 | 15. | —.76 | 5.42 |
| 10 | .00241 | 11.6 | .285 | 20. | —.51 | 5.67 |
| 12 | .00252 | 7.0 | .300 | 30. | —.06 | 5.86 |
| 14 | .00247 | 4.7 | .330 | 40. | .00 | 5.74 |
| 16 | .00234 | 3.5 | .37 | 50. | —.19 | 5.29 |
| | | | | 60. | —.70 | 4.65 |
| | | | | 70. | —.76 | 4.01 |
| | | | | 80. | —.70 | 3.25 |
| | | | | 90. | —.45 | 2.23 |
| | | | | 100. | .43 | .43 |

Radius L.E. = .57
Radius T.E. = .43

### *U.S.A. 27*

M.I.T., March 3, 1920

| i | $K_y$ | L/D | C.P. | Distance from L.E. | Lower Ordi-nate | Upper Ordi-nate |
|---|---|---|---|---|---|---|
| —6 | —.00047 | —2.5 | | 0.00 | 1.77 | 1.77 |
| —4 | .00012 | 1.1 | | 1.25 | .50 | 3.80 |
| —2 | .00062 | 8.0 | 56.5 | 2.50 | .33 | 5.07 |
| 0 | .00099 | 13.6 | 43.3 | 5.0 | .17 | 6.93 |
| 2 | .00140 | 16.1 | 38.3 | 7.5 | .10 | 8.22 |
| 4 | .00179 | 15.7 | 36.0 | 10. | 0.00 | 9.17 |
| 6 | .00213 | 14.6 | 34.7 | 15. | .10 | 10.50 |
| 8 | .00251 | 13.4 | 34.0 | 20. | .35 | 11.33 |
| 10 | .00284 | 12.0 | 33.3 | 30. | .95 | 11.90 |
| 12 | .00319 | 10.8 | 33.0 | 40. | 1.17 | 11.57 |
| 14 | .00346 | 9.9 | 33.3 | 50. | .80 | 10.77 |
| 16 | .00359 | 8.6 | 33.7 | 60. | .25 | 9.52 |
| 18 | .00350 | 6.8 | 34.0 | 70. | .10 | 8.00 |
| | | | | 77. | 0.00 | —— |
| | | | | 80. | .05 | 6.03 |
| | | | | 90. | .15 | 3.65 |
| | | | | 100. | .67 | .67 |

Radius L.E. = 1.06
Radius T.E. = .21

## *TABLE III (Continued)*

### *U.S.A.T.S. 5*

M.I.T., February, 1919

| Characteristics | | | | Ordinates | | |
|---|---|---|---|---|---|---|
| i | $k_y$ | L/D | C.P. | Distance from L.E. | Lower Ordi-nate | Upper Ordi-nate |
| —8 | .00017 | 1.8 | | 0.0 | 2.50 | 2.50 |
| —6 | .00051 | 5.8 | .91 | 2.5 | — .80 | 5.69 |
| —4 | .00088 | 9.6 | .62 | 5.0 | —1.80 | 7.50 |
| —2 | .00126 | 11.8 | .50 | 10. | —3.00 | 10.10 |
| 0 | .00164 | 12.6 | .44 | 15. | —3.40 | 12.00 |
| 2 | .00200 | 12.8 | .40 | 20. | —3.50 | 13.30 |
| 4 | .00237 | 12.2 | .38 | 30. | —2.90 | 14.60 |
| 6 | .00271 | 11.3 | .36 | 40. | —1.50 | 14.68 |
| 8 | .00303 | 10.5 | .355 | 50. | — .60 | 13.80 |
| 10 | .00336 | 9.9 | .35 | 60. | — .40 | 12.40 |
| 12 | .00362 | 9.3 | .34 | 70. | — .20 | 10.40 |
| 14 | .00385 | 8.5 | .33 | 80. | .00 | 7.76 |
| 16 | .00379 | 7.3 | .32 | 90. | .00 | 5.00 |
| | | | | 100. | 1.00 | 1.00 |

Radius L.E. = 3.00
Radius T.E. = 1.00

### *R.A.F. 15*

M.I.T., December 30, 1919

| i | $k_y$ | L/D | C.P. | Distance from L.E. | Lower Ordinate | Upper Ordinate |
|---|---|---|---|---|---|---|
| —2 | —.00004 | | | 0.00 | .27 | .27 |
| 0 | .00033 | 8.1 | .443 | 1.25 | — .47 | 1.86 |
| 2 | .00085 | 16.6 | .367 | 2.5 | — .70 | 2.78 |
| 4 | .00122 | 16.5 | .327 | 5.0 | — .93 | 3.95 |
| 6 | .00160 | 15.4 | .310 | 7.5 | —1.00 | 4.62 |
| 8 | .00194 | 13.4 | .300 | 10. | —1.00 | 5.05 |
| 10 | .00228 | 11.9 | .297 | 15. | — .78 | 5.53 |
| 12 | .00260 | 10.7 | .290 | 20. | — .50 | 5.75 |
| 14 | .00240 | 5.0 | .350 | 30. | — .06 | 5.83 |
| | | | | 40. | — .03 | 5.63 |
| | | | | 50. | — .20 | 5.30 |
| | | | | 60. | — .50 | 4.80 |
| | | | | 70. | — .67 | 4.13 |
| | | | | 80. | — .58 | 3.26 |
| | | | | 90. | — .35 | 2.23 |
| | | | | 95. | — .16 | 1.74 |
| | | | | 100. | .42 | .42 |

Radius L.E. = .45
Radius T.E. = .42

29. *Characteristics of Military Airplanes*—In Table IV are given data on most of the best airplanes of various types of both American and foreign design. The main purpose of the table is to present the structural characteristics and loadings of these airplanes and show the performance obtained. The horsepower loading is somewhat misleading, except in comparing airplanes with the same type of engine, because the loading is based on the rated power of the engines rather than their maximum, and the margin between the rated and maximum horsepowers varies considerably with different engines. Furthermore, even with a particular engine, the type of propeller that is used makes a very appreciable difference in the performance of the airplane. It is only by eliminating differences in their power plants through the use of a similar propeller and engine that the performance of two airplanes can be fairly compared. In columns 7-9 where two values are given the second value is for the lower wing.

30. *Structural Weight Analysis of Airplanes*—Table V is a tabulation of weights for several different types of airplanes which will aid in preparing preliminary weight estimates for new designs. The total weight is divided into several main groups, such as power plant, fuel and oil, etc., and each group is expressed as a percentage of the total weight.

P = Total weight in lbs. of engine or engines, including carburetors, ignition system, self-starter, radiator and connections, water, propeller and propeller hub.

F = Weight of the capacity of gasoline plus oil.

U = Weight of crew and all movable equipment such as bombs and racks, ammunition, baggage, guns, wireless, sights ,and camera.

B = Weight of fuselage complete, with all it contains except *P*, *F*, and *U*. This weight includes structure, doped covering, cowling, tanks, engine mounting, seats, controls, instruments, gun mounts, tail skid, etc. In case of multi-engined airplanes, it includes weight of cowling and tanks.

W = Total weight of main wings, including ailerons, interplane struts, wiring, fittings, center panel and center panel bracing, drag bracing tip skids, etc.

C = Total weight of landing gear—everything beyond fuselage and wings.

E = Total weight of empennage, including bracing exterior to the fuselage.

31. *Load on Wings and Weight Per Horsepower*—The loadings for different types of airplanes of American, English, French, Italian and German design are given in Table IV and the table on page 44 and may serve as an approximate guide in the selection of the wing area or of the size of engine to use.

TABLE 4 CHARACTERISTICS OF MILITARY AIRPLANES

| No. | Name of Airplane & Engine | Type | Function | Weight Empty * | Useful Load | Gross Weight | Span | Chord | Aspect Ratio | Gap | Horse-power | Lbs./sq.ft. | Lbs./H.P. | Ground Speed | Service Ceiling |
|---|---|---|---|---|---|---|---|---|---|---|---|---|---|---|---|
|  | 1 | 2 | 3 | 4 | 5 | 6 | 7 | 8 | 9 | 10 | 11 | 12 | 13 | 14 | 15 |
| 1 | Thomas-Morse, Hispano "H" | MB-3 | 1-Pursuit | 1505 | 590 | 2095 | 26.0 | 5.25 | 4.95 | 4.58 | 300 | 8.3 | 7.0 | 152 | 23,700 |
| 2 | *Ordnance, Hispano "H" | D | 1-Pursuit | 1775 | 655 | 2430 | 30.0 | 5.00 | 6.00 | 4.33 | 300 | 9.3 | 8.1 | 147 | 22,000 |
| 3 | Verville, Hispano "H" | VCP-1 | 1-Pursuit | 1980 | 640 | 2620 | 32.0 | 4.75 (mean) | 6.75 | 4.5 | 300 | 9.7 | 8.75 | 150 | 22,400 |
| 4 | Loening Monoplane, Hispano "H" | M-8 | 1-Pursuit | 1665 | 975 | 2640 | 32.75 | 7.00 | 4.68 |  | 300 | 12.3 | 8.8 | 143.5 | 18,500 |
| 5 | Pomilio, Liberty "8" | FVL-8 | 1-Pursuit | 1725 | 560 | 2285 | 26.67 | 5.50 | 4.9 | 5.25 | 290 | 8.05 | 7.9 | 133 |  |
| 6 | S.E.5A (Amer.) Hispano "E" |  | 1-Pursuit | 1485 | 575 | 2060 | 26.75 | 5.0 | [illegible] | 4.83 | 180 | 8.4 | 11.4 | 122 | 20,400 |
| 7 | Spad Herbemont, Hispano | 200-2 | 2-Pursuit<br>1-Pursuit | 1865<br>1865 | 1015<br>790 | 2880<br>2655 | 31.9<br>28.41 | 5.74<br>5.41 | 5.55<br>5.25 | 5.10 | 300 | 8.9<br>8.2 | 9.6<br>8.85 | 142<br>146 | 23,000<br>28,200*** |
| 8 | Spad, Hispano 8-BEC | 13C-1 | 1-Pursuit | 1255 | 560 | 1815 | 26.25 | 4.92 | 5.35 |  | 220 | 8.45 | 8.25 | 138 | 22,300 |
| 9 | Nieuport, Gnome Monovalve | 28 | 1-Pursuit | 960 | 575 | 1535 | 26.5<br>25.5 | 4.25<br>3.28 | 6.25<br>7.75 | 3.92 | 150 | 8.9 | 10.2 |  |  |
| 10 | *Nieuport, Hispano 8-FB | 29C-1 | 1-Pursuit | 1680 | 745 | 2425 | 31.83 |  |  |  | 300 | 8.35 | 8.1 | 140 | 24,900 |
| 11 | Spad, Hispano 8-AB | 7C-1 | 1-Pursuit | 1100 | 450 | 1550 | 25.75<br>25.0 | 5.17<br>4.58 | 5.0<br>5.45 | 3.58 | 200 | 8.0 | 7.2 | 124 | 21,000 |
| 12 | Gourdou Monoplane Hispano "E" | C-1 | 1-Pursuit | 1440 | 410 | 1850 | 29.53 | 6.17 | 4.8 |  | 180 | 10.3 | 10.2 | 152 |  |
| 13 | Morane Monopl. Monos. Gnome | AIC1 | 1-Pursuit | 925 | 500 | 1425 | 27.75 | 5.33 | 5.2 |  | 150 | 9.8 | 9.5 | 140 | 27,000 |
| 14 | S.V.A., Isotta Fraschini V-6 |  | 1-Pursuit | 1590 | 720 | 2310 | 30.0 | 5.05 | 5.95 | 5.86 | 260 | 8.8 | 8.9 | 147 |  |
| 15 | S.V.A., - S.P.A. Engine |  | 1-Pursuit | 1445 | 530 | 1975 | 24.21 | 5.25 | 5.0 |  | 220 | 8.5 | 8.98 | 143 |  |
| 16 | Ansaldo, - S.P.A. Engine | 1 | 1-Pursuit | 1470 | 495 | 1965 | 25.0 |  |  |  | 220 | 8.7 | 8.9 | 137 |  |
| 17 | *Sopwith, Hispano 8Fb | Dolphin C-1 | 1-Pursuit | 1515 | 790 | 2305 | 32.5 | 4.50 | 7.2 | 4.25 | 300 | 8.75 | 7.7 | 139** | 24,600 |
| 18 | Sopwith, Clerget | Camel | 1-Pursuit | 975 | 510 | 1485 | 28.0 | 4.50 | 6.2 | 4.50 | 130 | 6.45 | 11.4 | 119 | 19,000 |
| 19 | Sopwith, BR-2 | Salamander | 1-Armored French Fighter | 1845 | 670 | 2515 | 30.67 | 5.00 | 6.13 |  | 200 | 9.4 | 12.6 | 126 | 13,000 |
| 20 | Sopwith BR-2 | Snipe 7.F.1. | 1-Pursuit | 1210 | 755 | 1965 | 30.0 | 5.00 | 6.0 | 4.25 | 200 | 7.2 | 9.8 | 124.5 | 20,000 |
| 21 | Martinsyde Rolls R. Falcon |  | 1-Pursuit | 1790 | 535 | 2325 | 31.0 |  |  |  | 220 | 6.9 | 10.5 | 138** | 22,800 |
| 22 | *Nieuport, ABC Radial | Night Hawk | 1-Pursuit | 1365 | 735 | 2100 | 28.0 | 5.25 | 5.35 | 4.5 | 320 | 7.75 | 6.6 | 152 | 23,000 |
| 23 | Pfalz, Mercedes | G.141 | 1-Pursuit | 1530 | 525 | 2055 | 30.88<br>26.64 | 5.44<br>3.92 | 5.7<br>6.8 | 4.67 | 160 | 8.55 | 12.8 | 112 |  |
| 24 | Fokker, Mercedes | D-7 | 1-Pursuit | 1560 | 445 | 2005 | 29.29<br>22.92 | 5.25<br>3.98 | 5.6<br>5.75 | 4.53 | 180 | 8.1 | 11.1 | 117 |  |
| 25 | *U.S.X. B1 - A, Hispano "H" |  | 2-Night Observation | 2010 | 985 | 2995 | 39.33 | 5.50 | 7.15 | 5.42 | 300 | 7.4 | 10.0 | 124 | 20,900 |
| 26 | LePere, Liberty "12" | U.S.A. C-11 | 2-Fighter & Observation | 2560 | 1185 | 3745 | 41.50 | 5.48 | 7.6 | 5.05 | 400 | 9.05 | 9.35 | 135 | 20,200 |
| 27 | U.S.A. D-4 |  | 2-Observation | 2390 | 1190 | 3580 | 42.58 | 5.50 | 7.75 | 5.5 | 400 | 8.15 | 8.95 | 125 | 20,000 |
| 28 | LePere Triplane, 2-Liberty "12s" | U.S.A. C-11 | 3-Army Corps Observation | 5455 | 3120 | 8575 | 54.5 | 5.73 | 9.5 | 5.17 | 800 | 9.88 | 10.7 | 112 | 15,800 |
| 29 | Breguet, Liberty "12" | 14-A2 | 2-Day Bomber & Reconn. | 2390 | 1380 | 3770 | 47.25<br>40.67 | 6.50<br>6.25 | 7.2<br>6.5 | 5.58 | 400 | 7.15 | 9.4 | 130 | 21,800 |
| 30 | *Breguet, Renault 12Kd | 17-C2 | 2-Day Bomber & Reconn. | 2660 | 1410 | 4070 | 44.85 |  |  |  | 450 | 8.5 | 9.05 | 137 | 19,700 |
| 31 | Spad, Hispano 8-BEC | 11-A2 | 2-Observation | 1485 | 825 | 2310 | 36.67 | 4.58 | 8.0 |  | 220 | 7.15 | 10.5 | 115 | 21,000 |
| 32 | Salmson, Salmson C.U.9Z | 4-AB2 | 2-Armored French Fighter | 3255 | 1155 | 4410 | 49.9 | 6.8 | 7.35 |  | 230 | 7.25 | 19.1 | 92.5 |  |
| 33 | Salmson, Salmson C.U.9Z | 2-A2 | 2-Observation | 1675 | 1125 | 2800 | 38.5 | 5.51 | 7.0 | 5.58 | 230 | 7.0 | 12.2 | 119 | 20,500 |
| 34 | *Morane Saulnier, Liberty "12" | 22-C2 | 2-Observation | 2440 | 1245 | 3905 |  |  |  |  | 400 | 8.6 | 9.75 | 134.5 | 19,600 |
| 35 | Rumpler, Mercedes | G.117 | 2-Observation & Fighter | 2440 | 1000 | 3440 | 41.5<br>40.0 | 5.67<br>4.33 | 7.3<br>9.2 | 6.00 | 260 | 9.45 | 13.2 | 110 |  |
| 36 | L.V.G., Benz | C-6 | 2-Observation & Fighter | 2090 | 950 | 3040 | 42.75<br>40.71 | 5.25<br>4.65 | 8.15<br>8.75 | 4.75 | 220 | 8.55 | 13.8 |  |  |
| 37 | *Halberstadt, Benz | C-5 | 2-Observation & Fighter | 2060 | 950 | 3010 | 42.08 | 5.33 | 7.9 | 5.08 | 220 |  | 13.7 |  |  |
| 38 | Halberstadt, Mercedes | CL-2 | 2-Fighter | 1755 | 815 | 2570 | 35.27<br>34.92 | 5.27<br>4.29 | 6.7<br>8.1 | 3.85 | 180 | 8.3 | 14.3 |  |  |
| 39 | U.S. D-9A, Liberty "12" |  | 2-Observation & Day Bomber | 2815 | 1505<br>2055 | 4320<br>4870 | 45.92 | 5.75 | 8.0 | 6.12 | 400 | 8.8<br>9.9 | 10.8<br>12.2 | 126<br>121.5 | 18,700<br>14,400 |
| 40 | Glen Martin, 2-Liberty "12s" |  | 3-Day Bomber | 6700 | 3525 | 10225 | 71.42 | 7.92 | 9.0 | 8.33 | 800 | 9.55 | 12.8 | 105 | 10,300 |
| 41 | Pomilio, Liberty "12" | PVL-12 | 2-Day Bomber | 2825 | 1725 | 4550 | 48.25 | 6.08 | 7.98 | 5.90 | 400 | 7.9 | 11.4 | 111 | 15,700 |
| 42 | Vickers Vimy, Maori Sunbeams | FB-27a | 3-Night Bomber | 6735 | 3565 | 10300 | 67.17 | 10.5 | 6.4 | 10.17 | 500 | 7.8 | 20.6 | 92 | 8,000 |
| 43 | *Vickers Vimy, 2-Liberty "12s" |  | 3-Night Bomber | 6700 | 5400 | 12100 | 67.17 | 10.5 | 6.4 | 10.17 | 800 | 9.15 | 15.1 | 110 | 10,500 |
| 44 | Handley-Page, 4 R.R. Eagle "8" | V-1500 | 7-Night Bomber | 15210 | 8490 | 24700 | 125.5 |  |  |  | 1440 | 8.55 | 17.1 | 102 |  |
| 45 | Handley-Page, 2-Liberty "12s" |  | 4-Night Bomber | 8720 | 5705 | 14425 | 100.0 | 10.0 | 10.0 | 11.0 | 800 | 8.7 | 18.0 | 94 | 7,400 |
| 46 | Caproni Biplane, 3-Liberty "12s" (Navy) | CA-5 | 4-Night Bomber | 7700 | 4650 | 12350 | 76.77 | 9.12 | 8.4 | 9.12 | 1050 | 6.7 | 11.7 | 103 |  |
| 47 | Caproni Biplane, 3-Fiats A-12 Bis | Ca-5 | 3-Night Bomber | 6930 | 4400 | 11330 | 76.77 | 9.12 | 8.4 | 9.12 | 900 | 8.0 | 12.6 | 100 |  |
| 48 | Caudron, 2-Salmson C.U.9Z | C-23 | Night Bomber | 5050 | 4020 | 9070 | 77.0<br>73.0 | 8.29<br>6.71 | 9.3<br>10.9 |  | 460 | 8.4 | 19.7 | 95 |  |
| 49 | A.E.G., 2-Mercedes | 4 | 3-Night Bomber | 5390 | 3080 | 8470 | 60.12<br>51.87 | 7.64<br>7.33 | 7.95<br>7.1 | 8.0 | 520 | 11.6 | 16.3 | 90 |  |
| 50 | Gotha, 2-Mercedes | G-5 | 4-Night Bomber | 6040 | 2720 | 8760 | 77.0<br>71.75 | 7.85 | 10.5<br>9.75 | 7.0 | 520 | 8.9 | 16.8 | 75** |  |
| 51 | Friedrichshaven, 2-Mercedes | G-3 | 4-Night Bomber | 5930 | 2715 | 8645 | 78.0 | 7.67 | 10.2 | 7.0 | 520 | 9.25 | 16.6 |  |  |
| 52 | *Vought, Hispano "E" | VE-7 | 2-Advanced Training | 1560 | 535 | 2095 | 34.11 | 4.62 | 7.35 | 4.67 | 180 | 7.35 | 11.6 | 114 | 17,000 |
| 53 | Ordnance, LeRhone | C | 1-Advanced Training | 835 | 280 | 1115 | 26.0 | 4.0<br>3.75 | 6.5<br>6.95 | 3.83 | 80 | 5.85 | 14.0 | 98 | 13,500 |
| 54 | Thomas-Morse, Gnome | S-4C | 1-Training | 960 | 395 | 1355 | 26.54 | 5.50<br>4.25 | 4.83<br>6.25 | 4.50 | 100 | 5.78 | 13.5 | 100 |  |
| 55 | Standard, LeRhone | E-1 | 1-Training | 830 | 315 | 1145 | 24.0 | 3.50 | 6.85 | 4.00 | 80 | 7.5 | 14.3 | 100 | 14,800 (Absolute) |
| 56 | Curtiss, Hispano "A" or "I" | JN-6HB<br>JN-4H | 2-Training | 1795<br>1595 | 890<br>550 | 2685<br>2145 | 43.67<br>34.71 | 4.94 | 8.8<br>7.0 | 5.17<br>5.17 | 150 | 7.6<br>6.1 | 17.9<br>14.3 | 79<br>98 | 5,700<br>8,000 |
| 57 | Curtiss, Curtiss O.X.5 | JN-4 D2 | 2-Training | 1525 | 490 | 2015 | 43.67<br>34.71 | 4.94 | 8.8<br>7.0 | 5.17 | 90 | 5.7 | 22.4 | 75 | 6,500 |
| 58 | *Avro, LeRhone | 504-K | 2-Training | 1230 | 600 | 1830 | 36.0 | 4.83 | 7.45 | 5.50 | 110 | 5.55 | 16.6 | 90 |  |
| 59 | Nieuport, LeRhone | 24 | 1-Advanced Training | 780 | 425 | 1205 | 27.0<br>25.42 | 4.84<br>2.88 | 5.95<br>11.0 | 3.92 | 110 | 7.5 | 10.9 | 108 | 22,500 |

* Includes water
* These airplanes are of especial merit.
** Speed at 10,000 feet.
*** Stripped of all unnecessary equipment.

*TABLE V—WEIGHT ANALYSIS OF AIRPLANES*

| Group | Type | Gross Weight | Weight Empty (No Water) | % | P | % | F | % | U | % | B | % | W | % | C | % | E | % |
|---|---|---|---|---|---|---|---|---|---|---|---|---|---|---|---|---|---|---|
| 1 | 2 | 3 | 4 | 5 | 6 | 7 | 8 | 9 | 10 | 11 | 12 | 13 | 14 | 15 | 16 | 17 | 18 | 19 |
| 1 | Curtiss JN-4D2. | 2016 | 1475 | 73.2 | 537 | 26.6 | 151 | 7.5 | 340 | 16.9 | 430 | 21.3 | 407 | 20.2 | 91 | 4.5 | 60 | 3.0 |
| 2 | Vought VE-7... | 2095 | 1479 | 70.6 | 679 | 32.4 | 196 | 9.4 | 350 | 16.7 | 418 | 20.0 | 315 | 15.0 | 95 | 4.5 | 52 | 2.5 |
| 3 | Orenco "C" .... | 1117 | 835 | 74.7 | 285 | 25.5 | 121 | 10.8 | 161 | 14.4 | 301 | 26.9 | 150 | 13.4 | 77 | 6.9 | 21.5 | 1.9 |
| 4 | Orenco "D" .... | 2432 | 1660 | 68.3 | 886 | 36.5 | 339 | 13.9 | 317 | 13.0 | 438 | 18.0 | 307 | 12.6 | 106 | 4.4 | 39 | 1.6 |
| 5 | Thomas-Morse MB-3 ....... | 2094 | 1409 | 67.3 | 860 | 41.1 | 271 | 12.9 | 317 | 15.1 | 283 | 13.5 | 250 | 11.9 | 70 | 3.4 | 43.5 | 2.1 |
| 6 | Verville VCP-1. | 2617 | 1872 | 71.7 | 926 | 35.4 | 315 | 12.1 | 322 | 12.3 | 542 | 20.7 | 381 | 14.5 | 90 | 3.4 | 41 | 1.6 |
| 7 | Loening M-8 ... | 2639 | 1565 | 59.3 | 837 | 31.7 | 333 | 12.6 | 643 | 24.4 | 417 | 15.8 | 300 | 11.4 | 65 | 2.5 | 44 | 1.7 |
| 8 | Pomilio FVL-8.. | 2284 | 1636 | 71.6 | 849 | 37.2 | 241 | 10.5 | 317 | 13.9 | 482 | 21.1 | 245 | 10.7 | 115 | 5.0 | 34.5 | 1.5 |
| 9 | U.S. XB-1A ... | 2994 | 1905 | 63.7 | 879 | 29.3 | 341 | 11.4 | 643 | 21.5 | 524 | 17.5 | 455 | 15.2 | 112 | 3.7 | 40 | 1.3 |
| 10 | Fokker D-7 .... | 2005 | 1470 | 73.4 | 860 | 42.9 | 195 | 9.7 | 249 | 12.4 | 268 | 13.4 | 287 | 14.3 | 110 | 5.5 | 36 | 1.8 |
| 11 | Pomilio BVL-12 | 4552 | 2697 | 59.2 | 1164 | 25.6 | 767 | 16.8 | 961 | 21.1 | 737 | 16.2 | 676 | 14.8 | 175 | 3.8 | 72 | 1.6 |
| 12 | U.S. D-9A ..... | 4872 | 2661 | 54.6 | 1195 | 24.5 | 933 | 19.1 | 1124 | 23.1 | 767 | 15.7 | 625 | 12.8 | 145 | 3.0 | 82.5 | 1.7 |
| 13 | LePere Biplane U.S.A. C-11 .. | 3746 | 2465 | 65.8 | 1156 | 30.9 | 484 | 12.9 | 700 | 18.7 | 627 | 16.7 | 578 | 15.4 | 135 | 3.6 | 65 | 1.7 |
| 14 | U.S.A. D-4 .... | 3582 | 2260 | 63.1 | 1179 | 32.9 | 457 | 12.7 | 734 | 20.5 | 438 | 12.2 | 551 | 15.4 | 147 | 4.1 | 76 | 2.1 |
| 15 | LePere Triplane U.S.A. O-11 .. | 8577 | 5239 | 61.6 | 2406 | 28.1 | 1792 | 20.9 | 1330 | 15.5 | 1433 | 16.7 | 1240 | 14.5 | 277 | 3.2 | 98.5 | 1.2 |
| 16 | Glenn Martin .. | 10225 | 6486 | 63.2 | 2337 | 22.8 | 1415 | 13.8 | 2108 | 20.6 | 2316 | 22.6 | 1604 | 15.6 | 285 | 2.8 | 160 | 1.6 |
| 17 | Handley-Page .. | 14425 | 8421 | 58.4 | 2462 | 17.1 | 2500 | 17.3 | 3204 | 22.2 | 2849 | 19.7 | 2550 | 17.7 | 660 | 4.6 | 200 | 1.4 |
| 18 | U.S. G.A.X. ... | 9813 | 7348 | 74.8 | 2469 | 25.2 | 735 | 7.5 | 1504 | 15.3 | 2987* | 30.5 | 1622 | 16.5 | 358 | 3.6 | 138 | 1.4 |

*This fuselage weight includes 2460 lbs. of armor on nacelles and fuselage.

| No. of Machines | Type | Weight per sq. ft. of Wing Area | | Weight Per H.P. | |
|---|---|---|---|---|---|
| | | Range | Average | Range | Average |
| 9 | One-place Pursuit...... | 6.2— 9.2 | 8.3 | 6.8—11.7 | 9.2 |
| 8 | Two-place Pursuit and Reconnaissance ........ | 6.9— 9.4 | 8.2 | 9.0—14.8 | 11.8 |
| 6 | Heavy Bombers ....... | 7.5—11.6 | 9.4 | 11.4—18.7 | 15.9 |

32. *Aspect Ratio, and Shape of Wing Tips*—The aspect ratio of airplanes is affected by four main factors: considerations of visibility, structural weight, and aerodynamic efficiency and stability. In general, a high aspect ratio is favorable as regards visibility and aerodynamic efficiency and stability, but unfavorable from the structural point of view. The relative importance of these different factors varies with the type of airplane. In a commercial airplane, for instance, visibility in landing is even more important than in a military type, but visibility in other directions is not so essential. If a low aspect ratio is decided upon, care should be taken to select a wing with as small a center of pressure travel as possible, in order to secure good stability. In monoplane construction, and particularly in internally braced wings, low aspect ratio is important, as it is in single bay airplanes which are close to the economical limit for single bay construction. With suitable wing tips the effect of aspect ratios greater than 5.0 on aerodynamic efficiency is slight. It will be noticed from Table V that the MB-3, one of the fastest airplanes built, has an aspect ratio of only 4.95. For an extreme low limit that might still give satisfactory results a value of 3.5 is suggested. Another reason for a very low aspect ratio and consequently short span is the necessity, with certain types of both military and commercial airplanes, of landing in restricted areas such as mountainous regions or roads. Short spans also simplify the problem of housing an airplane. A method for increasing the effective aspect ratio of a wing of a given span and area is to taper it in plan form. It is not the average chord, but rather the chord at a distance from the tip equal to about .8 of the chord at that point, that determines the aspect ratio.

Wind tunnel tests on the effect of the shape of wing tips show that the most efficient tip is in the form of a semi-ellipse. For the most part it is undesirable to have as long a tip as is necessitated by an ellipse, and a semi-circular shape is used to good effect. The chief disadvantage of either of these types is the difference in length between the front and rear spars. This makes it difficult to proportion the cantilever span and the inner bays economically. The rear spar will be weak at the outer strut point, and the front spar in the bay adjacent to the cantilever span. The wing with a square tip, just having the corners rounded off, remedies this rather serious defect. Another objection to the elliptical or circular tip is that with such a tip a considerable area is cut from an aileron just where it is most effective. One of the most efficient high speed airplanes, the Curtiss 18-B, has this square tip slightly modified. As has been suggested, the influence of the shape of the tip is more pronounced with low aspect ratios. Since but a slight aerodynamic advantage is obtained by elliptical or circular tips, especially when the aspect ratio of the wing is greater than 5.0, the decided structural advantages of the square or modified square tips shown in Fig. 26 make these latter types the more satisfactory. A tip that has recently come into favor and which has many advantages is one that is absolutely square in plan view. The full depth of the wing section is maintained out to the tip, and balsa fairing, with a radius at each point along the chord of the wing equal to half the wing ordinate at that point, rounds off the wing. Such a tip greatly simplifies the construction, since no special short ribs are used at the tip and the leading edge does not have to be bent around the tip. Such a tip reduces the high speed between one and two miles per hour which is negligible in many types of airplanes. The raked tip has in the past been largely used because of its supposed efficiency and the fact that it afforded aileron area where it was most needed. On the other hand, the structural disadvantages of the elliptical or circular tip are accentuated in a raked tip. There is also a backward movement of the center of pressure which makes these disadvantages even more pronounced. For these reasons the raked tip is not recommended.

33. *Location of Lift Wires in Inner Bay*—It was at one time customary in a staggered biplane to keep the lift wires in the plane of the struts. To counteract the large drag loads put into the drag trusses by the backward component of the stresses in the lift wires with this arrangement, external drag wires, carried forward to the nose of the fuselage, were used. The location of the wires shown in Fig. 27 is recommended as being superior to the old arrangement. The double, front lift wire is kept in a vertical plane or brought back slightly. The rear lift wires are split. One is carried forward and attached to the fuselage either at the point of attachment of the front wire or at the bulkhead to the rear of this point. The other rear wire remains in the plane of the struts or is secured to a bulkhead so located as to give the wire the desired backward slope.

In the high incidence condition when the lift stresses in the front truss are large and the anti-drag forces on the wings are a maximum, the backward component of the stress in the front wire and in the rearmost wire will produce a drag tension in the front spar that will relieve the heavy compression there. The forward component of the stress in the forward rear wire will partially neutralize this, but since the wings are tending to move forward slightly the stress in this wire will be lessened. On the other hand, for the low incidence condition the drag truss will deflect backward somewhat, thus relieving the stress in the rearmost wire and putting it in the other rear wire which on account of its large forward component reduces the compression in the rear truss.

One reason for not splitting the front wires and bringing one of them forward is that to attach a wire to an engine bulkhead would subject it to severe vibration. Especially in the case of streamline wires this

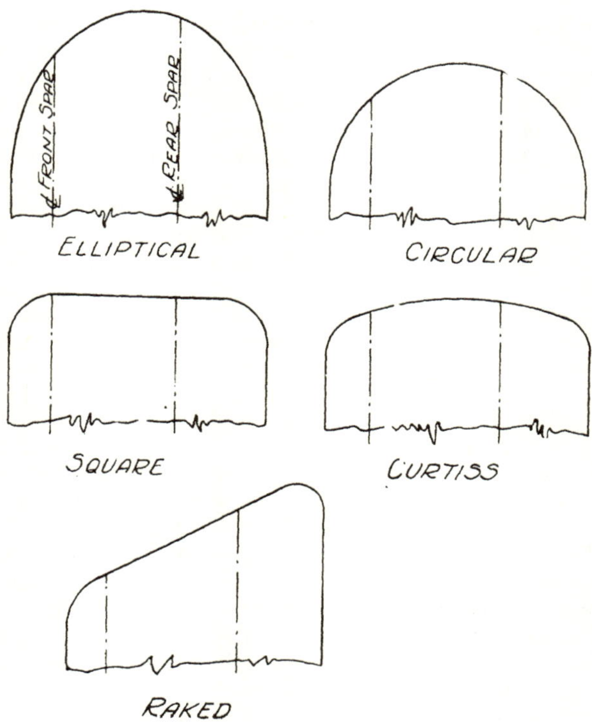


Fig. 26. Wing Tips

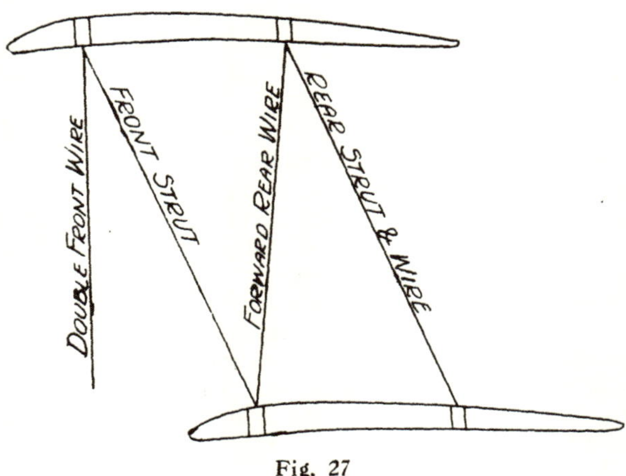


Fig. 27

would be injurious. Furthermore, locating a front wire at such point would result in a large drag compression in the front spar when it was most heavily stressed. The practice of using external drag wires is not recommended. They make the wing structure indeterminate, and are unnecessary. The drag truss is amply strong enough to care for any drag loads, and, in fact, usually does carry the entire anti-drag load which in a pursuit airplane is greater than the drag. However, if external drag bracing is eliminated, more care must be used in the design of the wing anchorage fittings than is necessary when drag wires are present. The center section struts, wires, and fittings must also be of ample strength.

34. *Location of Interplane Struts*—The struts should be so placed that the factors of safety of the spars at or near the strut points and in the span will be approximately equal. In an airplane of several bays the length of the panels decreases from the outer span toward the fuselage or center section. The purpose of this arrangement is to reduce the bending moment in the spars as the compression increases so as to keep the combined stresses nearly equal for the several panels. If there is a difference in the factors of safety for the panels, the inner panels should have a higher factor as they are more vital than the outer panels. It should be remembered that too great a strut spacing gives a flat slope to the flying wires with a consequent increase of the compression in the spars in addition to greater deflections. The combination of increased compression and deflection produces large secondary stresses.

As the maximum speed of an airplane increases, the elimination of interplane bracing becomes more and more important. In a comparatively low speed airplane an extra bay might well be used. This on a high-powered pursuit airplane would add far more to the structural resistance than would be saved in the weight of the members. In each case the problem must be solved by the application of the principle of equivalent weight (see Art. 99). The resistance and weight of extra interplane struts and wires must be balanced against the increased weight of the spars that is necessitated by the greater stresses which are the result of a reduction in the interplane bracing.

From a careful investigation of this subject in which the effect of aspect ratio, gap-chord ratio, stagger, and continuity of the spars were considered in connection with the monoplane, and the one, two and three bay biplane, the following deductions were arrived at and the values in Table VI determined.

I. When the spars are continuous at the center, the bays tend to increase in proportion to the total length, and the cantilever to decrease. This effect of the continuity of the spars is the most important factor in determining the location of the strut points in any given type of airplane.

II. A decrease in slope of the lift wires or monoplane struts will increase the proportionate length of the cantilevers. It is through this change of slope that a change in the aspect ratio or the gap-chord ratio affects the economic location of the strut points. In any case within the limits of practical designing the effect is not great, and where the spars are continuous over the center and there is no center span it is negligible.

III. Where the spars are continuous over the second joint from the left, the economical location of the struts is that which will make the factors of safety at this joint and the first joint from the left equal. If this second joint is not at the center line of the airplane, the proportion of the bay to the right of it and the bay to the left of it may be changed considerably without appreciably changing the strength of this joint. The proportions may be chosen so that the bay to the right or the bay to the left of this joint will be of the same strength as the joint, or both bays may be made stronger.

IV. Staggering the wings of a biplane does not materially, if at all, change the economical location of the strut points.

V. Owing to the absence of direct compression in the spars of the lower wing, it is the upper wing which governs the design. A change in the design of the lower wing will cause a change in the stress in the outer strut. This will affect the direct compressive stress in the bay 1—2 in the same manner as a change in the slope of the lift wires. The effect on the economical location of the strut is negligible for any likely difference in design of the two wings.

VI. If the rear spar be made longer than the front spar, the factor of safety at joint 1 of the rear spar will be decreased, and at joint 2 and in the bay 1—2 will be increased. The probability of using this type of construction was considered to some extent in recommending proportions. It is suggested that, if the rear spar is made the longer, the bays be proportioned on the basis of a mean spar with a length equal to that of the front spar plus one-third of the difference in the lengths of the two spars.

VII. The physical properties of the spar section have an influence in the economical location of the struts. The effect is small with small aspect ratios, but becomes appreciable with large aspect ratios such as are found in three bay biplanes.

In all cases the following data were assumed:

R.A.F. 15 wing section.

50 in. chord.

Front spar 13 per cent of chord back from leading edge.

Rear spar 67 per cent of chord back from leading edge.

Net wing loading 7.5 lbs. per sq. ft.

Location of center of pressure for high incidence condition at 29 per cent of chord.

Location of center of pressure for low incidence condition at 50 per cent of chord.

Square wing tips.

Decrease in tip loading as indicated in Fig. 34.

Spar Properties

| | Front | Rear |
|---|---|---|
| Area ........ | 2.15 | 2.08 |
| I ........... | 1.94 | 1.12 |
| I/y ......... | 1.34 | 0.99 |
| $\rho$ ........... | 0.95 | 0.735 |

Sections of these spars are shown in Fig. 28.

*TABLE VI—RECOMMENDED PROPORTIONS*

| | Canti-lever | Bay 1-2 | ½ Bay 2-3 | |
|---|---|---|---|---|
| Monoplane: | | | | |
| Without center span— | | | | |
| Hinged at center | | | | |
| Aspect ratio 4.0 .......... | 34½ | 65½ | 0 | |
| Aspect ratio 5.0 .......... | 35½ | 64½ | 0 | |
| Continuous at center | | | | |
| Aspect ratio 4.0 and 5.0... | 32½ | 67½ | 0 | |
| With center span— | | | | |
| Continuous throughout | | | | |
| Aspect ratio 5.0 ......... | 25 | 53½ | 21½ | |
| Biplane—Single Bay: | | | | |
| Without center span— | | | | |
| Hinged at center | | | | |
| Aspect ratio 4.0 and 6.0... | 36½ | 63½ | 0 | |
| Continuous at center | | | | |
| Aspect ratio 4.0 and 5.0... | 32½ | 67½ | 0 | |
| With center span of 30 in.— | | | | |
| Hinged at joint 2 | | | | |
| Aspect ratio 4.6 ......... | 36 | 64 | 0 | |
| Aspect ratio 6.6 ......... | 37 | 63 | 0 | |
| Continuous at joint 2 | | | | |
| Aspect ratio 5.6 ......... | 34 | 66 | 0 | |
| With center span greater than 30 in.— | | | | |
| Continuous throughout | | | | |
| Aspect ratio 4.0 and 5.3... | 25 | 51½ | 23½ | |
| Lower wing only— | | | | |
| Hinged at joint 2 | | | | |
| Aspect ratio 6.6 ......... | 32½ | 67½ | 0 | |
| Biplane—Two Bay: | | | Entire | Entire Bay 3-4 |
| With center span of 30 in.— | | | | |
| Hinged at joint 3 | | | | |
| Aspect ratio 7.0 ......... | 24½ | 45½ | 30 | |
| Biplane—Three Bay: | | | | |
| With center span of 30 in.— | | | | |
| Hinged at joint 4 | | | | |
| Aspect ratio 9.0 ......... | 20 | 40 | 25 | 15 |

*Note*:—The length of the cantilever in the above table is the actual cantilever length of the spars, not the loaded length used in sand test.

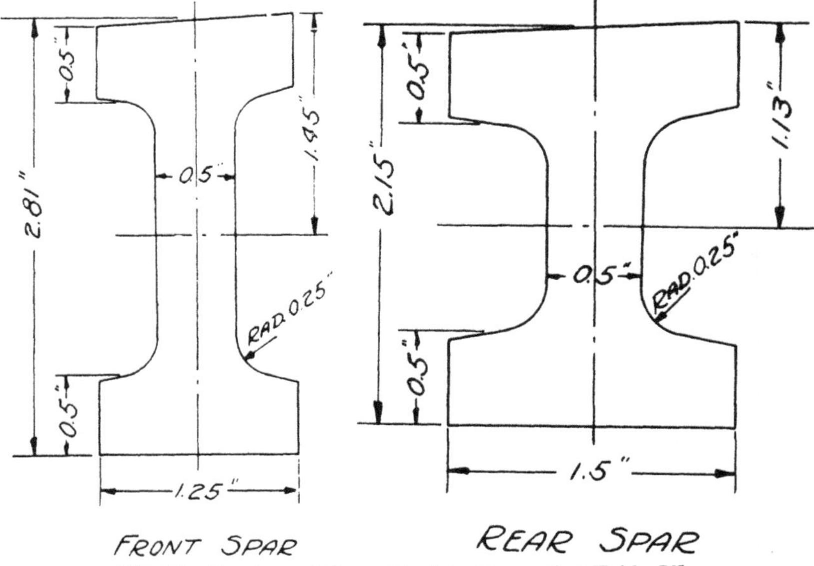


**Fig. 28. Sections of Spars Used in Computing Table VI**

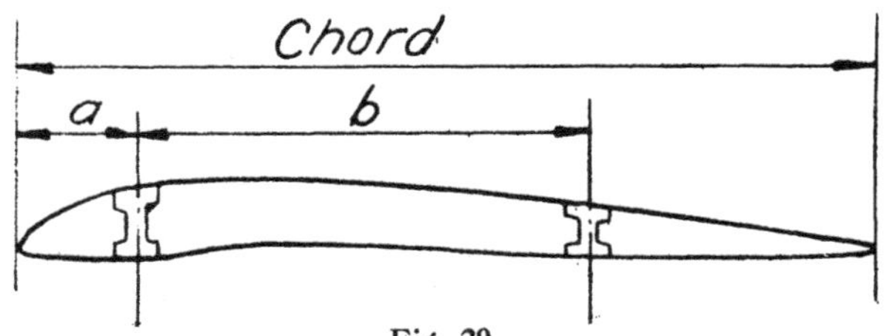


**Fig. 29**

35. *Location of Spars*—In the usual type of wing construction two spars are employed. They are generally located so as to receive about 70 per cent of the total load on the wing either at high or at low incidence flying conditions. Fig. 29 represents a wing section with two spars whose location will be determined under the assumption that each spar receives 70 per cent of the load. The center of pressure is assumed to be at .29 of the chord from the leading edge at high incidence and .50 of the chord from the leading edge at low incidence.

Low incidence condition:

$$\text{Load on rear spar} = \frac{.50L - a}{b} \times P = .70\,P$$

$$.50\,L - a = .70\,b$$

$$a + .70\,b = .50\,L \qquad (1)$$

High incidence condition:

$$\text{Load on front spar} = \frac{a + b - .29\,L}{b} \times P = .70\,P$$

$$a + b - .29\,L = .70\,b$$

$$(2) \qquad a + .30\,b = .29\,L$$

Solving (1) and (2)

$$a = .14\,L$$
$$b = .53\,L$$

An approximate location of the spars is then:

Front spar—1/7 of chord from leading edge.

Rear spar—2/3 of chord from leading edge.

These values are suitable for types of airplanes in common use at present. For very large airplanes the flying conditions will be such that the center of pressure will not go as far back as .50 of the chord. The maximum percentage of the load carried by the rear spar will therefore become less. Hence, there will be a tendency to move both spars forward somewhat in the wing section, the rear spar perhaps more than the front. This will help to equalize the loads on the spars. However, the rear spar depth will increase, and that of the front spar decrease, as the spars are brought forward. It may be noted that, although the overhang of the ribs at the trailing edge becomes large, the more forward location of the center of pressure at low incidence reduces the load on the rear part of the wing.

It is suggested that reasonable limits for the position of the front spar are 10 to 15 per cent of the chord from the leading edge, and for the rear spar, 60 to 70 per cent, the lower values being adapted to heavy airplanes.

36. *Torque of Engine*—With an airplane using two or more engines the torque is of some importance. In Fig. 30 let $T$ = torque of engine, $R$ = reaction on engine bearers, and $d$ = distance between bearers in inches. Let $P$ = horsepower developed by engine at "$n$" number of revolutions per minute.

$$T = \frac{33000\,P}{2\,\pi\,n}$$

$$R = \frac{T}{d} = \frac{12 \times 33000\,P}{2\,\pi\,nd} = 63000\,\frac{P}{nd} \text{ (in pounds)}$$

A simple rule for determining the directions of the reactions is that the direction of rotation of the couple, $Rd$, is the same as that of the engine shaft.

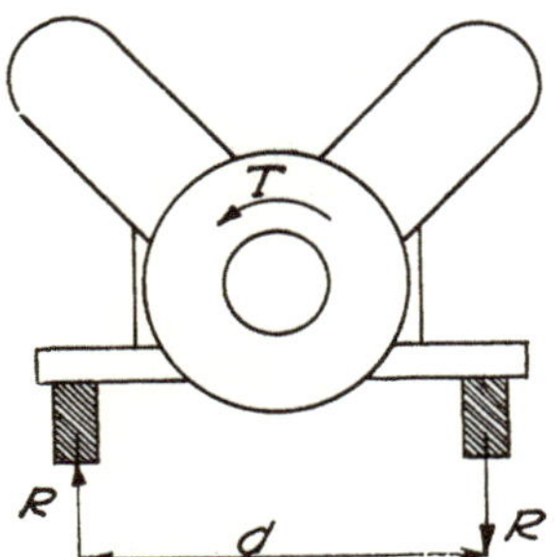


Fig. 30. Reactions from Engine Torque

37. *Method of Least Work*—The solution of stresses by statics can be accomplished only when there are no more than three unknown conditions. The existence of more than three unknowns makes it necessary to employ some method involving the elastic properties of the material. Such a method is the well-known principle of least work which states that "the deformation of any elastic body under the action of a balanced system of external forces will be such that the work done in causing the deformation, or the resilience of the body, is a minimum." This principle is explained in detail in Art. 116.

38. *Properties of Spar Sections*—In the case of most spars the top and bottom surfaces are seldom at right angles to the vertical axis. The cross-section of a typical spar is shown in Fig. 31 by the dotted lines. The computations of the properties of such a cross-section may be greatly simplified by assuming the section to be rectangular as indicated by the heavy full lines. The depth, *d*, is the center depth of the true section. The moment of inertia of the spar section will equal

$$I = \frac{wd^3 - c\,(d_1^3 + d_2^3)}{12}$$

When computing the section modulus either the value $y$ or $y^1$ should be used in the expression $I/y$ and not the value of $d/2$.

The area may be taken as the difference between that of the main rectangle and the areas of the shaded rectangles. $A = w \cdot d - (d_1 + d_2)c$.

The computation of the intensity of horizontal shear on any horizontal plane of the beam requires a property of the cross-section which is usually designated by the expression $Q/b\,I$.

Q = moment about the neutral axis of the area on the side of the plane (on which the horizontal shear is to be computed) away from the neutral axis.

b = width of the section at the plane under consideration.

I = moment of inertia of the whole cross-section.

The application of this formula is given in Art. 20. In computing the statical moment $Q$, the spar may be treated as rectangular, the same as in calculating the area or moment of inertia.

If it is desirable to take into account the fillets in the routing in computing the spar properties, this may be readily done by roughly computing the area of one fillet assuming it to be a triangle. The moment of inertia of the fillets equals their area times the sum of the squares of the distances from their centers of gravity to the vertical axis. Even without this correction the method given will have an error less than 1 per cent.

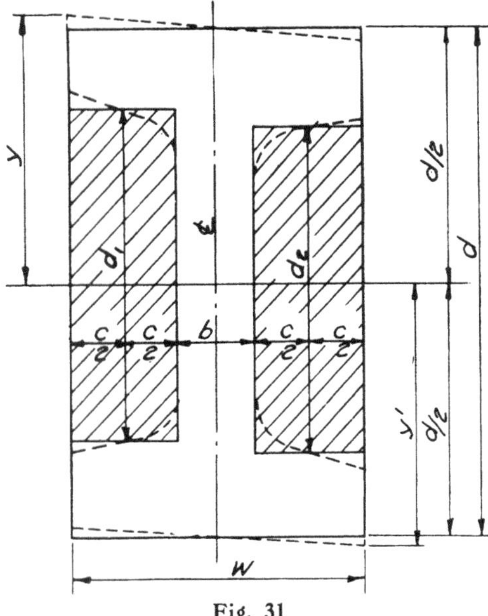


Fig. 31

39. *Eccentricity of Fittings.*—Fig. 32 illustrates a typical wing fitting on an upper spar. It is evident upon a little consideration that the downward pull from the lift wires is carried by bolt *A*. The true point of support for the spar is therefore at point *c* and not at the center line of the strut. The forces acting along the strut and the lift wire have eccentricities about point *c*, and consequently moments are produced in the spar. These moments should be considered in the design of the spar unless the fitting is so designed that the eccentricities are so small as to be negligible. Cases 4 and 5 of the three-moment equations treat this con-

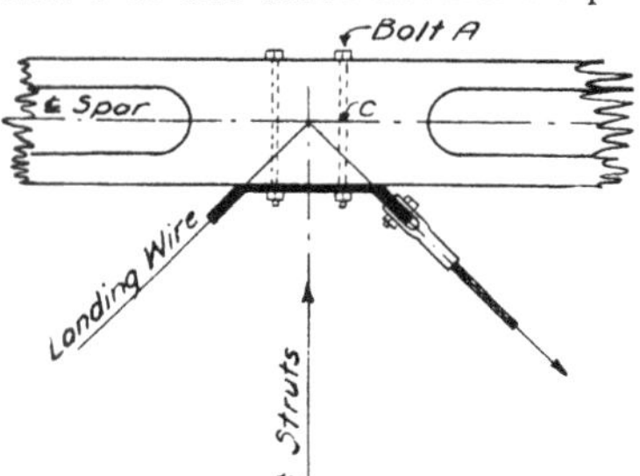


Fig. 32. Eccentric Wing Fitting

dition of moments at the panel points (see Art. 172). For the lower spar the reaction point is on the center line of the strut.

40. *Maximum Load on Wings*—Accelerometer tests have shown that the most severe loads on the wings occur in pulling sharply out of a steep dive. For a brief period of time the airplane is moving at a high, though decreasing velocity, in a path of decreasing radius of curvature. During the dive the angle of incidence is small, or even negative, which means that the center of pressure is far back on the wing. As soon as the airplane deviates from a straight path the angle of incidence begins to increase and the center of pressure to move forward. The maximum dynamic loads, due to the centrifugal force, occur at a high angle of incidence, close to the angle giving the maximum lift coefficient for the aerofoil. The greatest dynamic load that has yet been recorded by an accelerometer in a test is 4.2. It is possible that a load factor of 5.0 would be obtained in actual combat. The following discussion has reference to fast pursuit airplanes, and assumes that 4.2 is the maximum dynamic load and that it occurs at a high incidence when the center of pressure is about 30 to 31 per cent back from the leading edge. With an ordinary design in which the front spar is located at 12 per cent and the rear spar at 67 per cent of the chord, the front truss will carry 2/3 and the rear, 1/3 of the total load on the wings. Using these proportions the maximum load that can come on the front truss is 2.8 $W$ and on the rear truss 1.4 $W$, where $W$ is the gross weight of the airplane minus the weight of the wings. The necessary factor of safety is then arrived at as follows: If the resultant load $W$ is applied so that the front truss carries 2/3 $W$ and the rear 1/3 $W$ then the factor to produce loads of 2.8 W and 1.4 W is $2.8W/{}^2/_3W = 4.2$. To this is applied a material factor or true safety factor of about 2.0, making the total load factor 8.4. For the standard static test the required factor for pursuit airplanes is 8.5.

It is evident that the most severe condition for the front truss occurs when high incidence is combined with high speed, as in pulling out of a dive or in zooming. The corresponding condition for the rear truss will be either at high incidence, or in a straight dive when the resultant lift on the wings is zero or very small. It cannot occur in level flight at high speed, when the center of pressure is far back, because no dynamic effect can exist unless the airplane is traveling in a curved path in a vertical plane. With the high speed condition not more than 1.0 $W$ can come on the rear truss, and usually not more than .8 $W$ which would correspond with a center of pressure located at about 55 per cent of the chord. It is probable that, except for internally braced monoplanes, the critical flight condition for the rear truss is the same as for the front truss. The case of a straight dive is discussed in Art. 43. In addition to these conditions of flight there is that of reversed loading, which is met in sharply entering a dive at high speed, or with pursuit airplanes in various maneuvers or in actual upside down flight. This condition is of importance chiefly in the case of braced or semi-braced monoplanes in which it becomes one of the limiting conditions. There are certain other types of structures for which it would also be a critical condition. The load factor for reversed loading should be about .4 that for direct loading, and in no case less than 2.0. The question of proper load factor will be taken up in the appendix to which reference can be made.

41. *High and Low Incidence Conditions*—As previously stated, the maximum dynamic load occurs at an angle of incidence close to that for maximum $K_y$. For the purpose of analysis this angle will be taken as that at which the center of pressure reaches its most forward position, except for aerofoils with which $K_y$ becomes maximum before the center of pressure reaches this position. In such cases the angle at maximum $K_y$ is assumed, and the location of the center of pressure for this angle is employed in computations. The values of the angle of incidence and of the center of pressure travel are obtained from model tests of the particular aerofoil used. Should no tests be available the center of pressure may be taken at .3 of the chord from the leading edge, and the angle of incidence at 14 degs. In determining the direction of the resultant air load on an aerofoil the $L/D$ at the proper angle of incidence must be known. This is obtained from model tests, and should then have suitable correction factors, such as the factors for full scale, aspect ratio and biplane effect, applied to it. As seen from Fig. 35 this resultant air force will have a forward or anti-drag component which may be much more severe in certain cases than the backward drag forces in other conditions of flight. This is a fact not generally appreciated since, as a rule, the drag wires in drag trusses are double, while the anti-drag wires are single and since, although external drag wires are commonly used, provision is seldom made for external anti-drag wires. Incidentally, this goes to prove that external drag wires are usually unnecessary, as the drift truss is amply strong enough to carry all drag loads either positive or negative. Care should be taken, however, to have the incidence wires and struts in the center section and particularly the wing anchorage on each wing capable of transmitting the drag from the wings into the fuselage. The resistance of the struts, wires and fittings should be calculated and applied as concentrated drag forces at panel points. The velocity for which this resistance is found is determined by the dynamic or load factor suitable for the type of airplane under consideration. For pursuit airplanes this may be taken as 4 to 5 and for large bombers as 1¾. The velocity, $V = \sqrt{F \times W / K_y \times A}$, $F$ being the dynamic factor, $K_y$ the lift coefficient at the specific angle of incidence, $W$ the gross weight of the airplane and $A$ the net wing area. For approximate computations the resistance of the interplane bracing may be neglected, and to compensate for this the value of $L/D$ from model tests used uncorrected. It is evident that to underestimate the drag forces is conservative in the high incidence condition. The reverse, of course, is true for the low incidence condition. This distribution of load between wings and along the span is discussed in Arts. 44 to 46.

In the case of airplanes with a center fuselage and side nacelles, or with two fuselages and a center nacelle, the resistance of the nacelle should be considered as a concentrated drag load. For neither high nor low incidence need the weight of nacelles be considered to have a drag component. For both of these conditions the weights of nacelles should be taken as concentrated loads applied to the wing cellule and properly distributed between the trusses. As just suggested, in an analysis of a wing cellule the fuselage or fuselages which carry the tail surfaces should always be regarded as reaction points and the forces due to the nacelles as applied loads.

For the low incidence condition the location of the center of pressure is determined by the high speed of the airplane at the ground. Knowing this speed the $K_y$ can be computed from the formula, $K_y = W/AV^2$. Then from the curve for the center of pressure travel of the aerofoil the location of the center of pressure, for the angle of incidence corresponding to the $K_y$ calculated above, is found. As before, the $L/D$ at the same angle of incidence is obtained from the model test and suitably corrected for full scale, etc. The resistance of the interplane bracing at maximum ground speed is then added to the distributed drag load determined by resolving the resultant air force on the aerofoil into components normal and parallel to the wing chord. In the low incidence condition also, the approximation of neglecting the resistance of the interplane bracing and using uncorrected values of $L/D$ can be made, though it may not be on the safe side. In case characteristic curves of the aerofoil are not at hand the center of pressure may be taken as between .45 and .70 of the chord from the leading edge for thin and thick aerofoils, respectively, and the $L/D$ as 6.

Although the low incidence condition is not a limiting one in itself, yet, because it can be readily and accurately calculated, it serves as a means for proportioning the rear truss. The data upon which the computation of the stresses in a straight dive is based are not quite so definitely known. But the main reason why the rear truss is not designed solely for this condition is due to the indeterminate character of the cellule under the load that is produced in a dive. In a military airplane, especially one subjected to severe fire, the rear truss must be of sufficient strength to carry the entire weight of the airplane when the front truss is destroyed, with a safety factor of 2 to 3 depending on the size of the airplane.

The center of pressure for the reversed loading condition is well forward. Its exact position has not been determined for most aerofoils, but it may be taken at .25 of the chord from the leading edge. Information as to the angle of incidence and the $L/D$ for this condition is meager. It is probably safe to assume that these factors are the same as in the high incidence condition, with the exception that the angle of attack is negative. Aerofoils at negative angles are much less efficient than at positive angles. To counteract this, the angle of incidence is very large.

42. *Static Test Conditions for the Wing Cellule*—Standard static testing as conducted by the U. S. Government at McCook Field has been changed so that instead of attempting to combine two conditions of flight in a single test, separate tests are made for the conditions of high

and low incidences, and in certain cases, such as braced monoplanes, the cellule is tested for reversed loading also. As the front truss is most severely stressed in the test for the high incidence condition and the rear truss in the test for the low incidence condition, by carefully localizing the failure and repairing the break, the same cellule may be used for both tests. No attempt is made to test for the conditions of a straight dive. Since the tests are arranged so as to reproduce the conditions of flight as nearly as possible no separate stress analysis for static test need be made. The location of the center of pressure is determined as has been described above for the purposes of stress analysis. The distribution of load between the wings and along the span will be the same as, or equivalent to, the distribution of load for flight conditions. Since in static test the load acts vertically, the drag load is obtained by inclining the wing chord. The angle of inclination of the wings is such that the load will act in the same direction with respect to the wing chord as the resultant of the air forces on the cellule. These forces are made up of the combined lift and drag forces on the aerofoils and the resistance of the struts, wires, and fittings. In general, this angle of inclination will be about 7 degs. With the wings in reversed position for test, the leading edge will be up for the low incidence and down for the high incidence condition.

In static tests on airplanes with side or center nacelles, the resultant of the loads caused by the nacelle, which are made up of its resistance and weight, are reproduced by a cable attached to the cellule in a suitable manner.

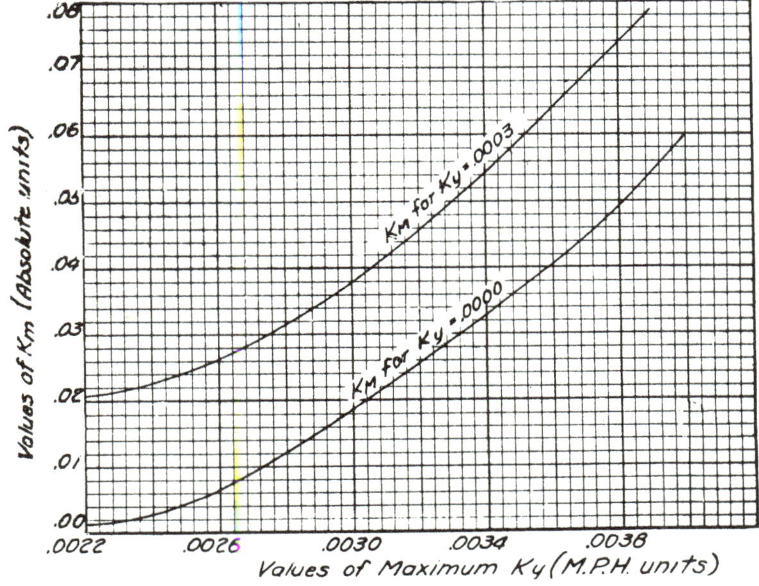

43. *Diving*—An airplane is assumed to dive at an angle close to the angle of zero lift, and to reach its limiting velocity at which the total resistance equals the weight of the airplane. Owing to the character of the distribution of pressure on a wing at very small angles, shown in Fig. 33, there is a large moment which tends to put the airplane into an even steeper dive. This moment is counteracted by downward forces on the tail sufficient to produce equilibrium. The forces acting on the wings are, therefore, downward loads on the front lift truss and upward loads on the rear lift truss. The upward loads are greater than the downward by the amount of the tail load. In addition to these there is the drag of the wings, struts, and wires.

From wind tunnel tests on aerofoils, moment coefficients for the wing section used can be obtained. The total diving moment then equals $M = K_m \rho c S V^2$, $D = W = K_d \rho S V^2$. From these relations $M = K_m c W / K_d$.

W = gross weight of airplane in lbs.
$\rho$ = density of air = .07608 lbs. per cu. ft.
S = total area of wings in sq. ft.
c = chord length in ft.
V = velocity in ft. per sec.
$K_m$ = moment coefficient in absolute units.
$K_d$ = coefficient of resistance of entire airplane in absolute units.

Values of $K_d$ based on full flight performance data are listed below for various airplanes. From these, the value of $K_d$ for any airplane may be estimated by comparing the airplane in question with the types in the table.

| Airplane | $K_d$ in absolute units |
|---|---|
| Martin Bomber | .0375 |
| JN-4 D2 | .0323 |
| DH-4 | .0297 |
| U.S.D.-9A Bomber | .0297 |
| Fokker D-7 | .0272 |
| VE-7 | .0249 |

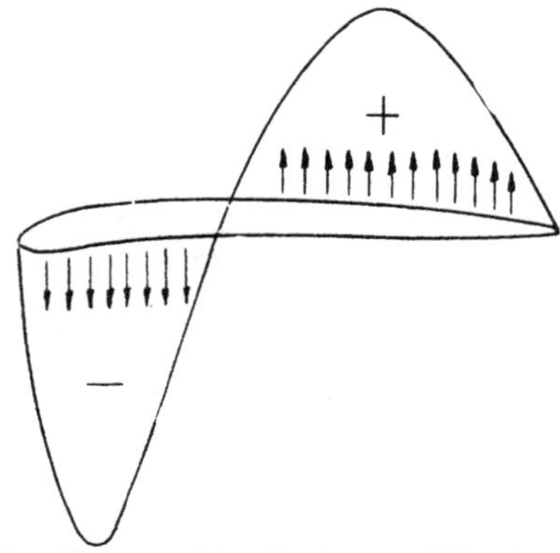


**Fig. 33. Pressure Distribution on Wing in a Dive**

| | |
|---|---|
| Ordnance D .................. | .0242 |
| SE-5 ........................ | .0236 |
| MB-3 ........................ | .0212 |
| U.S. XB-1A .................. | .0206 |
| V.C.P.-1 .................... | .0200 |

Curves of values of $K_m$ are given for $K_y = .0000$ and $.0003$ (M.P.H. units). These curves are averages of the $K_m$, at these values of $K_y$ of a large number of different aerofoils with maximum $K_y$'s ranging from .0022 to .0039. Although in individual cases the $K_m$ of an aerofoil may not fall exactly on the curves, the error will be small, especially in the lower portion of the curves which is the most important part since on airplanes that will dive very steeply, speed sections, such as the R.A.F. 15, with low maximum $K_y$'s will be used. Furthermore, in case an aerofoil having a high maximum $K_y$ is employed, the $K_d$ of the airplane will be large.

When the diving moment, given by the formula $M = K_m cW/K_d$, has been computed, the load $P$ on the tail plane may be found. If it is assumed that the centers of resistance and of thrust are at the center of gravity of the airplane, which is reasonable and conservative, $P = M/L$, $L$ being the distance from the center of gravity of the airplane to the resultant of the load on the tail plane. The load on the trusses may now be calculated as follows:

$F_1$ = load on front truss.
$F_2$ = load on rear truss.
$L_1$ = distance from front truss to the resultant of the load on the tail plane.
$L_2$ = distance between front and rear trusses.

Taking moments about the front truss,

$$L_1 P = L_2 F_2$$

$$F_2 = \frac{L_1 P}{L_2} = F_1 + P$$

With $F_1$ and $F_2$ known, the loads per inch run on the trusses can be computed and the stresses determined in the usual manner. It should be remembered that the loads on the front truss act downward and those on the rear truss upward. In the usual type of wing construction in which incidence wires are used, the latter will be severely stressed, and will tend to neutralize the truss stresses, thus considerably increasing the factor of safety. The bending moments on the spars produced by the distributed load and the stresses due to them will not, however, be affected by the incidence wires. A factor of safety of 1¾, neglecting the incidence wires is ample for this condition of diving.

In the internally braced monoplane there are, of course, no incidence wires tending to equalize the truss loads, and if such an airplane can be made to dive at an angle of zero lift much more severe stresses will be induced than in the case of an ordinary biplane, unless the wing is deep enough to permit effective internal bracing between the spars.

In addition to the stresses caused by the up and down loads on the lift trusses, there are large drag stresses which occur in a limiting dive. The proportion of the total drag that is made up of the drag of the wings and the resistance of the interplane struts, wires, and fittings varies from 45 to 60 per cent of the gross weight of the airplane. In an extremely clean wing structure with a speed aerofoil, a minimum of interplane bracing, and with streamline wires, the lower percentage is applicable. It is seldom that the wing drag exceeds 60 per cent, unless the wing loading is very light. For the U.S.D-4 it amounts to 55 per cent. To obtain the net drag on the wings, their weight, which may usually be assumed as 15 per cent of the total weight of the airplane, must be subtracted, leaving 30 to 45 per cent of the gross weight. If side or center nacelles are used, their drag must be added to this. Data relative to coefficients of resistance of nacelles, especially engine nacelles, are unsatisfactory. It may be said that in a limiting dive the resistance of an engine nacelle averages about 30 per cent of its weight, and if the nacelle carries a gas tank, also, 20 per cent of its weight. The weight of the nacelle acts vertically downward, and that component parallel to the line of flight must be subtracted from the nacelle resistance to give the net drag, or rather anti-drag.

44. *Distribution of Load Between Wings*—Reliable information on this subject is meager. The quantitative effect of various factors such as stagger, gap-chord ratio and angle of incidence has not been determined. Several tendencies may be pointed out, however.

At both high and low incidence conditions positive stagger decreases and negative stagger increases the relative lift on the lower wing of a biplane. At very high angles of incidence the lower wing becomes more effective than the upper wing. This is because the upper wing reaches its burble point sooner than the lower and so loses its efficiency.

In a biplane with a gap-chord ratio between 0.8 and 1.2, and no stagger, the load per sq. ft. on the lower wing may be taken as 90 per cent of the load per sq. ft. on the upper wing for the high incidence condition, and as 95 per cent for the low incidence condition.

When there is 50 per cent positive stagger, these values should be decreased to 80 per cent and 75 per cent for high and low incidence, respectively. For 50 per cent negative stagger they increase to 110 per cent and 125 per cent.

With a triplane the loads per sq. ft. on the middle and lower wings may be taken as 75 per cent and 85 per cent, respectively, of the load per sq. ft. on the upper wing in the high incidence condition. For the low incidence condition both of these values become 80 per cent of the load per sq. ft. on the upper wing.

In a standard static test the load would be distributed in accordance with the proportions given above.

45. *Distribution of Load Along Span and Chord*—The following assumptions regarding the location of the center of pressure will be made as representing a close average for most wing sections of the general character of the R.A.F. 15. The location is given in per cent of the chord from the leading edge. In the case of a few wing sections the

center of pressure movement is much greater than is indicated below:

Low incidence ............. = 50 per cent of chord
High incidence ............. = 30 per cent of chord
Static test (direct loading)... = 30 and 50 per cent of chord
Static test (reversed loading). = 25 per cent of chord

For unusual wing sections in which the center of pressure travel is much greater or less than is indicated by the first two of these values, proper percentages obtained from wind tunnel tests should be used.

In static tests the load along the span is constant. The effect of rounded tips and of reduced loading on tips is allowed for by decreasing the loaded length on the cantilever span so that the cantilever moment will be the same as the moment of the loading treated in Art 46.

46. *Distribution of Pressure Along Wing Tip*—No extensive series of experiments have been carried out with a view to determining the variation in pressure on the wing tip. Experiments conducted at the Royal Aircraft Factory resulted in the curve given in Fig. 34, which is a fair representation of the diminution of pressure at the tip.

In order to simplify computations the distribution is taken as shown by the dotted line. At a distance from the wing tip equal to the chord the load per inch of run is reduced to 0.9 of the maximum and at a point 1/2 of the chord length from the tip the load is reduced to 0.7 of the maximum. Whenever the cantilever length is less than the chord, the maximum load is carried out to the strut point to facilitate the calculations of the moments and shears on the spars. Since the load to be carried by a wing is a definite amount, it is evident that the maximum load per inch of run is greater when allowance is made for the decrease toward the tip than when the load is uniform over the entire span.

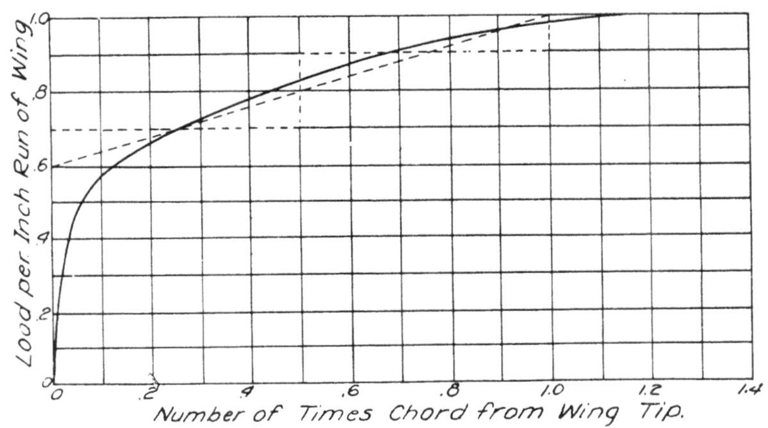


**Fig. 34. Distribution of Load on Wing Tip**

47. *Distribution of Load Between Spars*—For the condition of loading under consideration, the location of the center of pressure, and hence

of the resultant force on the wing, is first determined. The proportion of the total load carried by the wing that goes into each spar is inversely proportional to the respective distances of the spars from the resultant force. From the gross load supported by the wing must be subtracted the weight of the wing cellule, in order to get the net load on the wing. For a small airplane with a low safety factor the weight of the wing cellule may be as low as .9 lbs. per sq. ft. of wing surface, while with large airplanes of the bombing type the weight will seldom be less than 1.5 lbs. per sq. ft. For a usual design of a medium-size airplane this weight may be taken as about 1.1 lbs. per sq. ft.

The loading on the wing tip is reduced as described in Art. 46. The running loads on the front and rear spars are then calculated by the following formulas:

$$W = .7w_r l_r + .9w_r l_1 + 1.0w_r l + .7w_f l_f + .9w_f l_1 + 1.0w_f l$$

$$w_f = w_r \cdot p_f / p_r$$

$W$ = total net load on wing.
$w_f$ = normal running load on front spar.
$w_r$ = normal running load on rear spar.
$p_f$ = per cent of W on front spar.
$p_r$ = per cent of W on rear spar.
$l$ = length of front and rear spars with normal load.
$l_1$ = length of front and rear spars with .9 normal load.
$l_f$ = length of front spar with .7 normal load.
$l_r$ = length of rear spar with .7 normal load.
$l_1 + l_f$ = cantilever length of front spar.
$l_1 + l_r$ = cantilever length of rear spar.

As a rule, no reduction of load on the rear spar due to cutouts at the fuselage is allowed for. This method will give a correct distribution of load with square or only slightly rounded wing tips. If the wing tip is given a large rake, the load on the rear spar given by this formula is low because with a raked tip the center of pressure moves back. Yet, except in extreme conditions, this method will give satisfactory results.

No aileron loads, in addition to the normal load on the rear spar, are considered. Ailerons would be used but little under the flight conditions which produce maximum stresses in the rear spars. An exception is made in the case of a balanced aileron, part of which projects beyond

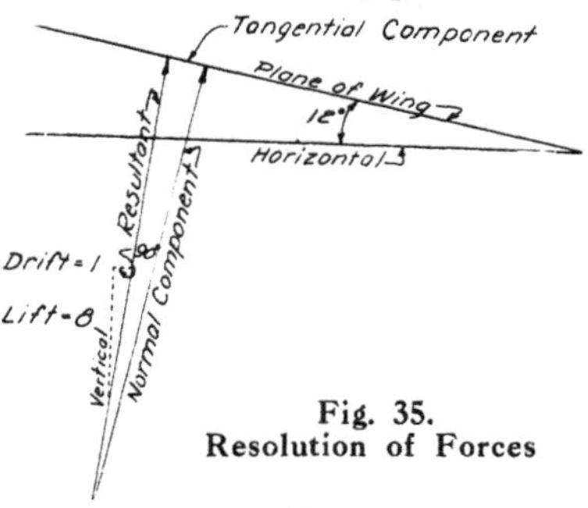


**Fig. 35.**
**Resolution of Forces**

the contour of the wing proper. On this portion of such an aileron a loading equal to .7 of the normal wing loading is assumed. All of this load is carried by the rear spar.

48. *Resolution of Lift and Drag Forces*—The resultant pressure on a wing is composed of a vertical force, called "lift," and a horizontal force, called "drag." The ratio of lift to drag ($L/D$) determines the inclination of the resultant force. Fig. 35 is a resolution for the high incidence condition in which the angle of incidence is 12 degs. positive and $L/D$ equals 8. The resultant is resolved into a component normal to the wing chord and a component parallel to the wing chord. The normal component produces stresses in the lift trusses while the tangential component produces stresses in the drag trusses. The drag is always laid off parallel to the direction of the wind and the lift perpendicular to this direction.

The vertical component of the resultant equals the weight of the airplane, but for designing purposes it is sufficiently accurate to take the resultant itself as equal to the weight. In the sand test loading the resultant is vertical, and, when the angle of inclination of the wing chord is 7 degs., the component normal to the wing chord is .992 of the resultant and the component parallel to the chord .122 of the resultant.

49. *Moments, Shears, and Reactions*—When the running loads on the spars have been determined, the next step is the calculation of the moments on them at their supports. If a spar is continuous over several supports or if the loading is not uniform, the necessary three moment equations can be found in Art. 172. Care should be taken not to assume a spar continuous unless it is truly so. All properly constructed splices can be considered continuous as well as joints made with fairly long metal sleeves. There are few types of wing anchorage fittings, however, that justify the assumption of continuity, even though they may theoretically do so before the close fit of the parts is destroyed by the wear and tear of service. From the moments at the supports the shears and reactions of the spars may be computed by the methods of Art. 19. If in the truss under consideration there are no moments from eccentric strut loads or wire pulls, then the next step is the calculation of the bending moment in each bay at the point of zero shear.

50. *Truss Stresses*—Both lift and drag trusses are assumed to be pin-jointed structures for the calculation of the truss stresses. No account is taken of the effect of deformation of the truss under load. All truss loads are considered as concentrated loads applied at the panel points. With a lift truss these loads are the reactions from the spars, calculated as described above. When any weights, such as engine nacelles, are located on the wing their effect in relieving the lift stresses must be allowed for. The presence of incidence wires connecting the front and rear trusses renders them indeterminate. Therefore, except in considering cases in which lift wires are shot away or in which the load on one truss is up and on the other down, as in a steep nose dive, the effect of incidence wires is neglected, unless, as in a static test tensiometer measurements have been made of the stresses in them. Either

graphical or analytical methods can be used in calculating truss stresses. If care is taken it is believed that the graphical method is somewhat more satisfactory.

51. *Eccentric Strut Loads and Wire Pulls*—After the lift truss stresses have been determined the effect of eccentric struts loads and wire pulls can be computed by means of the three moment equations in Art. 172. This effect is most serious in single bay airplanes with the spars hinged at the fuselage or center section. If the spars are continuous over two or more points of support the stresses from eccentric loads will not largely alter those produced by the direct uniform load, especially within the spans. With eccentric wire pulls, the moments at the strut points due to the wire eccentricities change from positive to negative. However, in calculating the fiber stresses at these points, the stress due to the maximum moment should be combined with that due to the maximum compression, since the maximum values of each type of stress occur simultaneously. In a single bay airplane or in the case of unusually large eccentricities the reactions produced by the eccentricities should be computed, and the reactions from the direct air loads modified accordingly before obtaining the final values for the lift truss stresses. After the moments and shears from the direct air loading and the eccentric loads are combined, the points of zero shear should be determined, and then the bending moments at these points calculated.

52. *Deflection Moments*—When the stresses in the lift and drag trusses and the moments and shears from the direct air load have been calculated, and after any modification in them necessitated by strut or wire eccentricities has been made, the deflections of the spars, at the final points of zero shear within the bays in which the spars are under compression, should be computed from the formulas for deflection in Art. 172. It is unnecessary to obtain separately the deflection due to eccentric wire pulls, for if the moments and shears at the strut points due to eccentric wire pulls are combined with the moments and shears from the air load, and these combined values are substituted in the formula for deflection due to uniform load, the deflection obtained will be the resultant of the deflection produced by the eccentric wire pulls and the uniform load. If, however, there are present eccentric strut loads also, the deflections caused by them should be computed separately. In order to obtain the deflection it is necessary to know the moment of inertia of the spar, and hence if it has not been designed some approximate size must be assumed for the preliminary calculations. When the final moment of inertia is known the deflection should be corrected. The product of this deflection and the compression in the span gives the deflection moment which is added to the bending moment in the span, previously obtained. Because the panels of the drag truss are only half as long as the panels in the lift truss there will be a change in the compression near the center of the span. The maximum compression is the one to use in finding the deflection moment; i. e., if the stress in the drag truss is tension, then the smaller

value of the tension should be subtracted from the lift compression, while if the drag stress is compression the larger value should be added to the lift stress.

53. *Stress Table*—The tabulation of spar properties, moments, compressions, unit stresses, and factors of safety given in Table XI is in the most satisfactory form for use. The values of *Pd* in column 8 represent the product of the compression and the deflection each calculated for unit load factor. Since this deflection moment is the product of these two factors, each of which vary as the load factor, it varies as the square of the load factor. Hence, the factors of safety of spars in the spans under compression must be determined by the method of trial. Before the deflection moment as given in column 8 can be added to the main moment in column 7 it must be multiplied by the factor of safety for the spar in the bay under consideration. By doing this the fact that the deflection moment varies as the square of the load factor is taken into account.

54. *Unit Fiber Stresses*—In airplane design the stresses employed are based on ultimate strengths rather than elastic limit fiber stresses. There are two main reasons for this. First, airplane structures are very frequently subjected to static tests which are carried to destruction. They are designed to have a definite load factor under these conditions. Second, airplane structures are unique in that the extreme loads imposed upon them are of very short duration. In wood, especially, little injury is done by exceeding the elastic limit or even approaching the ultimate strength, provided the load is removed within a few seconds.

Tables XXXIX and XL of the appendix gives the physical properties of steels and various non-ferrous metals. These tables were compiled from standard government specifications. Ultimate stresses for steel columns and struts can be obtained from the parabolic formula given in Art. 24. This formula is regarded as more satisfactory than the Rankine or the various straight-line formulas used in structural work. This is particularly true for high strength steels and for materials on which few data regarding column strength are available. Attention should be called to the fact that with ductile materials the ultimate compressive strength for struts will not be greater than the yield point of the material.

In thin walled tubes the ultimate strength may be further limited by so-called "crinkling action," by which is meant local crumpling of the tube wall. This can be avoided by using tubes in which the ratio of wall thickness to tube diameter is greater than about .06. When the ratio is less than this the crinkling stress falls below the yield point of the material and the design may be limited by this stress rather than by the ultimate strength computed from the parabolic formula. The curves in Fig. 36 are suggested by Mr. Barling as giving a reasonable reduction in ultimate strength due to crinkling action. In connection with thin walled tubes it is recommended that when the tubes are to be welded, the gage should not be less than No. 20 (.035 in.) although skilled welders can weld No. 22 gage material satisfactorily.

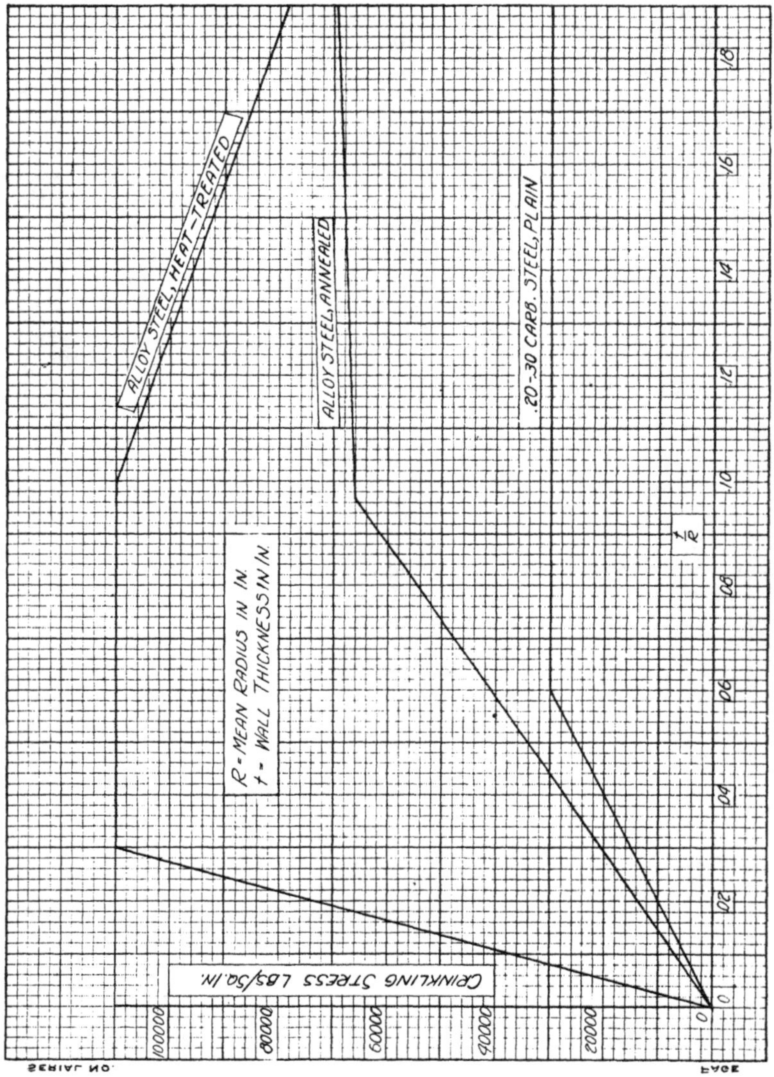

Fig. 36. Effect of Crinkling Stresses

A similar reduction in ultimate torsional stress because of thin walled tubes is treated in Art. 149 and a curve showing this reduction for medium carbon steel is given in Fig. 153. For alloy steels the curve would probably be similar, with all the values proportionally increased.

In the appendix, Art. 174, a rather complete discussion is given of stresses in airplane woods including the usual method for calculating the ultimate allowable stress in spars carrying both axial and lateral loads.

For high ratios of $L/\rho$, i. e., greater than about 90, this formula will not give safe results as it does not make sufficient allowance for the column action which is present when $L/\rho$ is large. Experimental work on combined bending and compression is now in progress. Until it is completed the following discussion and methods are suggested and it is believed will give satisfactory results.

In a spar subjected to lateral and axial loads the deflection may be regarded as being made up of two parts, a primary deflection produced by the lateral load and a series of secondary deflections due to what may be termed column action. The first type of deflection can be computed for any condition of loading and for continuous as well as simple beams by means of the formulas and methods given in Arts. 172 and 173. The rest of this discussion, however, applies to simple beams. The compression in the spar at any point has an eccentricity equal to the deflection of the spar at that point. Thus the spar has imposed upon it a bending moment which causes the first secondary deflection, as it may be called. In a similar manner bending moments are produced by the compression acting through this secondary deflection, and another deflection occurs. This last deflection in turn causes further deflection, but each deflection is smaller than the preceding one and the ratio between successive deflections remains constant. These deflections, therefore, form a series in geometric progression and may be readily summed up, once the geometric ratio has been determined. An example will be worked out illustrating the method.

Assume a strut loaded as shown in Fig. 37. The primary deflection of this strut under the lateral load of 2.0 lbs. per in. =

$$\delta_1 = \frac{5\ wL^3}{384\ EI} = \frac{5 \times 200 \times 100^3}{384 \times 1{,}600{,}000 \times 2.67} = .61 \text{ in.}$$

The maximum bending moment caused by this deflection $= m_1 = 2000 \times .61 = 1220$ in. lbs. Since the error in assuming the elastic curve of the strut to be parabolic is negligible, the bending moment on the strut due to the deflection has a parabolic variation and is the same as the bending moment produced by a uniform load. The value of this hypothetical load is determined by the equation 1220 in. lbs. $= \dfrac{wL^2}{8} = \dfrac{w \times 100^2}{8}$ or $w = .975$ lbs. per in. The first secondary deflection due to this uniform load can be calculated from the primary deflection by direct proportion:

$$\delta_2 = \delta_1 \times \frac{w_2}{w_1} = .61 \times \frac{.975}{2.00} = .297 \text{ in.}$$

In a similar manner this secondary deflection produces bending moment along the beam which results in a second secondary deflection.

$$m = 2000 \times .297 = 594 \text{ in. lbs.} = \frac{w_3 \times 100^2}{8}\ ;\ w_3 = .475\ ;\ \delta_3 = .297 \times \frac{.475}{.975}$$

$= .145$ in. The ratios of $\delta_2$ to $\delta_1$ and $\delta_3$ to $\delta_2$ are equal: $.297/.61 = .145/.297 = .487$, which is the geometric constant. The moment due to the entire series of deflection $= m_n = 1220$ in. lbs. $(1.00+.487+.487^2+$

$$.487^3 + 487^{n-1}) = \frac{1 \times m_1}{(1.00-r)} = \frac{1220}{(1.00-.487)} = 2380 \text{ in. lbs. for a load}$$

factor of unity. As explained in Art. 53 this deflection moment varies as the square of the load factor and must therefore be multiplied by the load factor before it can be added to the direct moment caused by the uniform load. This latter moment equals $\frac{2.0 \times 100^2}{8} = 2500$ in. lbs.

Assume a load factor of 2.75.

$$M_r = (2500 + 2.75 \times 2380) = 9040 \text{ in. lbs.}$$

$$f_b = \frac{M_t \cdot y}{I} = \frac{9040 \times 1}{2.67} = 3390 \text{ lbs. per sq. in.}$$

$$f_c = \frac{P}{A} = \frac{2000}{8} = 250 \text{ lbs. per sq. in.}$$

$$f_t = 3640 \text{ lbs. per sq. in.}$$

If 10,300 lbs. per sq. in. equals the modulus of rupture of spruce and 5,500 lbs. per sq. in. its ultimate compressive strength, the safety factor for this beam is calculated as follows:

$$f_a = \frac{3390}{3640} \cdot (10{,}300 - 5{,}500) + 5{,}500 = 9{,}960$$

$$F.S. = \frac{9960}{3640} = 2.74$$

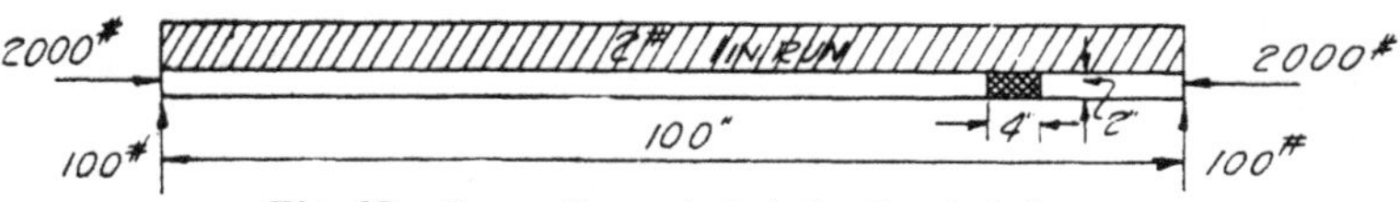


**Fig. 37. Laterally and Axially Loaded Struts**

It should be noted that in computing the ultimate allowable stress when the total deflection is known the ultimate compressive strength is not reduced to allow for column action but is its full value. In order to determine the geometric constant it is unnecessary to compute more than the primary and first secondary deflections. Cases will arise in which the value of this geometric constant will be greater than unity. The series then diverges instead of converging, indicating an indefinite deflection. This means that the beam is in an unstable condition, and either its section must be increased or the imposed load decreased.

One valid objection to this general method is that it will not give entirely correct ultimate loads because the elastic properties of the material do not remain constant when the elastic limit has once been passed. The deflections tend to increase faster than the formula would show.

With a beam continuous over one or more supports the correct primary deflection is calculated by formula, but it is impracticable to compute exactly the secondary deflections resulting from column action. If, however, in determining these secondary deflections the spar is assumed to be a simple beam the same procedure that is given above may be used. This assumption will make the deflections somewhat greater than they should be, which will offset the tendency of the deflections to be larger than calculated after the elastic limit has been passed.

# CHAPTER III

## Wing Stress Analysis

55. *Preliminary Data*—The VE-7 Training airplane has been selected for analysis, and the trusses, wing plans, and spar sections are shown in Figs. 38 to 41 inclusive. The data which can be taken directly from the plans are:

| | |
|---|---|
| Chord .............. | 55½ in. |
| Gap ................. | 56 in. |
| Gap-chord ratio ....... | 1.01 |
| Span ................. | 410.6 in. |
| Aspect ratio .......... | $\frac{410.6}{55.5} = 7.4$ |

The areas of the wings were determined by carefully laying out the wings to a scale of 1 in. to 1 ft. and dividing the total area up into smaller sections whose area could be readily computed.

| | |
|---|---|
| Area of upper wing... | 151.0 sq. ft. |
| Area of lower wing... | 138.8 sq. ft. |
| Total area of wings... | 289.8 sq. ft. |

The weight of the airplane fully loaded is 1937 lbs.

The analysis will be carried through for the high incidence condition only, as this case will bring out the methods of stress analysis. The other conditions are repetitions of this one with differences of loadings.

56. *Location of Center of Pressure*—The center of pressure for high incidence will be assumed to be at 29 per cent of the chord from the leading edge, or $.29 \times 55.5 = 16.10$ in. In the same manner, for high speed the center of pressure is 50 per cent or 27.75 in. from the leading edge.

57. *Distribution of Loads Between Spars*—The front spar is 8.75 in. from the leading edge. For high incidence the center of pressure is 7.35 in. (16.10—8.75 = 7.35) from the front spar. Since the spars are 27.375 in. apart, the proportion of load carried by the rear spar equals 7.35/27.375=.269. The front spar will receive the remainder, or 1.000 — .269 = .731 of the total load. The distribution for low incidence is computed in the same way.

*DISTRIBUTION BETWEEN SPARS*

| Loading | % Load on Front Spar | % Load on Rear Spar |
|---|---|---|
| High incidence ......... | 73 | 27 |
| Low incidence .......... | 30 | 70 |

58. *Distribution of Load Between Wings*—The proportion of load carried by the upper and lower wings has been changed since the present analysis was completed. Therefore, the old proportions will be used instead of the later values recommended in Art. 44.

A = 151.0 sq. ft. B = 138.8 sq. ft.

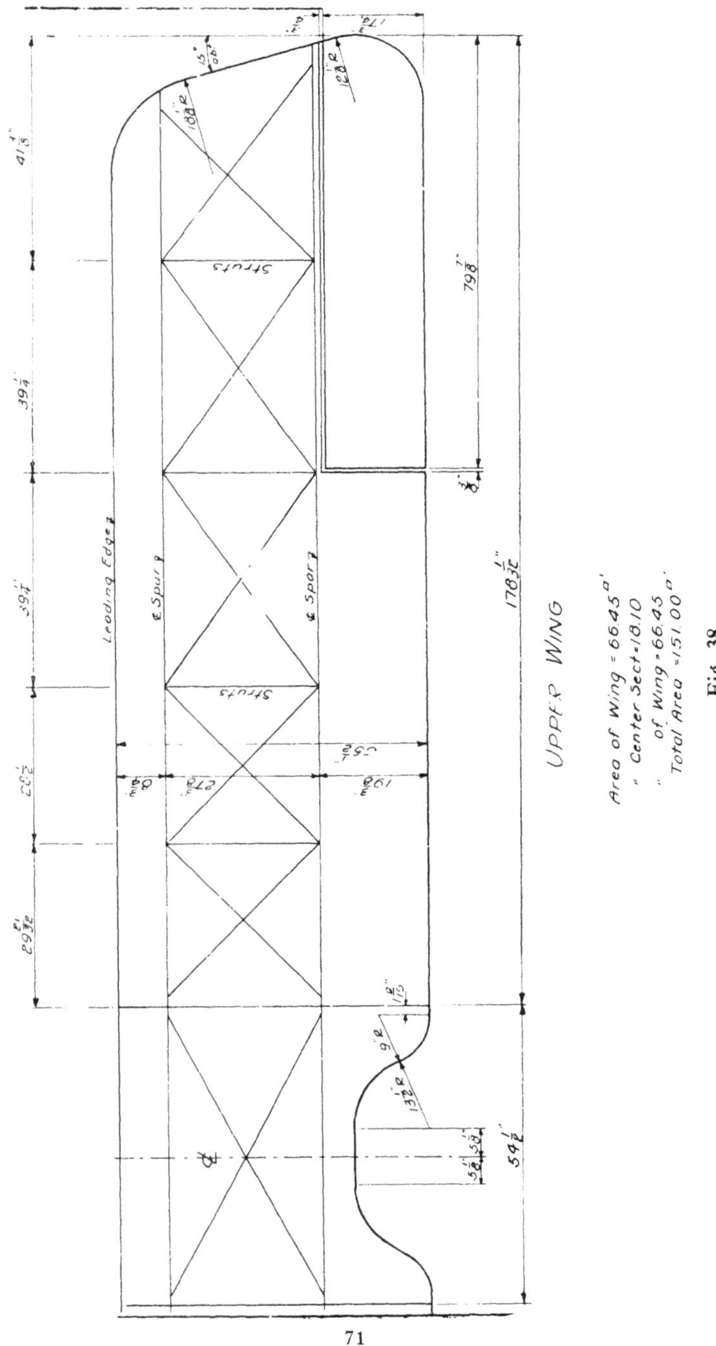
Leading Edge
℄ Spar
℄ Spar
Struts
Struts
UPPER WING
Area of Wing = 66.45 □'
" Center Sect = 18.10
" of Wing = 66.45
" Total Area = 151.00 □'

Fig. 38

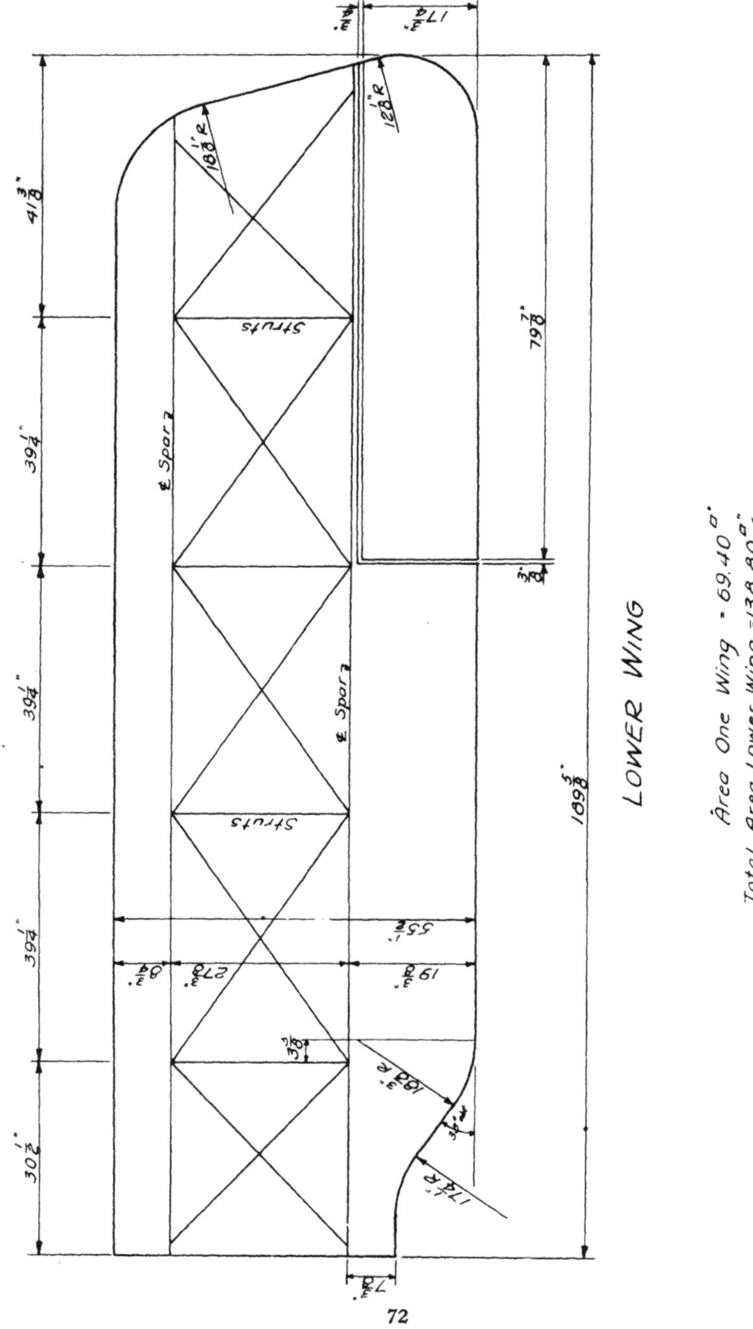
LOWER WING
Area One Wing = 69.40 □'
Total Area Lower Wing = 138.80 □'
Struts
Struts
℄ Spar
℄ Spar

Fig. 39

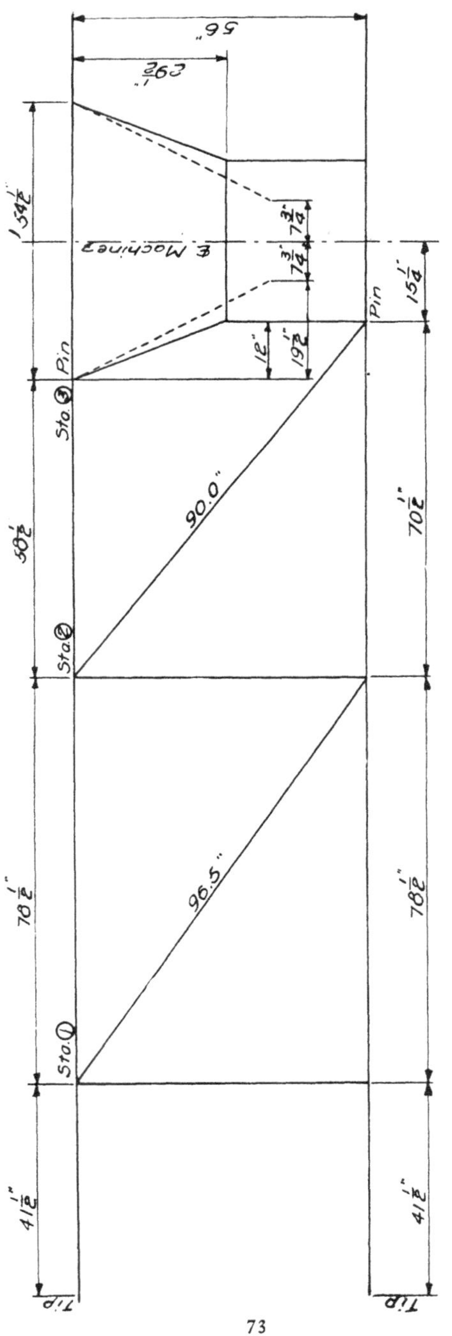


Fig. 40. Lift Truss

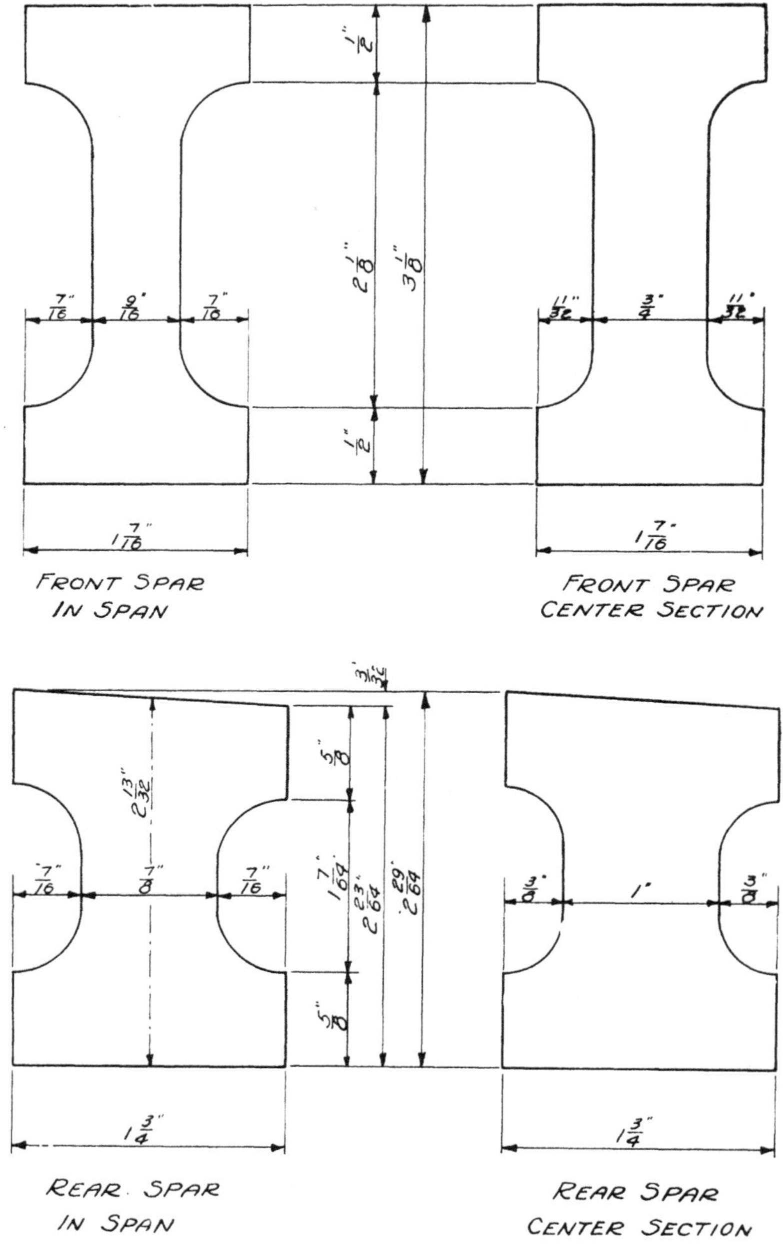


Fig. 41. Spar Sections

The weight of the wings will be taken as 0.9 lbs. per sq. ft. in determining the value of the net weight of the airplane, $W$. Then $W$ equals $1937 - 289.8 \times .9 = 1937 - 261 = 1676$ lbs.

Substituting these values in the formula—

$$\frac{11}{9} \times 151.0\,w + 138.8\,w = 1676 \text{ lbs.}$$

$$w = 5.19 \text{ lbs. per sq. ft. on lower wing.}$$

$$\frac{11}{9}\,w = 6.34 \text{ lbs. per sq. ft. on upper wing.}$$

59. *Load on Wings*—The load on the upper wing is

$151.0 \times 6.34 = 956$ lbs. (net)
$151.0 \times (6.34 + .90) = 1093$ lbs. (gross)

The load on the lower wing is

$138.8 \times 5.19 = 720$ lbs. (net)
$138.8 \times (5.19 + .90) = 845$ lbs. (gross)

60. *Distribution of Pressure on Wing Tips*—Fig. 42 is a plan view of the wing and gives the distribution of load per inch of run. As explained in Art. 46, the maximum load is carried out to the strut points for ease of computation.

61. *Load per Inch of Run*—The loads for each spar are given in Fig. 43. The method of calculating these loads will be illustrated by taking the upper wing spars as examples. Let $w$ = maximum load per inch of run on rear spar. By reference to Fig. 43 it will be seen that at the strut point the load drops to $.9w$ and extends for 14 in. toward the wing tip. Then the load drops to $.7w$ and continues to the end of the spar for a distance of 26 in. At high incidence the rear spar carries 27 per cent of the total load, and the front spar 73 per cent. Hence the front spar carries $73/27 = 2.70$ times the load on the rear spar. The maximum load per inch of run for the front spar is therefore $2.7w$. As with the rear spar, the load drops to .9 of the maximum at the strut point, or to $.9 \times 2.70w = 2.43\,w$, and at 14 in. out from the strut point it drops to $.7 \times 2.70w = 1.89w$. Since only one-half of the wing is shown in Fig. 43 the sum of these uniform loads must equal one-half the load on the wing.

| | | |
|---|---|---|
| Rear Spar | 164 × 1.0 w = | 164.0 w |
| | 14 × .9 w = | 12.6 w |
| | 26 × .7 w = | 18.2 w |
| Front Spar | 164 × 2.7 w = | 442.8 w |
| | 14 × 2.43 w = | 34.0 w |
| | 18 × 1.89 w = | 34.0 w |
| Total | .................. | 705.6 w |

From Art. 59 the net load on upper wing is 956 lbs.

$$\text{Then} \quad 705.6\,w = \frac{956}{2} = 478 \text{ lbs.}$$

$$w = .68 \text{ lbs. per in. of run}$$

Substituting this value in the above expressions for loads per inch of run the true values may be obtained. The loads for the lower wing were obtained in the same manner.

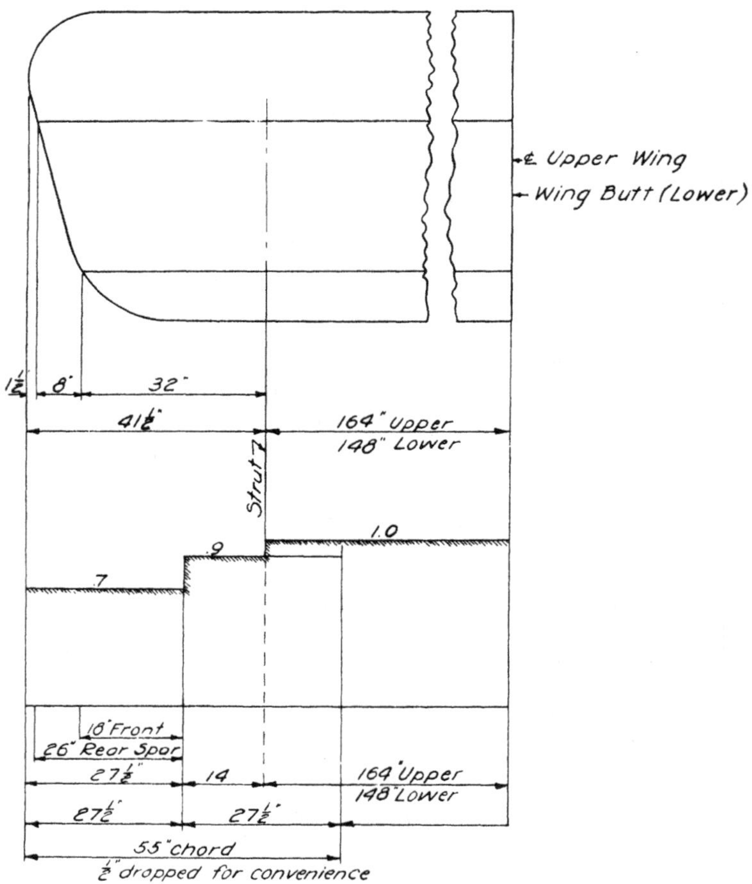


Fig. 42. Distribution of Load on Wing Tip

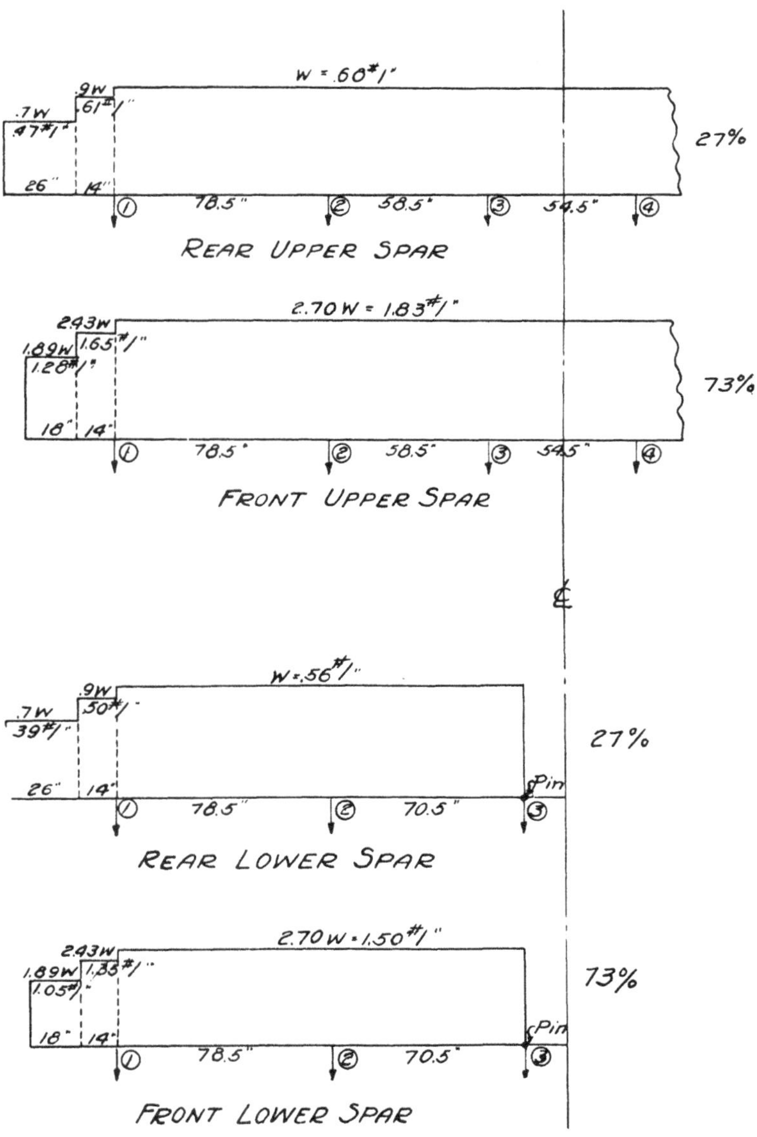


Fig. 43. Running Loads on Spars

62. *Moments and Shears in Front Upper Spar*—The connection of the wing spars to the center section spars is shown in Fig. 44. This joint is capable of transmitting bending moment, and therefore the spars in the upper wing will be assumed as continuous across the center section. In the lower wing, however, the spars are pinned at the fuselage. The strut points are assigned station numbers in Fig. 45. The outer strut point is station 1, intermediate strut point, station 2, and so forth. The complete calculations for moments and shears are given below. The handling of the three-moment equation is explained in Art. 22. The important relation between the shears and moments in a span (discussed in Art. 19) is used frequently in this work and should be well understood.

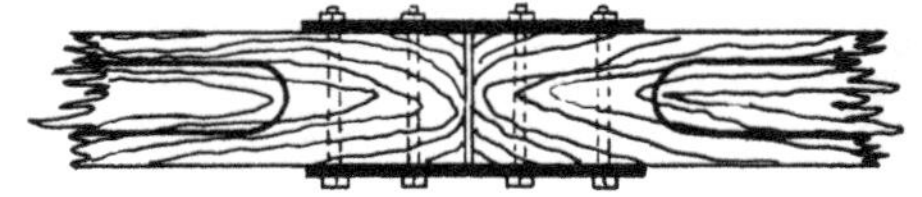

Fig. 44. Joint in Upper Spar

Front upper spar:

$m_1 = (18 \times 1.28) \times 23 + (14 \times 1.65) \times 7 = +692$ in. lbs.

Apply the three-moment equation, Case 2, to spans 1—2 and 2—3.

$$(78.5 \times 690) + 2\, m_2 (78.5 + 58.5) + 58.5\, m_3 = \frac{1.83}{4} \left[ (78.5)^3 + (58.5)^3 \right]$$

(1) $274\, m_2 + 58.5\, m_3 = 258740$

Apply the three-moment equation, Case 2, to spans 2—3 and 3—4, noting that $m_4 = m_3$ because of symmetry about the center line of the airplane:

$$58.5\, m_2 + 2\, m_3\, (58.5 + 54.5) + 54.5\, m_3 = \frac{1.83}{4} \left[ (58.5)^3 + (54.5)^3 \right]$$

(2) $58.5\, m_2 + 280.5\, m_3 = 165650$

Eliminate $m_2$ between (1) and (2). Divide (1) by 274 and (2) by 58.5.

(3) $m_2 + .2135\, m_3 = 944.3$

(4) $m_2 + 4.7949\, m_3 = 2831.1$

Subtracting (3) from (4)

$4.5814\, m_3 = 1886.8$

$m_3 = +412.0$

$m_2 + .2135 \times 412.0 = 944.3$

$m_2 = 944.3 - 88.0 = +856.3$

$s_{-1} = -18 \times 1.28 - 14 \times 1.65 = -46.1$

$$78.5\, s_1 + 692 + \frac{1.83 \times (78.5)^2}{2} = m_2 = +856.3$$

$$s_1 = -\frac{5474}{78.5} = -69.7$$

$$s_{-2} = -(1.83 \times 78.5) - (-69.7) = -143.7 + 69.7 = -74.0$$

$$58.5\, s_2 + 856.3 + \frac{1.83 \times (58.5)^2}{2} = +412.0$$

$$s_2 = -\frac{3575}{58.5} = -61.1$$

$$s_{-3} = -(1.83 \times 58.5) - (-61.1) = -107.1 + 61.1 = -46.0$$

$$s_3 = s_{-4} = \frac{18.3 \times 54.5}{3} = -49.9$$

$$r_1 = s_1 + s_{-1} = -69.7 - 46.1 = -115.8$$
$$r_2 = s_2 + s_{-2} = -61.1 - 74.0 = -135.1$$
$$r_3 = s_3 + s_{-3} = -49.9 - 46.0 = -\ 95.9$$

Total............ = 346.8 lbs.

Total load on half spar = 23.0+23.1+300.6 = 346.7 lbs. (check)
The maximum moments in the spans occur at the points of zero shear.
Consider span 1—2

Distance from station 1 to point of zero shear $= \frac{69.7}{1.83} = 38.1$ in.

$$m_{1-2} = +692 + \frac{1.83 \times (38.1)^2}{2} - (69.7 \times 38.1) = -638 \text{ in. lbs.}$$

Considering span 2—3, the distance from station 2 to point of zero shear equals $\frac{61.1}{1.83} = 33.4$ in.

$$m_{2-3} = +856 - (61.1 \times 33.4) + \frac{1.83 \times (33.4)^2}{2} = -164 \text{ in. lbs.}$$

Considering the span 2—3, the point of zero shear occurs at the center of the span.

$$m_{3-3} = +412 - 49.9 \times 27.25 + \frac{1.83 \times (27.25)^2}{2} = -268 \text{ in. lbs.}$$

Fig. 45 gives the curves of the moments and shears. It will be noted that the points of maximum moment occur where the shear equals zero or changes from plus to minus.

63. *Moments, Shears and Reactions*—The computations for all spars follow the example given in Art. 62. Table 7 is a summary of the

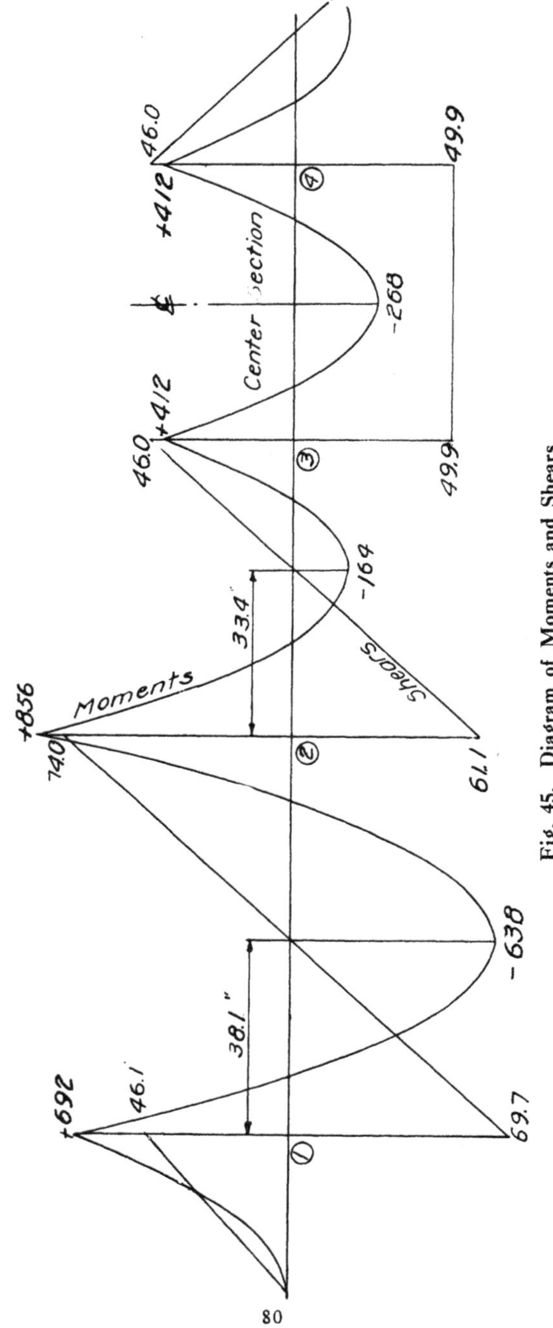


Fig. 45. Diagram of Moments and Shears

results for the four spars. They do not include the effect of eccentric loading, which will be treated in a later article.

*TABLE VII*

*MOMENTS, SHEARS AND REACTIONS*

| | Front Lower | Rear Lower | Front Upper | Rear Upper |
|---|---|---|---|---|
| $r_1$ | 92.4 lbs. | 39.3 lbs. | 115.8 lbs. | 48.8 lbs. |
| $r_2$ | 128.8 | 45.9 | 135.1 | 47.2 |
| $r_3$ | 40.1 | 15.4 | 95.9 | 36.4 |
| $s_{-1}$ | — 37.8 | — 17.1 | — 46.1 | — 20.7 |
| $s_1$ | — 54.6 | — 22.2 | — 69.7 | — 28.1 |
| $s_{-2}$ | — 63.1 | — 21.8 | — 74.0 | — 25.3 |
| $s_2$ | — 65.7 | — 24.1 | — 61.1 | — 21.9 |
| $s_{-3}$ | — 40.1 | — 15.4 | — 46.0 | — 17.9 |
| $s_3$ | 0.0 | 0.0 | — 49.9 | — 18.5 |
| $m_1$ | +567 in. lbs. | +323 in. lbs. | +692 in. lbs. | +390 in. lbs. |
| $m_2$ | +900 | +307 | +856 | +278 |
| $m_3$ | 0 | 0 | +412 | +161 |
| $m_{1-2}$ | —427 | | —638 | |
| $m_{2-3}$ | —540 | | —164 | |
| $m_{3-3}$ | 0 | | —268 | |

64. *Front Lift Truss*—The solution of the front lift truss will be given in detail. The rear truss is analyzed in exactly the same way. Fig. 46 shows the outline of the truss in vertical projection, together with the loads at the panel points and the graphical solution for the stresses. The true plane of the truss is inclined to the vertical plane on account of the stagger, but a little thought will show that when the panel loads are vertical the stresses in the spars are correct when the vertical projection of the truss is used. However, the stagger and slopes of wires will change the actual stresses in the struts and wires. The analytical and graphical methods are explained in Arts. 6 to 11 and 12 to 15, respectively. The analytical computations are given below.

Stress in strut 1-1 = 92 lbs. compression (Method of Joints)
Stress in spar 1-2, lower wing = 0 lbs. compression (Method of Joints)

The stress in spar 1-2, upper wing, is found by the method of moments, taking moments about station 2 of the lower wing.

$$\text{Stress in spar 1-2, upper wing} = \frac{208 \times 78.5}{56.0} = 292 \text{ lbs. compression.}$$

The stress in wire 1 upper to 2 lower is found by the method of shears, using the principle that the vertical component of the stress in the wire equals the shear. This is true because the chords of the truss are parallel and horizontal. The shear equals the loads to the left of a section, or 92+116 = 208 lbs.

$$\text{Projected stress in wire} = \frac{96.5}{56} \times 208 = 358 \text{ lbs. tension.}$$

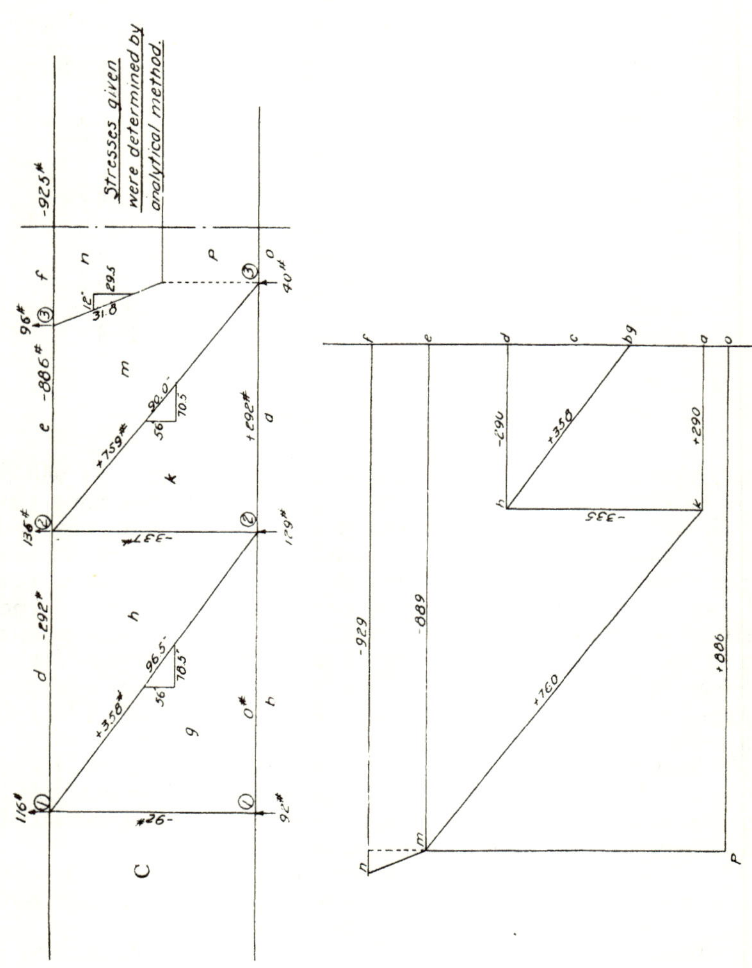


**Fig. 46. Stress Diagram of Front Truss for High Incidence Conditions**

The stress in strut 2—2 is found by applying $\Sigma V = 0$ to joint 2 of lower wing.

Stress in strut = 92 + 116 + 129 = 337 lbs. compression.

The solution for stresses in members of panel 2—3 is similar to that just given for panel 1—2. Using the method of joints at station 3 of upper spar, the vertical component of stress in the cabane strut equals 96 lbs. This stress has a horizontal component of $\frac{12}{29.5} \times 96 = 39$ lbs., and since the strut is in tension this component will produce 40 lbs. additional compression in the center section spar. By the method of joints at station 3 the stress in center section spar equals 886+39 = 925 lbs. compression.

The stresses given on the truss diagram in Fig. 46 are those obtained by the analytical computations. It will be seen that they are in very close agreement with those obtained by the graphical method.

65. *Wing Spar Fittings*—The fittings for the front upper spar which transmit the stresses in the flying wires are shown in Fig. 47. These fittings are eccentric and very similar to the case taken up in Art. 38. The vertical bolts are the true reaction points for the spar, and the eccentricities of the struts and wires are given. The same type of fitting is used at all strut points on this airplane.

66. *Moments and Shears for Offset Strut Loads*—Fig. 48 gives the loading on the upper front spar from the strut loads above. These loads are the stresses in the struts as determined in Art. 64. Case 1, given in Art. 172 for continuous beams applies to this case of loading.

$m_1 = 92 \times 1.75 = +161$ in. lbs.

Apply the three-moment equation (Case 1) to spans 1—2 and 2—3,

$$161 \times 78.5 + 2\,m_2\,(78.5 + 56.75) + 56.75\,m_3 = \frac{337 \times 76.75}{78.5}\left[(78.5)^2 - (76.75)^2\right]$$

$$12640 + 270.5\,m_2 + 56.75\,m_3 = +89520$$

$$(1) \qquad 270.5\,m_2 + 56.75\,m_3 = +76880$$

Apply Case 1 to spans 2—3 and 3—4, noting that $m_4 = m_3$ because of symmetry of wings about the center line of the airplane.

$$56.75\,m_2 + 2\,m_3\,(56.75 + 54.5) + 54.5\,m_3 = 0$$

$$56.75\,m_2 + 277.0\,m_3 = 0$$

$$m_3 = -\frac{56.75\,m_2}{277.0}$$

Substituting this value of $m_3$ in equation (1):

$$270.5\,m_2 - \frac{(56.75)^2\,m_2}{277.0} = 76880$$

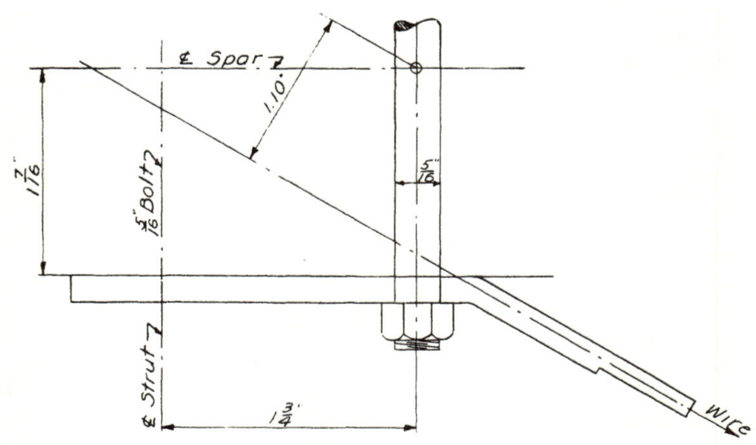


Fitting at Outer Strut Point ① Front Spar

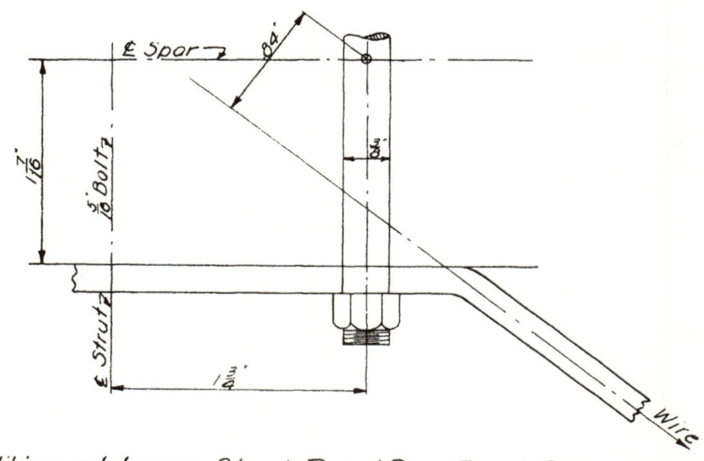


Fitting at Inner Strut Point ② Front Spar

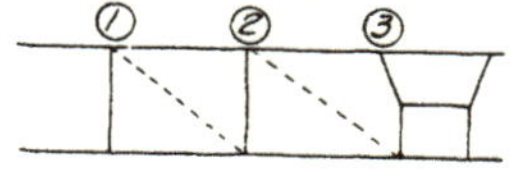


1/4" bolts at Sta. 3

**Fig. 47. Wing Spar Wire Fittings.**

$$m_2 = +\frac{76880}{258.9} = +296.9$$

$$m_3 = -\frac{56.75}{277.0} + 296.9 = -61$$

$$s_{-1} = -92 \text{ lbs.}$$

$$78.5\, s_1 + 161 + 337 \times 1.75 = +297$$

$$s_1 = -\frac{454}{78.5} = -5.8 \text{ lbs.}$$

$$s_{-2} = -337 - (-5.8) = -331.2 \text{ lbs.}$$

$$56.75\, s_2 + 296.9 = -61$$

$$s_2 = -\frac{357.9}{56.75} = -6.3$$

$$56.75\, s_{-3} - 61 = 296.9$$

$$s_{-3} = +6.3$$

$$s_3 = 0$$

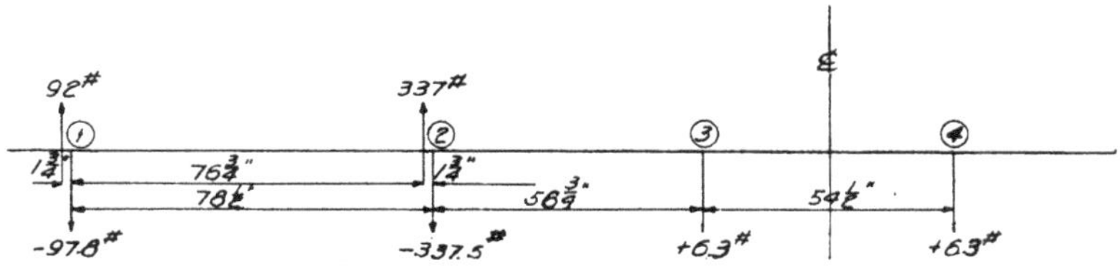

Fig. 48. Offset Strut Loads for Front Upper Spar

67. *Moments and Shears for Eccentric Wire Pulls*—The wire pulls and their eccentricities are shown in Fig. 49. The stresses in the wires were taken from Fig. 46. Case 5 given in Art. 172 applies to this condition of loading.

$$M_1 = 358 \times 1.10 = -394 \text{ in. lbs. (Applied moment)}$$

$$M_2 = 762 \times .84 = -640 \text{ in. lbs. (Applied moment)}$$

$$m_1 = 0$$

$$2\, m_2\, (78.5+56.75)+56.75\, m_3 = 394\times 78.5+2\times 640\times 56.75$$

(1) $270.5\, m_2+56.75\, m_3 = +103570$

$$56.75\, m_2+2\, m_3\, (56.75+54.5)+54.5\, m_3 = 640\times 56.75 = 36320$$

(2) $56.75\ m_2 + 277.0\ m_3 = +36320$

Eliminate $m_2$ between (1) and (2). Divide (1) by 270.5 and (2) by 56.75.

(3) $m_2 + .2098\ m_3 = +382.9$

(4) $m_2 + 4.882\ m_3 = +640.0$

$$4.672\ m_3 = 257.1$$

$$m_3 = + 55.0$$

$$m_2 + (.2098 \times 55.0) = +382.9$$

$$m_2 = +371.4$$

$$78.5\ s_1 - 394 = +371.4$$

$$s_1 = \frac{+765.4}{78.5} = +9.7$$

$$s_{-2} = -9.7$$

$$56.75\ s_2 + 371.4 - 640 = +55$$

$$s_2 = \frac{+323.6}{56.75} = +5.7$$

$$s_{-3} = -5.7$$

$$s_3 = 0$$

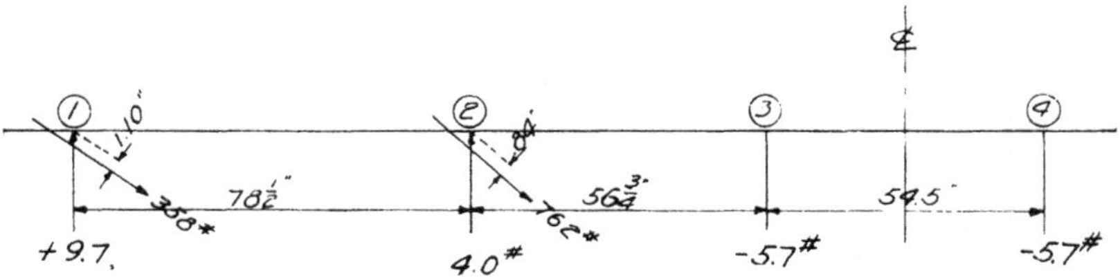


Fig. 49. Eccentric Wire Pulls for Front Upper Spar

68. *Summary of Moments, Shears and Reactions for Front Upper Spar*—Table VIII gives a summary which includes the effect of eccentric strut loads and wire pulls.

*TABLE VIII*

| | Distributed Load | Eccentric Strut Load | Eccentric Wire Pull | Resultant |
|---|---|---|---|---|
| $r_1$ | —115.8 lbs. | — 97.8 lbs. | +9.7 lbs. | —203.9 |
| $r_2$ | —135.1 | —337.5 | —4.0 | —476.6 |
| $r_3$ | — 95.9 | + 6.3 | —5.7 | — 95.3 |
| $s_{-1}$ | — 46.1 | — 92.0 | 0.0 | —138.0 |
| $s_{+1}$ | — 69.7 | — 5.8 | +9.7 | — 65.8 |
| $s_{-2}$ | — 74.0 | —331.2 | —9.7 | —414.9 |
| $s_{+2}$ | — 61.1 | — 6.3 | +5.7 | — 61.7 |
| $s_{-3}$ | — 46.0 | + 6.3 | —5.7 | — 45.4 |
| $s_{+3}$ | — 49.9 | 0.0 | 0.0 | — 49.9 |
| $m_1$ | +692 in. lbs. | +161 in. lbs. | 0 in. lbs. | +853 in. lbs. |
| $m_2$ | +856 | +297 | +371 | +1524 |
| $m_3$ | +412 | — 61 | + 55 | + 406 |
| $m_{1-2}$ | —638 | | | — 724 |
| $m_{2-3}$ | —164 | | | — 156 |
| $m_{3-4}$ | —268 | | | — 274 |

The maximum moment in each span is determined at the point of zero shear by the application of the formula expressing the relation between bending moments at different points in a span. In the following computations the bending moment at the point of zero shear is expressed in terms of the bending moment and vertical shear just to the right of the adjacent support. It should be particularly noted that the bending moment just to the right of the support is equal to the bending moment at the support changed by the moment of the eccentric wire pull. The necessary computations follow:

$$m_{1-2} = m_1 \pm \frac{s_1^2}{2\,w} = m_1 + s_1 x + \frac{wx^2}{2}$$

$$= (853—394) — \frac{65.8^2}{2 \times 1.83} = —724 \text{ in. lbs.}$$

$$m_{2-3} = (1524—640) — \frac{61.7^2}{2 \times 1.83} = —156 \text{ in. lbs.}$$

69. *Spar Properties*—The properties given in Table IX are computed by the exact method, but the approximate method is accurate enough. The sections of the spars in the spans are shown in Fig. 41.

### *TABLE IX*

### *SPAR PROPERTIES ABOUT HORIZONTAL AXIS*

| Spar | Location | Area | I | S | $\rho$ | Q/bI | Remarks |
|---|---|---|---|---|---|---|---|
| Front | In Span | 2.81 sq. in. | 3.12 | 2.00 | 1.055 in. | .763 | Routed |
| Front | Center Section | 3.14 | 3.17 | 2.03 | 1.005 | .597 | Routed |
| Front | Strut Points | 4.50 | 3.66 | 2.34 | — | .333 | Solid |
| Front | Strut Points | 3.23 | 2.63 | 1.68 | — | — | Solid—hole for 3/8 in. bolt |
| Front | Strut Points | 3.43 | 2.79 | 1.78 | — | — | Solid—hole for 5/16 in. bolt |
| Front | Strut Points | 3.62 | 2.95 | 1.88 | — | — | Solid—hole for 1/4 in. bolt |
| Rear | In Span | 3.38 | 1.97 | 1.58 | .764 | .675 | Routed |
| Rear | Center Section | 3.46 | 1.98 | 1.58 | .756 | .597 | Routed |
| Rear | Strut Points | 4.21 | 2.04 | 1.63 | — | .356 | Solid |
| Rear | Strut Points | 3.23 | 1.56 | 1.25 | — | — | Solid—hole for 3/8 in. bolt |
| Rear | Strut Points | 3.38 | 1.64 | 1.31 | — | — | Solid—hole for 5/16 in. bolt |
| Rear | Strut Points | 3.54 | 1.71 | 1.37 | — | — | Solid—hole for 1/4 in. bolt |

The holes were taken 1/32 in. greater than the diameter of bolt.
All holes are vertical.

70. *Deflection of Front Upper Spar*—The formula for deflection in a span with end moments and shears and a uniformly distributed load is given under Case 2A in Art. 172. In using this formula it should be borne in mind that $m_1$ is the moment just to the right of the support at the outer end of the span. As an example of the application of this formula the deflection will be computed for span 1—2. The formula is:

$$EIv = x\,(x-L)\left\{\frac{m_1}{2} + \frac{s_1}{6}(x+L) + \frac{w}{24}(x_2 + xL + L^2)\right\}$$

| | |
|---|---|
| Moment to left of support (1) | = +853 in. lbs. |
| Moment due to eccentric wire pull | = —394 |
| Moment to right of support (2) | = +459 in. lbs. |
| Shear to right of support (2) | = — 65.8 lbs. |
| w = 1.83 lbs. per in. | L = 78.5 in. |

$$\text{When shear equals zero, } x = \frac{65.8}{1.83} = 36.0 \text{ in.}$$

$$EIv = 36.0(36.0-78.5)\left\{\frac{459}{2} - \frac{65.8}{6}(36.0+78.5) + \frac{1.83}{24}(36.0^2 + 36.0\times 78.5 + 78.5^2)\right\}$$

$$EIv = +370{,}300$$

The value of $I = 3.12$ for the span may be taken from the table of properties in Art. 69 and $E = 1{,}600{,}000$ lbs. per sq. in. Then $v = \dfrac{370{,}300}{1{,}600{,}000 \times 3.12} = 0.0742$ in. Maximum moment in span $= m_1 \pm \dfrac{s_1^{\,2}}{2w}$ $= 459 - \dfrac{65.8^2}{2\times 1.83} = -723$ in. lbs. The sign of $\dfrac{s_1^{\,2}}{2w}$ is positive when $s_1$ is positive and negative when $s_1$ is negative.

71. *Drag*—The subject of drag and resolution of forces has been taken up in Art. 48. As this analysis is for high incidence, a value of 8.0 for the ratio of *L/D* will be taken with an angle of incidence of 12 degs. Fig. 50 gives the resolution of forces and it will be noted that the drag component acts forward, or in a negative direction.

72. *Drag Components of Struts and Wires*—Fig. 51 is a diagram of the struts at station 2 and the flying wires in panel 2—3. The loads shown in Fig. 52 are taken from Table VII and are sufficiently exact

for drag calculations. The components of stress may be obtained from the dimensions given in Figs. 51 and 53. The horizontal or drag component parallel to the chord = 11/56, or .196 times the vertical component of the stress in the member for both the struts and wires. A drag component acting forward is taken as negative, and one acting toward the rear as positive.

### *DRAG COMPONENTS AT STATION 1 UPPER WING*

| Member | Vertical Component | Horizontal Component |
|---|---|---|
| Front Strut... | 92 lbs. | .196 × 92 = —18 lbs. |
| Rear Strut.... | 39 lbs. | .196 × 39 = — 8 lbs. |
| Front Wire.... | 92 + 116 = 208 lbs. | .196 × 208 = +41 lbs. |
| Rear Wire.... | 39 + 49 = 88 lbs. | .196 × 88 = +17 lbs. |
| Resultant Drag ............................ | | +32 lbs. |

The total drag, then, for the struts and wires is a concentrated load of 32 lbs. acting towards the rear at the panel point of the drag truss.

The calculation of the drag component of the stresses in the wires in panel 2—3 is made more difficult because of the divided wires from each of the upper strut points at station 2. It is necessary to make use of the theory of least work, which gives a simple solution. Fig. 51 shows these wires, and the first step is to determine the true lengths of the lift wires and the components of their stresses, assuming the vertical components to equal unity. As an example of this analysis, the computations of the stresses in wires C and G will be given in detail.

Wire G

$(70.5)^2 = 4970.25$
$(54.0)^2 = 2916.00$
$(9.25)^2 = 85.56$
$(89.3)^2 = 7971.81$

Wire B

$(70.5)^2 = 4970.25$
$(54.0)^2 = 2916.00$
$(36.6)^2 = 1339.56$
$(96.0)^2 = 9225.81$

Wire D

$(70.5)^2 = 4970.25$
$(56.0)^2 = 3136.00$
$(11.0)^2 = 121.00$
$(90.7)^2 = 8227.25$

Wire C

$(70.5)^2 = 4970.25$
$(56.0)^2 = 3136.00$
$(38.4)^2 = 1474.56$
$(97.9)^2 = 9580.81$

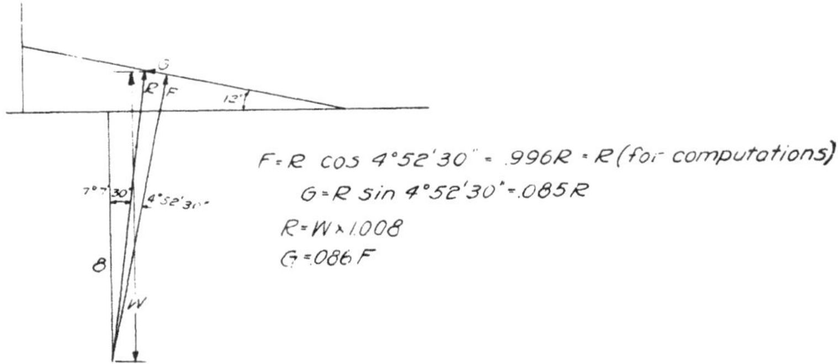


Fig. 50. Resolution of Forces at High Incidence Conditions

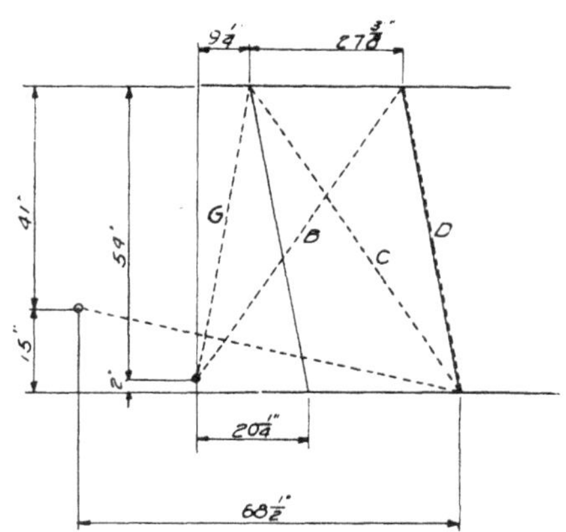


Fig. 51. Wiring in Inner Bay

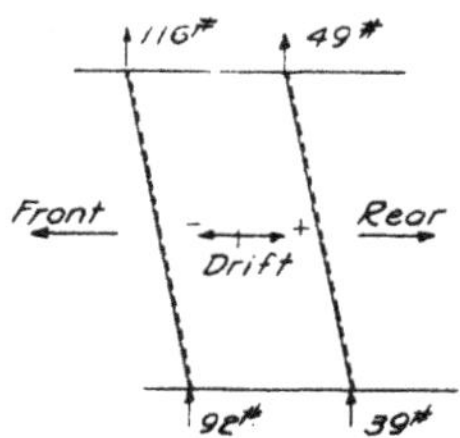


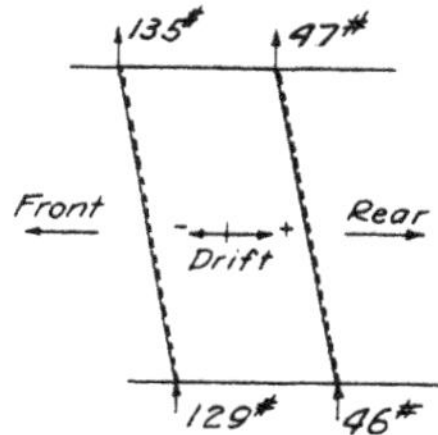


Fig. 52

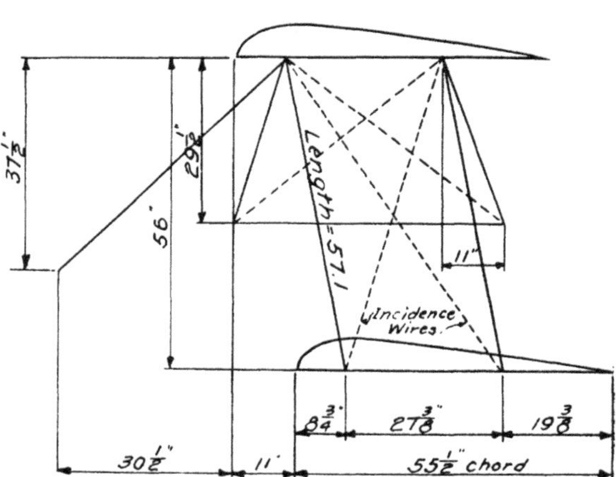


Fig. 53. Wire in Outer and Center Section

*TABLE IXa*

| Wire | Length | Drag Comp. | Stress/Vert. Comp. |
|---|---|---|---|
| G | 89.3 in. | $\frac{9.25}{89.3} = -.104$ | $\frac{89.3}{54.0} = 1.654$ |
| B | 96.0 in. | $\frac{36.6}{96.0} = -.382$ | $\frac{96.0}{54.0} = 1.779$ |
| C | 97.9 in. | $\frac{38.4}{97.9} = +.392$ | $\frac{97.9}{56.0} = 1.749$ |
| D | 90.7 in. | $\frac{11.0}{90.7} = +.121$ | $\frac{90.7}{56.0} = 1.620$ |

The assumption will be made that the vertical load carried by the two wires *G* and *C*, or *B* and *D* is divided between them in such proportions that the work performed by the stresses is a minimum.

For convenience let the vertical components of the stresses in the wires *G* and *C* be denoted by the letters *G* and *C* respectively. The total vertical component of the two wires may be taken from Table VII.

From Art. 63.

$R_1$ (front lower spar) = 92.4
$R_1$ (front upper spar) = 115.8
$R_2$ (front upper spar) = 135.1
$R_2$ (front lower spar) = 128.8

Total Vert. Comp. = 472.1 lbs.

Using the reaction given in Table VIII, $R_2 = 476.6$ lbs.

There is very little difference for this particular airplane, but in some cases the reactions obtained by the method used for Table VIII will be more accurate.

The stresses in the wires are: 1.654 *G* for wire *G*
1.749 *C* for wire *C*

$G + C = 472$ lbs.

Let *R* denote the total work performed in stressing the wires. The expression for the work is given in Art. 116.

$$R = \frac{(1.654G)^2 \times 89.3}{2\,AE} + \frac{(1.749C)^2 \times 97.9}{2\,AE}$$

The two wires will be of the same size, so that the expression 2 *AE* is the same for both terms.

$$R = \frac{1}{2\,AE}\left[244.3\,G^2 + 299.9\,C^2\right]$$

$$C = 472 - G$$

Let $$R = \frac{1}{2\,AE}\left[244.3\,G^2 + 299.9\,(222784 - 944\,G + G^2)\right]$$

Differentiate $R$ with respect to $G$ and put the result equal to zero.

$$\frac{dR}{dG} = 488.6\,G + 299.9\,(-944 + 2\,G) = 0$$

$$1088.4\,G = 299.9 \times 944$$

$$G = 259.9$$

$$C = 472 - 259.9 = 212.1$$

Stress in wire G = 1.654 × 259.9 = 430 lbs.
Stress in wire C = 1.749 × 212.1 = 371 lbs.
Drag component of wire *G* = —.104 × 430 = — 45
Drag component of wire *C* = .392 × 371 = +145

Resultant drag = 100

In the same way, the stress in wire *B* is found to be 141 lbs., and in wire *D* 165 lbs.

Drag component of wire *B* = —.382 × 141 = —54 lbs.
Drag component of wire D = .121 × 165 = +20 lbs.

Resultant drag = —34 lbs.

The drag component of the stress in the strut equals 11.0/56.0, or .196 times the vertical component.

Fig. 52 gives the reactions at the strut points at stations 1 and 2.
Drag of front strut = .196 × (92+116+129) = —66 lbs.
Drag of rear strut = .196 × (39 + 49 + 46) = —26 lbs.

Total drag = —92 lbs.

*SUMMARY OF DRAG COMPONENTS AT STATION 2*

| Member | Drag |
|---|---|
| Front strut | — 66 lbs. |
| Rear strut | — 26 lbs. |
| Wire *G* | — 45 lbs. |
| Wire *B* | — 54 lbs. |
| Wire *C* | +145 lbs. |
| Wire *D* | + 20 lbs. |
| Resultant drag | — 26 lbs. |

73. *Air Load Drag*—The drag from the air load on the upper wing equals .085 times the load on the wing, and acts forward as shown in Fig. 50. It will be sufficiently accurate to assume that the drag load is concentrated at the panel points of the drag truss and that the panel points carry the drag load from half of the adjacent spans. In the case of the cantilever portion of the wing the entire drag will be assumed as concentrated at the panel point located over the outer struts. The drag load between the outer struts = —.085 (1.83 + .68) = —.213 lbs. per inch of run.

74. *Loads on Drag Truss*—The loads at the panel points, as shown in Fig. 54, are made up of the drag from the air load, the thrusts of the struts, and the wire pulls.

Load at Point (*A*)

From cantilever = —.085 × 66.7 = — 5.7
From panel A—B = —.213 × 19.6 = — 4.2

—10
From struts and wires = +32
Total = +22 lbs.

Panel Point (*B*)

Load = —.213 × 39.25 = —8 lbs.

Panel Point (*C*)

$$\text{From panel} = -.213 \times \left[\frac{39.25 + 28.5}{2}\right] = -7$$

From struts and wires = —26

Total = —33 lbs.

Panel Point (*D*)

$$\text{Load} = -.213 \times \left[\frac{28.5+29.6}{2}\right] = -6 \text{ lbs.}$$

75. *Struts*—The dimensions of the struts are given in Table X. The properties of the sections were obtained with the aid of the following formulas:

$$\text{Area} = .71 \text{ bd}$$
$$\text{I (long axis)} = .042 \text{ bd}^3$$
$$\rho \text{ (long axis)} = .24 \text{ d}$$

These formulas were applied to the former U.S.A. standard strut section for values of fineness ratio between 2.0 and 5.0. The maximum diameter was located at 40 per cent of the long axis from the nose. Since the maximum diameter of the strut section of this airplane is at 38 per cent of the long axis from the nose it will be accurate enough to use these formulas. The actual length of the front struts is 53½ in. and of the rear struts 54½ in. All struts are tapered.

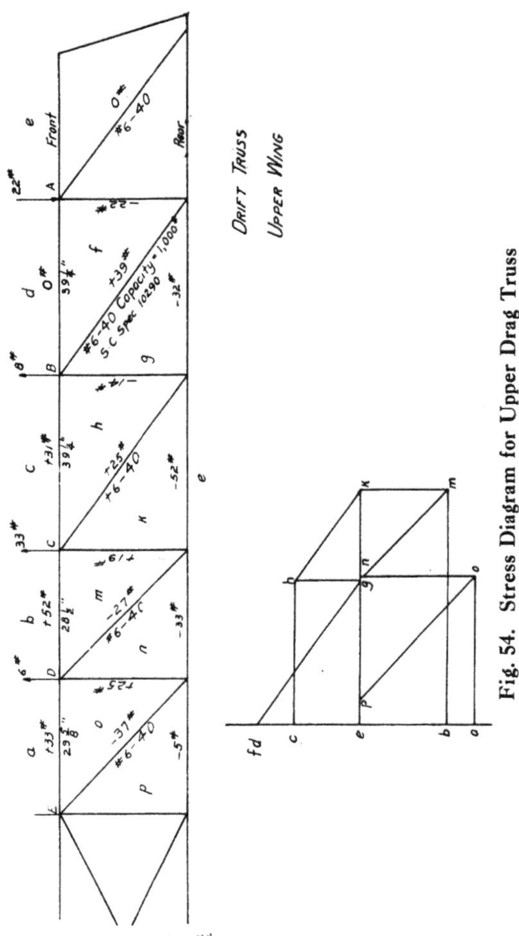


Fig. 54. Stress Diagram for Upper Drag Truss

## TABLE X
## DIMENSIONS OF STRUTS

| Struts | Average Length | Short Diameter "d" | Long Diameter "b" | Fineness Ratio b/d |
|---|---|---|---|---|
| Front and Rear Outer ....... | 54 in. | $1\,^1/_8$ in. | $4^1/_4$ in. | 3.78 |
| Front and Rear Intermediate.. | 54 in. | $1^{15}/_{32}$ in. | $4^9/_{16}$ in. | 3.12 |

## PROPERTIES OF STRUT SECTIONS

| Struts | I Long Axis | $\rho$ Long Axis | Area | $L/\rho$ |
|---|---|---|---|---|
| Front and Rear—Outer...... | .254 | .27 | 3.40 | 200 |
| Front and Rear—Intermediate | .608 | .35 | 4.77 | 154 |

As an example of strut calculations, the factor of safety for the front intermediate strut will be given. The strut will be assumed as pin-ended and the ultimate load computed from Euler's formula, using a value of $E = 1{,}600{,}000$ lbs. per sq. in.

$$\text{Ultimate}\ \frac{P}{A} = \frac{\pi^2 E}{(L/\rho)^2} = \frac{\pi^2 \times 1{,}600{,}000}{(154)^2} = 665 \text{ lbs. per sq. in.}$$

$$\text{Ultimate load} = 665 \times \text{Area} = 665 \times 4.77 = 3180 \text{ lbs.}$$

This load is for a strut of uniform cross-section. These struts being tapered, the ultimate load equals $.85 \times 3180 = 2700$ lbs.

From Fig. 46 the vertical components of the stress in the strut equal 337 lbs.

$$\text{Strut load} = \frac{57.1}{56.0} \times 337 = 344 \text{ lbs.}$$

$$\text{The factor of safety} = \frac{\text{Ultimate load}}{\text{Load}} = \frac{2700}{344} = 7.9$$

The front outer strut has a factor of safety of 12.1

76. *Stress Table*—Table XI gives the properties of the spar sections at each of the strut points, and at the point of maximum bending moment in each span. Reference should be made at this point to Fig. 41 in which sections of the front and rear spars are shown. The areas, section moduli, and radii of gyration are taken directly from Table IX.

In the determination of the ratio of slenderness, $L/\rho$, the unsupported length for intermediate points is considered as the distance between strut points, and for inner strut points as two-tenths of the sum of the adjacent spans.

The deflection at the point of zero shear in the outer span has already been calculated in Art. 70, and the same method may be employed in the determination of the deflection in the inner span.

The moments in the next column are taken from Table VIII and represent the resultant moments due to the distributed load, eccentric strut loads, and eccentric wire pulls. Reference should be made at this point to Art. 51 in regard to the condition of maximum bending moments at strut points. It will be noted that the spar section at the strut points is solid with a vertical bolt hole deducted. At a point five or six inches from strut points routing begins. The strength of the spar at this point as well as at the strut point should be investigated, taking into consideration the change in bending moment, the modification of the spar section, and the reduction in the allowable unit stresses due to the effect of routing.

The Moment *Pd* is due to the eccentric application of the direct stress and is obtained as the product of the resultant direct stress and the maximum deflection in the span under consideration. It should be noted that the eccentric moment varies as the square of the load factor since it is the product of two quantities, each of which varies directly as the load factor. The moment due to the distributed load and eccentric strut and wire loads varies directly as the load factor. Consequently it is necessary to multiply the eccentric moment by an assumed value of the load factor before computing the resultant moment. It may be necessary to make several trials before the assumed value of the load factor corresponds with the actual load factor.

The ultimate allowable stress is determined as suggested in Art. 174 for the case of laterally loaded columns. At the outer strut point the full compressive strength of the material is allowed in the calculation of the allowable fibre stress. In all other cases a reduction in the maximum compressive stress is made to compensate for column action as recommended in the article referred to above. The modulus of rupture is here assumed as 8000 lbs. per sq. in. for the routed sections. The correct value for the reduction of the modulus of rupture due to routing of the spar may be calculated as indicated in Art. 174.

## *SPECIAL FORMS OF LIFT TRUSSES*

77. *Clark Truss*—The special features of this truss are clearly brought out in Fig. 55. There is only one lift truss, the lift wire attaching to the rear upper spar. The wire may be attached to the fuselage at the lower front spar as shown in Fig. 55, or at any other convenient point. The stress analysis of this truss is similar to the one given in this chapter, but it should be noted that the lift load at point *A* has to be carried down through strut *AC* and up through the strut *CB* before it can be transferred to the lift wire. In the same manner the loads at point *C* and *D* have to be transmitted to point *B*.

For normal flight the rear upper spar is the only one carrying the lift compression, while the other three spars carry only bending moments, or moments and drift compression. This allows a saving in weight for the other spars, although increasing the size of the rear upper spar. In reversed, or upside down flights the front lower spar is the only one

## *UPPER FRONT SPAR*

## *TABLE XI*

| Station | Area | Sect. Mod. | $\rho$ | $L/\rho$ | Deflection d | Moments | | | Compression | | | $f_c$ | $f_b$ | $f_c+f_b$ | Ult. Allow. Stress | F.S. |
|---|---|---|---|---|---|---|---|---|---|---|---|---|---|---|---|---|
| | | | | | | Load | Pd | Total | Direct | Drag | Total | | | | | |
| 1 | 2 | 3 | 4 | 5 | 6 | 7 | 8 | 9 | 10 | 11 | 12 | 13 | 14 | 15 | 16 | 17 |
| 1 | 3.43 | 1.78 | | | | + 853 | | + 853 | —292 | 0 | —292 | 85 | 479 | 564 | 9,575 | 17.0 |
| 1—2 | 2.81 | 2.00 | 1.055 | 75 | 0.0742 | — 724 | —19.3 | — 965 | —292 | +32 | —260 | 93 | 482 | 575 | 7,160 | 12.5 |
| 2 | 3.23 | 1.68 | 1.005 | 27 | | +1524 | | +1524 | —886 | +52 | —834 | 258 | 908 | 1166 | 9,170 | 7.9 |
| 2—3 | 2.81 | 2.00 | 1.055 | 56 | 0.00171 | — 156 | — 1.4 | — 174 | —886 | +33 | —853 | 303 | 87 | 390 | 4,940 | 12.7 |
| 3 | 3.62 | 1.88 | 1.055 | 20 | | + 406 | | + 406 | —925 | +33 | —892 | 247 | 216 | 463 | 7,740 | 16.7 |

carrying lift compression. If the lift wire is carried forward still more than shown in the sketch it may be possible to relieve the compression in the rear upper spar by transferring the compression to the front upper spar. Trial computations will establish the correct location of the wire for the least weight of spars.

The drift loads from the thrusts of the struts and the wire pull must be determined before solving for the stresses in the drift trusses. These trusses are best adapted for large stagger for the reason that when the lift truss is vertical the drift loads from the struts and wires are a minimum.

The chief advantage of a single lift truss is the lessening of the resistance due to the small number of wires. In a pursuit airplane the single truss has the great disadvantage that if one wire is shot away the wings will collapse. The upper front and lower rear spars may be damaged without injuring the lift truss.

Fig. 60 illustrates the typical truss applied to multi-engined airplanes.

78. *S.P.A.D. Truss*—This truss is arranged to support the spars in the center of their span by the auxiliary struts at station 2 in Fig. 56. The portion of the truss in action when in normal flight is shown by the full lines in Fig. 57.

The loads $W_3$ and $W_4$ are equally divided between members 3 and 4. The vertical component of the stress in 3 is carried upward by the strut 5 and then down through the wire 6—4. The horizontal component of the stress in 3 produces compression in the lower spar 9—10. The remainder of the stress analysis is similar to the one in the first part of this chapter. In landing, wire 11 is in action and wire 6 is inactive.

This type is especially applicable in triplane construction, and in biplane construction where the span or loading is too great for an ecenomical design of a one-bay airplane, yet not large enough to warrant a two-bay airplane.

79. *Thomas-Morse Truss*—This form of truss, shown in Fig. 58, resembles somewhat that used on the S.P.A.D. airplane, but the lift and flying wires which cross at the inner interplane strut have no connection to the strut. It is evident that there is no tension on the wing hinges of the lower wing when in normal flight. The stress analysis presents no difficulties. It may be noted that the lower wing spans may be shot away without endangering the upper wing or truss system.

80. *S.V.A. Truss*—This airplane, Fig. 59, uses the familiar Warren truss in which the diagonals take either tension or compression. The diagonals are oval steel tubes with the ordinary crossed incidence wires. A drift wire (shown by the dotted line) leads from the rear upper spar to the nose of the fuselage. This type of truss permits an overhang on the upper wing without an increase in the strength of the spars. Rigidity and ease of assembly are two of the advantages of this form.

81. *Monoplane Wings*—Partly because of a number of accidents due to faulty designing, the monoplane has not been as widely used as it should have been. There are several important advantages inherent in this type of wing construction. Aerodynamically, a monoplane wing

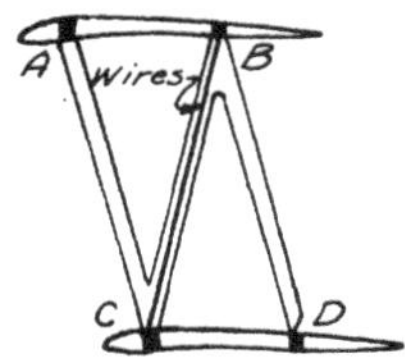


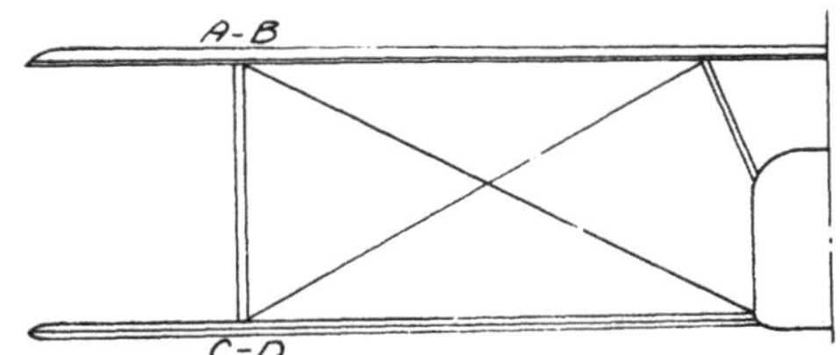


Fig. 55. Clark Truss

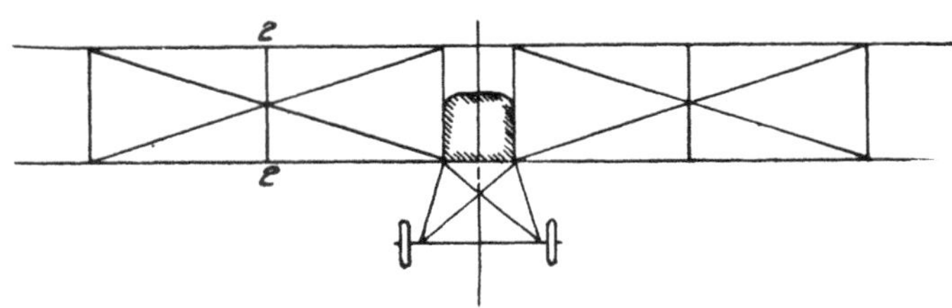


Fig. 56. S.P.A.D Truss

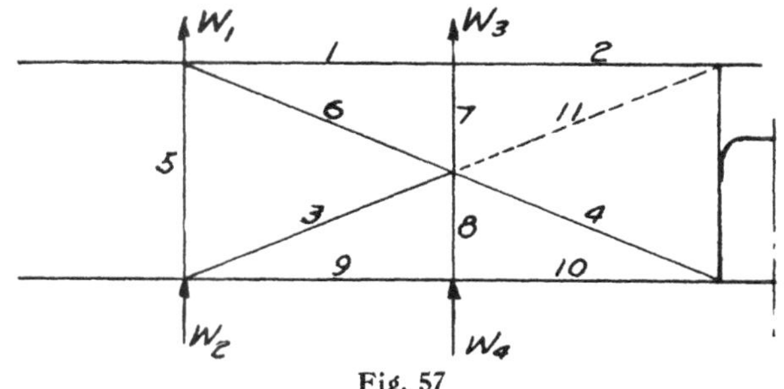


Fig. 57

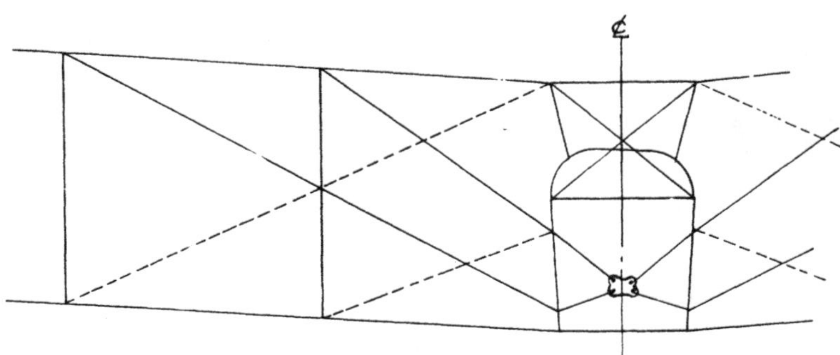


Fig. 58. Thomas-Morse Truss

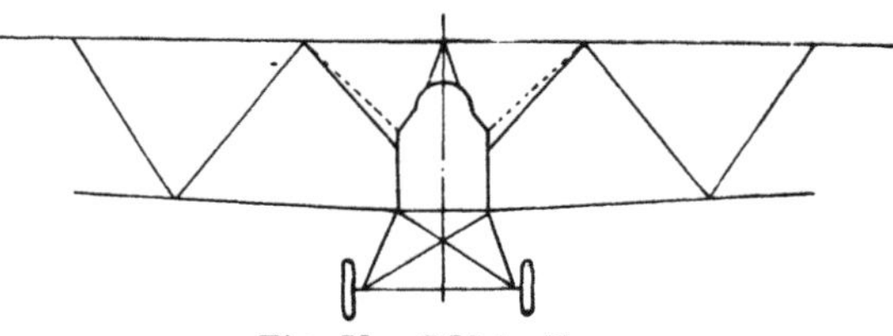

Fig. 59. S.V.A. Truss

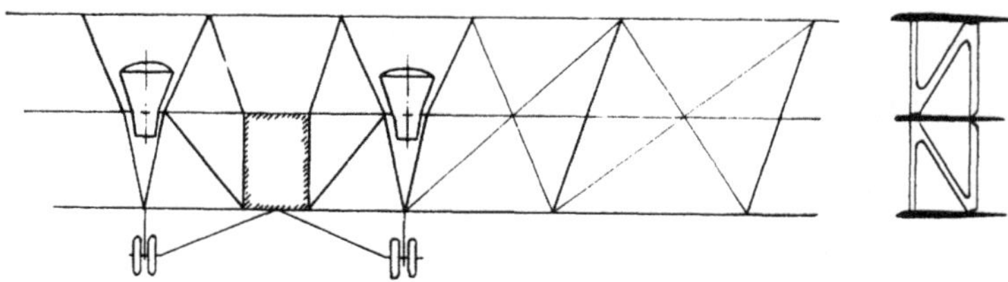

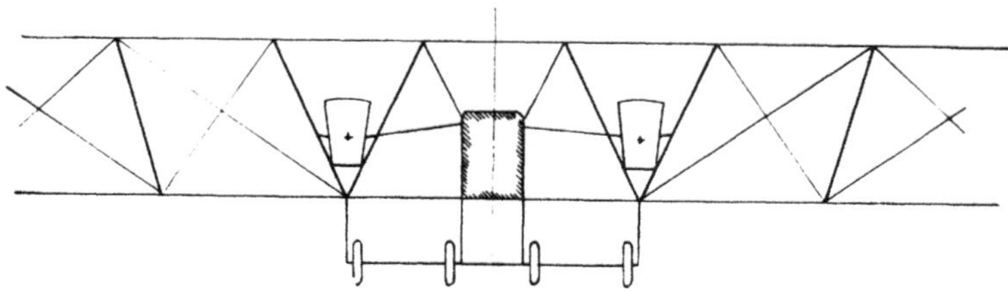

Fig. 60. Typical Wing Trusses of Multi-Engined Airplanes

is much more efficient than a biplane or triplane wing. From the standpoint of visibility a monoplane wing is excellent because there is no interference from a lower wing. If a pilot is so placed that his eye is on a level with the wing of a monoplane his blind angle due to the wings is very small. Even an externally braced monoplane cellule is much easier to assemble and true up than a biplane cellule, while an internally braced monoplane wing offers no rigging difficulties whatever. Monoplanes braced by complicated trussing are entirely obsolete. If external bracing is employed it should consist of struts or wires from the front and rear spars to the bottom of the fuselage. If struts are used, the reactions from a negative air load are carried by them in compression, and if the wings are braced by wires against the direct air loads only, then the spars, which must in this case be continuous over the fuselage, acting as cantilevers carry in bending all negative loads. A wing structure should have a factor of 3.0 to 3.5 under reversed load. If a design is well adapted to a wing with continuous spars the wire type of bracing is satisfactory as it offers a minimum of resistance. In comparing a wire braced wing with an internally braced one it should be noted that the former type when acting as a cantilever has only about 40 per cent of the load to carry that comes on the latter from the direct air load. For this reason one of the thin wing sections might be used on a wire braced wing if the greater efficiency of the thin wing is of much importance. For internally braced monoplanes the U.S.A. 27 wing section is excellent, combining as it does large spar depth, high lift and high efficiency. It will not prove as satisfactory on a strut braced cellule, as it is impossible to take full advantage of the spar depth afforded. Incidence wires in the plane of the struts on a braced wing should be omitted. If they are effective in carrying drag loads, they put such heavy secondary stresses in the struts that the latter fail at a low factor. All drag should be taken care of by the internal drag trussing. The use of both incidence wires and drag trussing is an example of the kind of an indeterminate structure that should always be avoided.

82. *Multi-Spar Construction*—Wing construction of this type may be used to meet the specifications of wing strength and wing shape. Aerodynamic considerations may require a wing section so thin that it would be impracticable to use the ordinary two-spar construction because of the uneconomical cross section of spar which strength requirements would necessitate.

## Distribution of Load Between Spars

In the stress analysis, the first problem that arises is that of the division among the spars of the distributed load. The problem will be discussed from two points of view: First, assuming that the deflections are equal and second, that the deflections may or may not be equal. In either case there are, as fixed conditions of the problem, the spar spacing, the spar stiffness denoted by the quantity $EI$, and the wing loading. There are then three known or fixed conditions. With the first assumption accepted, i.e., that the deflections of the spars are all equal, there is a redundancy of limiting conditions. The problem can be solved by using two of the above known conditions and neglecting the third. First, answer the question with the spar spacing fixed and the spar stiffness known. Since the deflections of all the spars are equal, it can be stated that $d = \frac{KW(L^n)}{EI} = d_1 = \frac{KW_1(L^n)}{EI_1}$, etc., and since $L$ and $E$ are equal, by writing $W/I = W_1/I_1$, etc., the proposition is developed that, no matter what the distribution of loading may be, the spars take a load in proportion to their moments of inertia.

In the same manner, by holding to the two conditions of wing loading and spar stiffness, the proposition may be set forth that no matter what the spar spacing may be, the load is divided among the spars in proportion to their moments of inertia.

Third, with the condition of loading and spar spacing as the two factors considered we may, by application of the "Three Moment Equation," solve for the distribution of the load among the spars. In the case of a wing built with rigid compression ribs closely spaced, so that the wing construction approaches so-called "unit" construction, the assumption that the deflections of the spars are equal will closely approximate the truth. Hence, with such a type of construction the proposition has justification that no matter what the condition of loading may be, the distribution will be in proportion to the moments of inertia of the spars.

In general it probably will not be economical to tie the spars together rigidly enough to obtain this unit construction, and in that case it is evident that manner of loading, spar spacing and spar stiffness are all vital factors, none of which may be neglected. The assumption that all the spars deflect equally may then be considered fallacious except in some special cases.

One of the most important of these is the case of a cantilever or internally braced wing. Here the deflection of the spars at all points along their length is in the same direction. By running a metal strap over the weaker or more heavily loaded trusses, such as the extreme front and rear trusses, and under the stronger, and generally less heavily loaded trusses in the deeper part of the wing section, or by otherwise suitably bracing between these trusses at several points, the deflection of all the trusses can be nearly equalized. Therefore, they will all carry loads closely proportional to their moments of inertia. In the construction of large monoplanes of the heavy bomber type this is an important consideration.

For another case, make no assumption as to the relative deflection of the spars. Take for a specific example the condition as shown by Fig. 61. Fig. 62 shows the method of distribution of the load between the spars. The wing is considered as being stiff enough to carry the load to the spars but not stiff enough to influence, through its rigidity, the distribution of the load to each spar. It is clear that spar spacing and condition of loading are the determining factors of the problem, and that the factor of spar stiffness does not affect the distribution under this assumption. In justification of this method of distribution of load between the spars the example in Fig. 63 is shown. Here a standard two-spar section is taken and the proportionate part of the load going into the front spar is computed by the method given in this paragraph, and the result compared with the value obtained by the standard method.

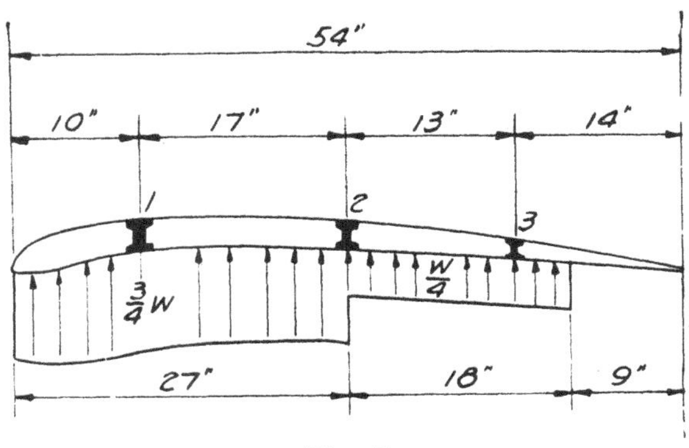


Fig. 61

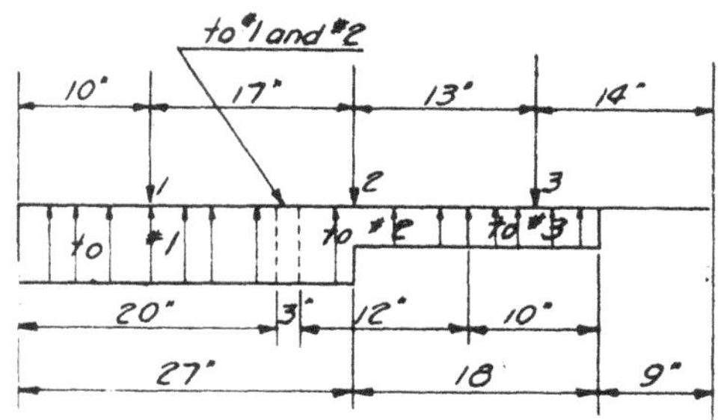


COMPUTATIONS OF REACTIONS

SPAR

1 $\quad 10 \times 2 \times \frac{W}{36} + \frac{W}{36} \times 3 \times \frac{5.5}{17} = .555\,W + .027\,W = .582\,W$

2 $\quad \frac{W}{72} \times 8 + \frac{W}{36} \times 4 + \frac{W}{36} \times 3 \times \frac{11.5}{17} = .222\,W + .056\,W = .278\,W$

3 $\quad 5 \times 2 \times \frac{W}{72} = .139\,W$

Fig. 62

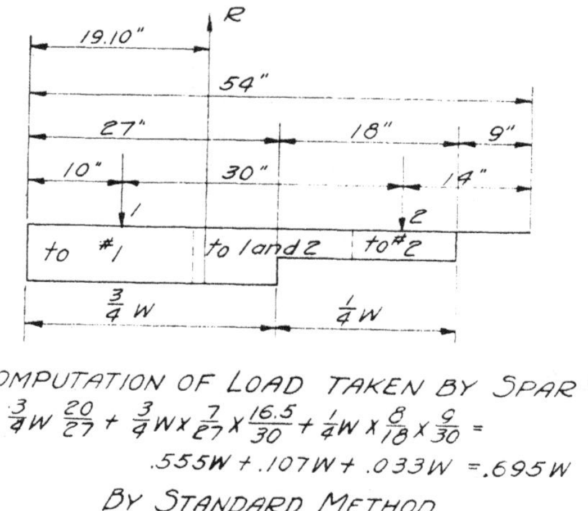


Fig. 63

## Division of Eccentric Moments and Shears Among the Spars

These moments and shears in the upper wing are caused by eccentric strut thrusts or lift wire pulls. The points of application of these forces are on the center lines of spars 1 and 3. The forces are in about the same proportion to each other as are the stresses in the front and rear lift trusses. These forces are applied at the strut points, and since at these points the spars are very rigidly connected, they all act together as a unit. If the load were uniformly distributed over the wing section at this strut point it would be divided among the spars in proportion to their moments of inertia. But, since these forces are applied directly at the center lines of the outside spars, the division of the load will not be in proportion to the moments of inertia, but will approach that division where each outside spar takes all the load applied directly to it. The true division probably lies between these two outside limits and, considering the magnitude of the moments and shears caused by these eccentric thrusts and wire pulls, the error caused by arbitrarily taking the average between the results obtained by these two limiting conceptions will not be outside the limits of precision of the complete analysis.

## Division of the Loads from the Lift Trusses

The loads from the lift trusses are also applied at the center lines of the outside spars. These loads are to be resolved with one component in the plane of the spars. With stiff connections between the spars at the strut points, and perhaps with the spars tied together at one or two places in mid span, the action is that of an eccentrically loaded column. The load on each spar may be computed by an adaptation of the formula $L = A_s (P/A + M \times y/I)$. Here $L$ is the proportionate part of the load taken by each spar; $A_s$ is the area of the spar under consideration; $P$ is the load whose effect is desired; $A$ is the area of all the spars; $I$ is the moment of inertia of all the spars about the center of gravity of all the spars; $y$ is the distance of the spar under consideration from the center of gravity of all the spars, and $M$ is the moment caused by the load $P$ which is eccentric with respect to the center of gravity of all the spars.

## Solution of the Drag Truss System

By assuming the drag truss to be pin-jointed and by neglecting the stresses brought into the spars because of the rigidity of the compression ribs, it is evident that there are no direct stresses from the drag truss in the intermediate spars. Fig. 64 represents a stress diagram for the drag truss computed under these assumptions. It is clear, however, that this solution can only approximate the truth and that it will not, except in a few cases, be of sufficient accuracy. By assuming stiff connections at the joints, the application of the theory of "Least Work" to the problem is justified. Fig. 65 shows the drag truss stress diagram with stresses computed by this method. For ease of computation, the assumption is made that the areas of the spars are equal. Referring to Fig. 65, and with particular reference to the stresses in the members cut by section *AA*, by taking moments about points 1 and 2 there can be obtained two independent equations involving three unknowns. The stress in the diagonal can be computed by the method of shear. The one equation of condition that is then required may be obtained by stating that the total work done by the members cut by the section *AA* will be a minimum. Fig. 65 shows the drag truss stresses in the spars as computed by this method.

Comparison of the results as given by Fig. 64 with those as given by Fig. 65 shows that the simpler method of Fig. 64 is conservative and is close enough for design purposes, but that for an accurate stress analysis the method of Fig. 65 is probably better.

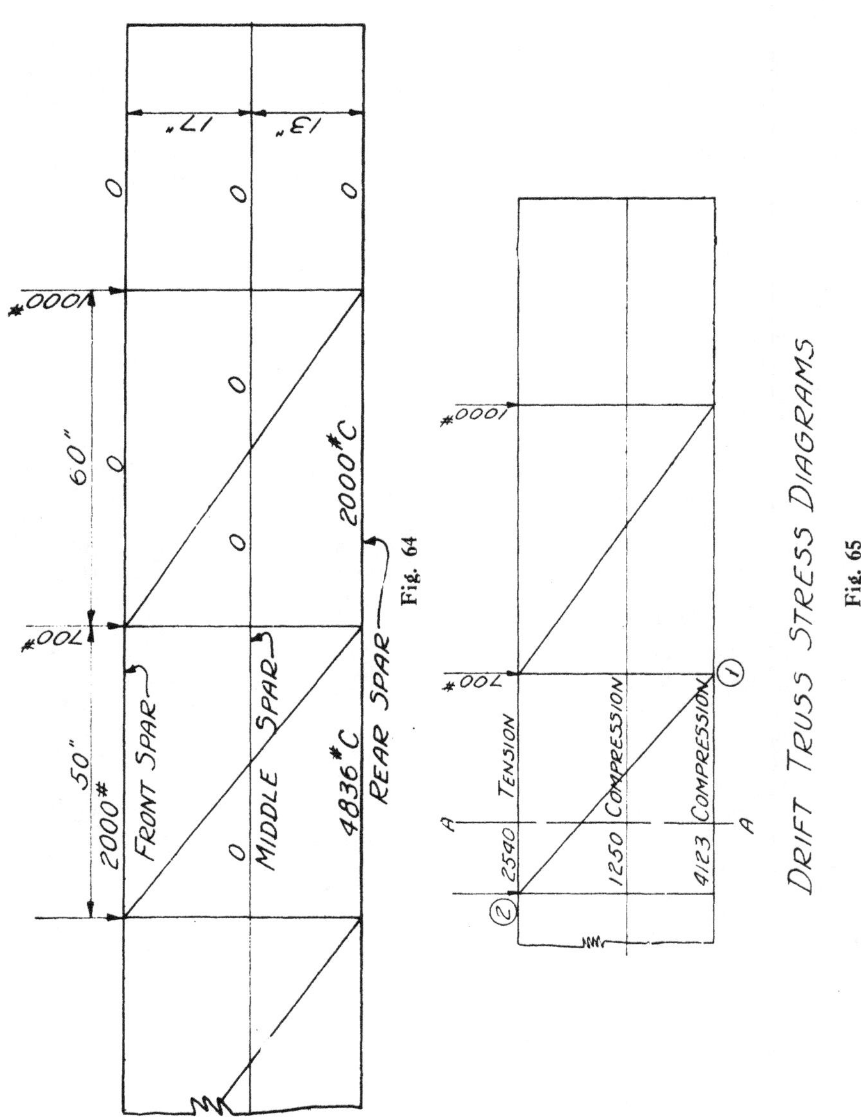


Fig. 64

DRIFT TRUSS STRESS DIAGRAMS

Fig. 65

## CHAPTER IV

# MISCELLANEOUS DESIGN

### Ribs

83. *Plywood Web Type*—For chord lengths greater than a certain value, the built-up truss rib is more economical than the plywood web rib. This value will vary for different types of wing sections, being less for the thicker types. For the sections represented by the R.A.F. 15, 95 to 100 inches chord length is approximately the maximum. Extensive testing and development work has shown that the rib section of the type shown in Fig. 66 is the best. For chords longer than 75 inches, the capstrips should be increased in size. The web thickness, however, is probably satisfactory for all chord lengths between 60 and 100 in. The function of the web in carrying shear and compression determines the direction of the face plies. Since the vertical members are in compression, and since only the material with the grain vertical is effective in compression, the best results are secured by having the face plies thin and placed with their grain vertical. The relatively thick core

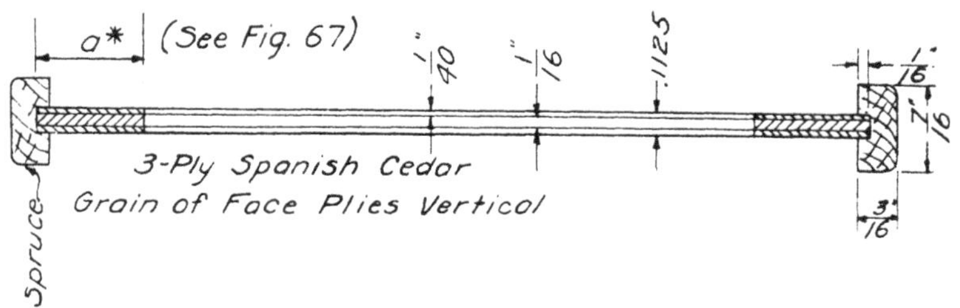


Fig. 66. Cross-section of Standard Rib

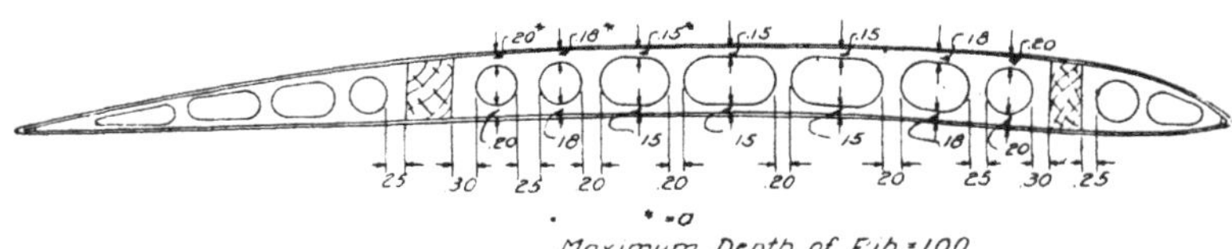


Fig. 67. Location of Cutouts in Terms of Maximum Rib Depth

used by spacing these plies far apart increases their moment of inertia. Numerous tests have demonstrated the superiority of an all-Spanish cedar web over a web with face plies of thin (1/70 to 1/100 in.) maple or birch veneer and a core of poplar or basswood, or a web in which both face plies and core are of some light wood such as pine, poplar, basswood, or spruce. The size, shape, and location of the web cutouts is another important matter. Fig. 67 gives what is perhaps nearly an ideal arrangement, but the presence of internal drift wires which must pass through the ribs may change this arrangement of the lightening holes somewhat, in order to secure clearance with all the ribs. The amount of material necessary in the web is dependent on the shear, hence the small cutouts near the spars. The ultimate strength of ribs of the type just described, for a chord length of 66 in., may be taken as 350 to 400 lbs. if the attachment of the rib to the spar is good.

84. *Attachment of Rib to Spar*—Since many rib failures are due to improper methods of securing the ribs to the spar a consideration of the various methods of doing this is important. The difficulty is caused in transferring the shear from the web to the spar. Unless some positive means is provided for accomplishing this, the shear is carried by the capstrips, and if these are not properly fastened to the spar they are very liable to pull free from the web. Incidentally, the use of light, cement-coated nails to connect the capstrip and web will not prevent this. In general, nails should be used only to hold the members in place during gluing. A direct, and excellent method of securing the web to the spar is shown in Fig. 68. The use of triangular blocks is best adapted to box spar construction. The same idea can be carried out on routed spars with plywood blocks, as shown in Fig. 69. However, the general tendency of construction is to reduce to a minimum extra parts and weight. The use of a tenoned web illustrated in No. 2, Fig. 70 overcomes this disadvantage and accomplishes the same results as the blocks. There are two objections to this method; the fact that the ribs must be assembled on the spar, and that a shrinkage of the spar might cause serious splitting stresses in the latter if the tenon fitted closely. Yet the method is very satisfactory from the point of view of low weight, simplicity, and strength, and has gained considerable favor. Nos. 1 and 3 of Fig. 70 illustrate two methods for preventing separation of the capstrips and web. The rib is taped and glued in No. 1, and in No. 3 the capstrip is held by very light steel clips secured to the spar by 2½ inch screws. The clips are used only on the top of the spars, and unless the overhangs at the nose and tail are rather large, only on the inside edge, as is indicated in the photograph. However, in the case of machines which are to be stunted, a steel clip should be used on the top of the front spar on the outside as well as on the inside edge. If tape is employed instead of steel clips, the ribs on such airplanes should be taped on the front side of the front spar also. For the reason that the screws have a tendency to split the spar flanges clips are not advised if the flange thickness is less than 3/4 in. In connection with these last two methods experiments have shown that the capstrips are strong enough

in straight shear to develop the full strength of the rib if they can be prevented from pulling away from the web. The manner in which the wings are assembled is important in deciding on the means of attaching the rib to the spar. Complete assembly of the ribs before slipping on the spar is not recommended, because of the clearances that are necessary and the danger of breaking the glue joint between the web and capstrips. Also, fittings and blocks have to be put on the spar after the ribs are in position if this method is followed, a feature that is often undesirable. In summary it may be said that methods employing little blocks or pieces of plywood are too laborious to be adapted to production. The added weight is also a serious item. The method involving tenoning of the web is difficult from the production standpoint. However, if labor is not an important consideration the results obtained are excellent. The two methods recommended as satisfactory in every way are those using the tape and glue, or the steel clip.

85. *Thick Sections*—In connection with a very thick wing section such as is found on the Fokker airplane it is suggested that several small vertical stiffeners of triangular section glued to the web between the spars add much to the stiffness of a thin veneer web, which relieved of its vertical compression can carry horizontal shear.

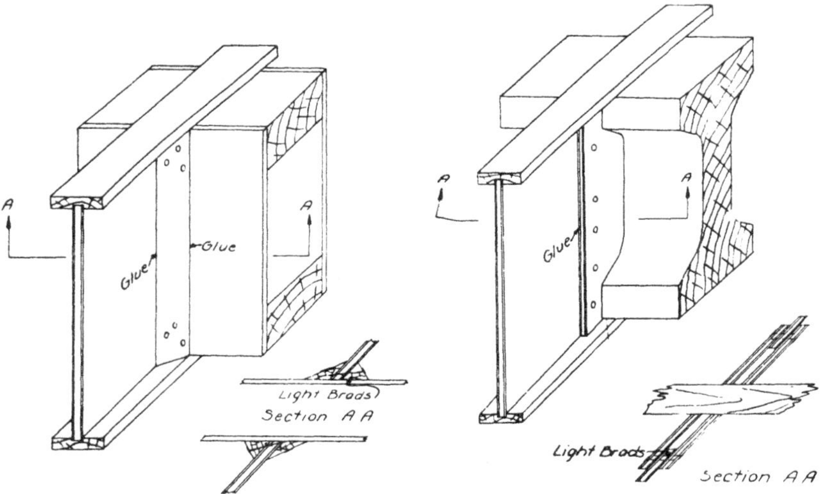


Fig. 68. Attachment of Ribs to Spar with Triangular Blocks.

Fig. 69. Attachment of Ribs to Spar with Plywood Blocks.

**Fig. 70. Three Methods of Attaching Rib to Spar**

86. *Split Capstrip*—Instead of a single capstrip, grooved a 1/16 in. to receive the web, good results have been obtained in some designs with a double or split capstrip, as shown in Fig. 71. Because of the large gluing surface the likelihood of separation of the capstrips and web is reduced. If a very thin web is used it is practically impossible to nail vertically through the web, as is done with the one piece capstrip. But with the split capstrip horizontal nails can be used very effectively. The latter type, however, is the more difficult to assemble, partly because of the extra piece, and partly because such a rib must be completely assembled and then slipped on the spar, which, as previously suggested, requires undesirable clearances.

87. *Compression Ribs*—In place of the usual double web compression rib, the type shown in section in Fig. 72 is suggested as being more satisfactory. The size of the ribbed part of the web is calculated by treating it as a column with round ends as far as lateral bending is concerned.

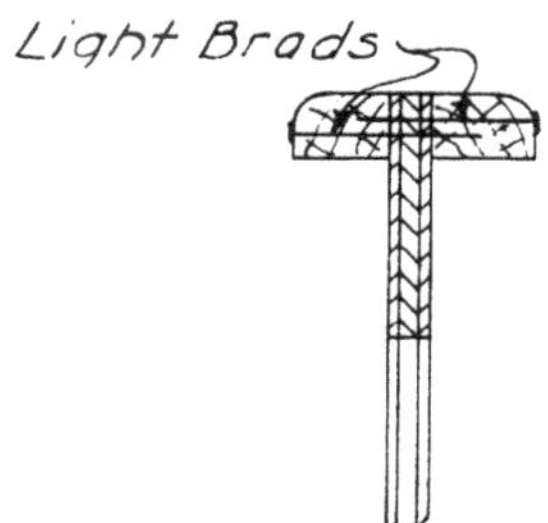


Fig. 71. Split Capstrip

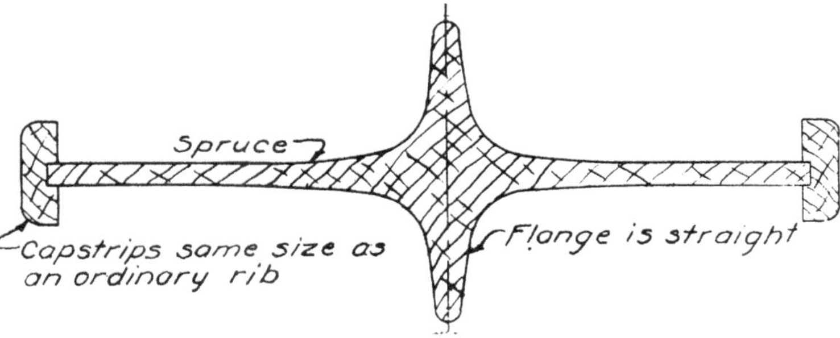


Fig. 72. Cruciform Compression Rib

88. *Truss Ribs*—The Forest Products Laboratory recently completed a series of tests on 15 ft. truss ribs of wood construction. Many different types of design were tried out, including Warren trusses and single and double Pratt trusses. The design found the most satisfactory as regards the strength-weight ratio, reliability, and simplicity of construction is the Warren truss, all-spruce rib illustrated in Fig. 73. With loading 2 of Fig. 76, which gives the most severe stresses, the average ultimate load for three ribs was 742 lbs.; with loading 1, 974 lbs. The average weight of these six ribs was 2.25 lbs. Failure seldom occurs in built-up ribs except at the joints, and the large majority of these failures take place at one or more of the eight joints adjacent to the spars. Attention should therefore be directed toward securing strength at such points. The manner in which the joints of the rib in Fig. 73 are taped and glued provides considerable residual strength even should the gluing of the wood surfaces be imperfect.

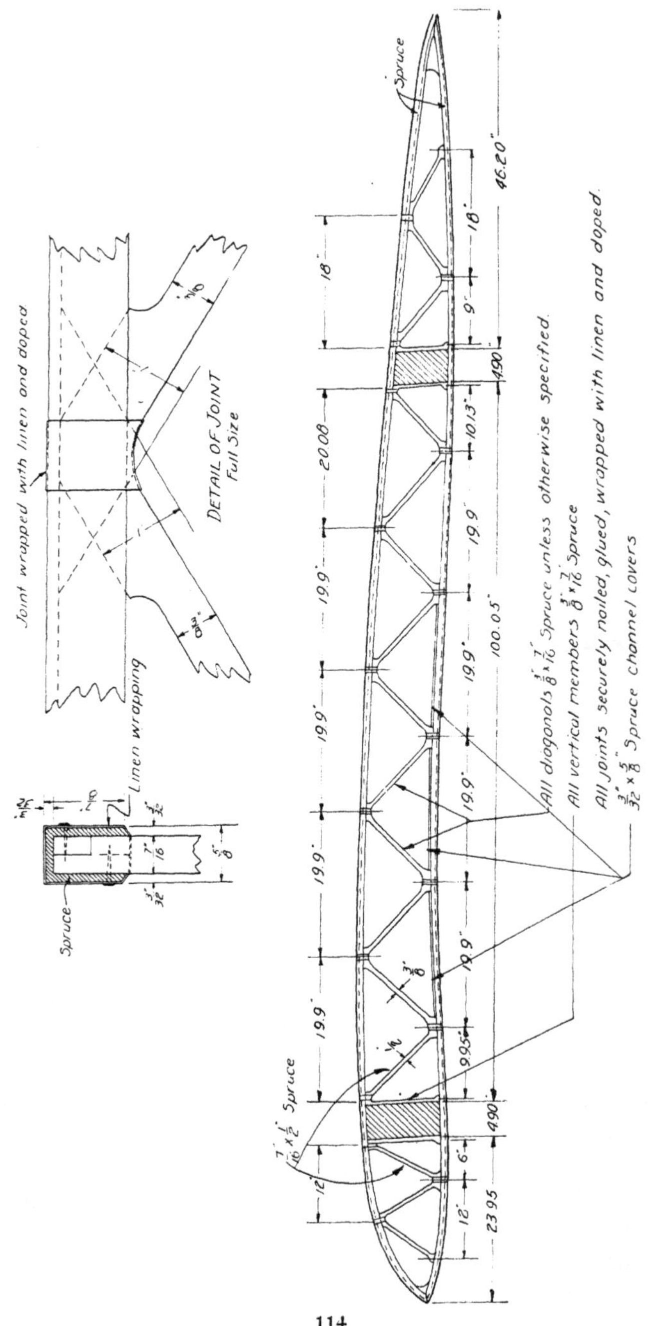


**Fig. 73. Built-up Truss Rib**

Fig. 74 is a modification of the construction discussed above. The principal difference is in the method of joining the web and chord members. The plywood gusset plates are effective. The chief objection to them is the labor entailed in screwing them to the chords and diagonals. This design is probably superior in reliability and ease of production to the rib in Fig. 73. Tests were made on three ribs of this design with 162 in. chord. The average weight of the ribs was 31 oz. and their average ultimate strength 1025 lbs. under a loading similar to loading 2, Fig. 76, but with the peak rounded off and with the load distributed between the upper and lower chord in the proportions of 1 to 2 respectively, instead of concentrated on the lower chord as is customary.

The type of construction shown in Fig. 75 is a recent development in rib design. Its two main objects are elimination of weakness at joints and ease of production. The first object is attained by making the chords and web integral through the use of plywood; the second object by the elimination of complicated joints of the type used either on the Forest Products Laboratory or the Barling ribs. It will be noted that the diagonal reinforcing strips are uniform in section throughout their length and are cut square on the ends. But two, or at most three, sizes of diagonals are required, and two of these sizes can be the same as those of the chord members. A saving in weight can be effected by making the upper chord, which in normal flight is in tension except in the overhangs, smaller than the lower chord which is in compression. With an especially large rib it would be worth while to cut down the size of the lower chord to that of the upper on both overhangs, which are in tension for normal flight. The best type of plywood is one with spruce faces and poplar core, the plies being of equal thickness. The grain of the face plies is horizontal. For ribs from 100 to 120 in. long each ply should be about 1/32 in. thick; for lengths between 120 and 150 in. 3/64 in. is a suitable thickness; for lengths from 150 to 180 in. each ply should be about 1/16 in. thick. Severe stresses are caused in the lower chord by the beam action of the chord between panel points. The stiffness of the chord can be readily increased by making the plywood leg of the chord section deeper. In calculating the stresses in a truss rib this secondary bending must be allowed for. Ribs of this plywood truss type are stiffer as a whole than ribs of other types, and their strength is less variable.

Seven reinforced plywood truss ribs of a 106 in. chord length have been tested with loading 1, Fig. 76. Some of them were constructed with the spruce-poplar plywood recommended, and the balance with an all birch plywood of the same thickness. The first type weighed 15 to 15½ oz. unvarnished, and carried an average ultimate load of 600 lbs. Those with birch plywood weighed 17 to 19 oz. unvarnished, and sustained an average maximum load of 800 lbs. The birch plywood ribs were appreciably stiffer than those with a spruce-poplar web. How-

Fig. 74. Truss Rib with Gusset Plates

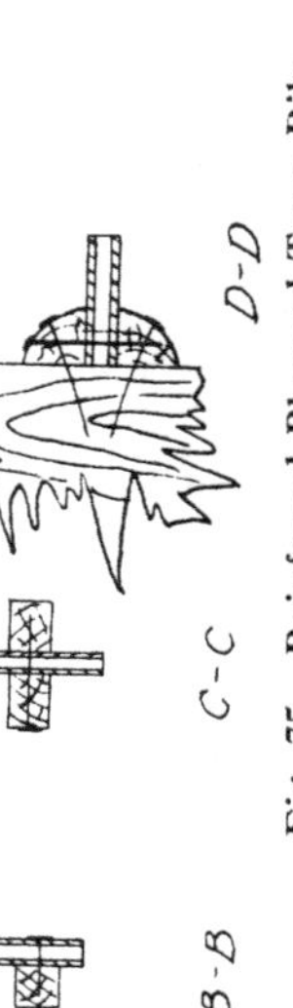


Fig. 75. Reinforced Plywood Truss Ribs

ever, the latter were considerably lighter and developed all the strength that could be utilized efficiently. Beyond a certain point additional strength is of no value. All the diagonal reinforcing was 3/16 x 3/16 in. spruce, except the diagonals adjacent to the spars on the inside, which were 3/16 x 1/4 in. The upper chord reinforcement was 3/16 x 3/16 in. spruce; the lower chord, 3/16 x 1/4 in. Triangular or half round strips glued and nailed to the web of the rib and to the spar after the rib has been finally located on the spar make a good connection, especially with box spar construction. If an "I" section spar is employed the taping shown in Fig. 70 is advisable.

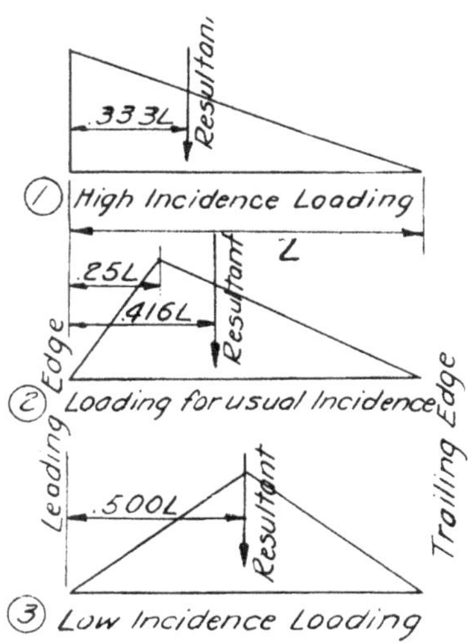


Fig. 76. Rib Loading for Static Test

89. *Rib Loading*—For purposes of design and static testing the three types of loading given in Fig. 76 have been largely used. Loading 2 is the most severe for members between the spars, while the other loadings limit the design in the case of the nose and tail portions of the rib, especially where the overhangs are unusually great. For static testing the total load is divided into equal loads spaced at 4 to 5 in., according to

the chord length. Where the type of sewing is such that all the load on the top surface of the aerofoil is transmitted to the lower chord of the rib, the entire load in a static test is applied to the lower chord. But if the fabric is sewed to the rib so that all the load on the top surface of the wing is carried by the upper chord, the load in static test is divided between the upper and lower chords, 50—70 per cent going on the upper chord. In design work provision should be made for reversed loading equal to half the direct loading. In calculating the ultimate load for a rib the factor required under sand test for the wing structure should be considered.

90. *Rib Spacing*—Wing loading and maximum speed are both factors in determining the spacing of ribs. For high powered pursuit airplanes a minimum spacing of 7 in. and a maximum of 10 in. are suitable. Within the slipstream and for perhaps a foot beyond, the minimum spacing should be employed. This may be gradually increased to the maximum up to a distance from the wing tip approximately equal to the chord length, beyond which the maximum may be used. For fast airplanes the maintenance of the true aerofoil section is important, which is one reason for the close rib spacing. In the case of two place airplanes with a high speed of 120 to 140 M.P.H., the maximum and minimum limits of spacing may be taken as 9 and 15 in. With heavy bombers the limits may be further increased to 15 and 20 in., depending on the size of the bomber. The maximum spacing of 20 in. should be suitable for airplanes with a gross weight of 20,000 to 25,000 lbs.

For larger airplanes it is possible that an even greater spacing can be used. Rib spacing on airplanes of large size is a function rather of the strength of fabric than of rib design. With built-up truss ribs it is generally easier to design strong, somewhat heavy ribs suitable for a large rib spacing, than lighter ribs of less strength that would be used with a smaller spacing. It may be well to note that it is not so much the normal wing loading which determines the maximum load to be carried by the ribs as it is the dynamic wing load. For instance, with small, high-powered airplanes the maximum dynamic loading may be 4 to 5 times the normal wing loading, while in the case of large bombers this dynamic loading would probably not exceed 1.75 times the normal loading.

With all types of airplanes one false rib should be used between each two main ribs from the leading edge to the front spar. This should be of such a character that the true aerofoil section will be maintained.

The use of thin veneer or plywood on the upper wing surface between the leading edge and the front spar, and preferably extending the entire length of the wing, is almost essential on airplanes of a speed greater than 120 M.P.H. For lower speed airplanes only the slipstream length need be covered. Mahogany plywood 3/64 in. thick has proved very satisfactory for this purpose. When no such reinforcing is used the rib spacing should be somewhat closer.

## Struts

For struts the forms in general use are the solid or hollow wood streamline type, and the elliptical or circular tubular type with which fairing is used. The properties of these various sections follow.

91. *Streamline Section*—The streamline strut section which has been adopted by the U. S. Air Service is shown in Fig. 77. Its dimensions are given in Table XII. This section was originally supplied to the National Physical Laboratory by the Royal Aircraft Factory. It has the lowest air resistance of any section so far developed, and is very good structurally.

*TABLE XII*

*DIMENSIONS OF STANDARD STREAMLINE SECTION*

| Distance from Leading Edge Decimal Part of "*L*" | Offsets from Center Line Decimal Part of "*D*" |
|---|---|
| 0.000 | 0.0000 |
| 0.005 | 0.0638 |
| 0.010 | 0.1000 |
| 0.020 | 0.1437 |
| 0.040 | 0.2150 |
| 0.070 | 0.3000 |
| 0.100 | 0.3637 |
| 0.150 | 0.4350 |
| 0.200 | 0.4763 |
| 0.250 | 0.4900 |
| 0.300 | 0.5000 |
| 0.350 | 0.4950 |
| 0.400 | 0.4912 |
| 0.500 | 0.4675 |
| 0.600 | 0.4225 |
| 0.700 | 0.3612 |
| 0.800 | 0.2850 |
| 0.900 | 0.1887 |
| 0.950 | 0.1237 |
| 0.980 | 0.0762 |
| 0.990 | 0.0500 |
| 1.000 | 0.0000 |

Radius of Nose = .12 D

The properties of the solid section may be accurately expressed in terms of the length and diameter.

Area, A = .730 LD

| | About long axis | About short axis |
|---|---|---|
| Moment of Inertia, I | $.0432\ LD^3$ | $.0432\ L^3D$ |
| Radius of Gyration, $\rho$ | .243 D | .243 L |

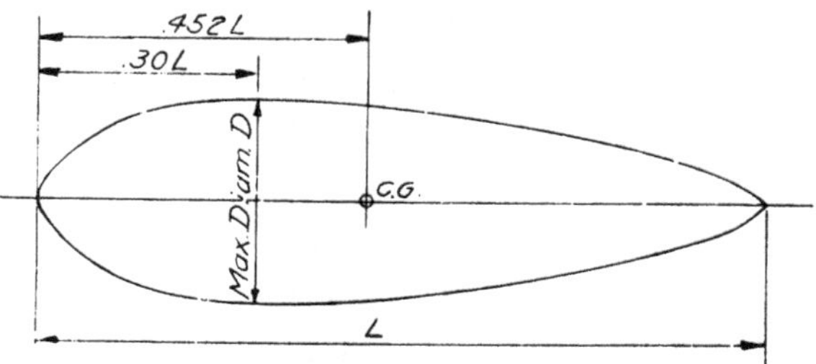


Fig. 77. Standard Strut Section

The properties of the hollow streamline section of this contour may be expressed in terms of the ratio of the mean length to the mean diameter. The mean length equals the outside length minus the thickness of the tube, and the mean diameter equals the maximum outside diameter minus the thickness. The ratio of these two distances may be defined as the mean fineness ratio to distinguish it from the true fineness ratio of the outside dimensions. The formulas apply only for mean fineness ratios between 2.0 and 4.5 and for a maximum ratio of wall thickness to outside diameter of 1/10. For greater wall thickness it would be more accurate to lay out the strut section to a large scale, and by scaling the ordinates to the inside of the section, compute the properties of the hollow portion which may be deducted from those of a solid section. However, for both streamline and elliptical sections the values of their properties, as given by the formulas in Art. 91-93 in which are used constants taken from the charts, become less than their true values as the ratio of $t/D$ becomes greater. It is, therefore, always safe to use the formulas.

Let L = mean length. D = mean diameter. R = mean fineness ratio = L/D. $K_I$ = moment of inertia constant. $K_\rho$ = radius of gyration constant. $K_A$ = area constant (also mean perimeter constant)

$$I = K_I \, tD^3 \qquad A = K_A \, td \qquad \rho = K_\rho \, D$$

$$K_A = 1.875\,R + .992$$

$$K_I = .290\,R + .054 \text{ (about long axis)}$$

$$K_\rho = .394 - \frac{.060}{R} \text{ (about long axis)}$$

These constants may be read off the diagram of Fig. 78.

92. *Hollow Circular Tubing*—The exact formulas for the properties of a hollow circular section are very simple and easy to apply.

Let d = outside diameter, $d_1$ = inside diameter, and t = thickness.

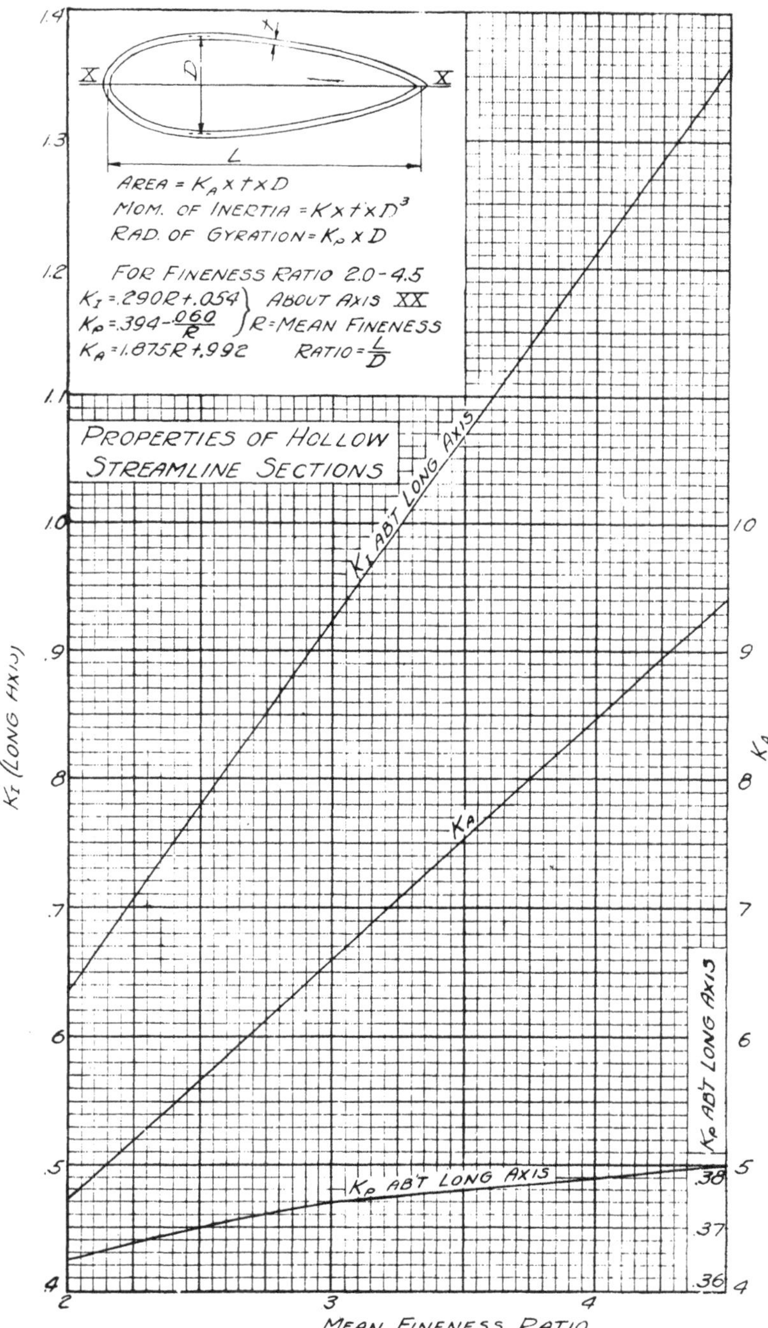


Fig. 78. Properties of Streamline Tubing

$$D = \text{mean diameter} = \frac{d + d_1}{2}$$

$$\text{Area, } A = \frac{\pi(d^2 - d_1^2)}{4} = \pi Dt$$

$$\text{Moment of Inertia, } I = \frac{\pi(d^4 - d_1^4)}{64}$$

$$\text{Section Modulus, } S = \frac{\pi(d^4 - d_1^4)}{32d}$$

$$\text{Radius of Gyration, } \rho = \frac{\sqrt{d^2 + d_1^2}}{4}$$

When the thickness, *t*, does not exceed 1/10 of the outside diameter, *d*, the following approximate formulas may be used with no appreciable error:

$$A = \pi Dt \text{ (exact)} \qquad I = .394\, D^3t \qquad \rho = .354\, D$$

93. *Hollow Elliptical Tubing*—In Fig. 79 is given a set of curves for the property constants for various mean fineness ratios. The mean fineness ratio is the same as that for hollow streamline sections. These constants give results of sufficient accuracy when the ratio of thickness to outside diameter is not greater than 1/10.

Area $= K_A\, Dt$
Moment of Inertia $= K_I\, D^3t$
Radius of Gyration $= K_\rho D$

When the ratio of thickness to outside diameter of hollow sections is as large as 1/10 the error of the various approximate formulas used above is about 1 per cent on the safe side.

94. *Design of Struts*—Interplane struts should be calculated as round ended columns. Euler's formula, $P = \pi^2 EI/L^2$, should be used for computing the strength of struts when the $L/\rho$ for the strut equals or exceeds these values: fir 90, spruce 100, steel 125. $E$ for spruce or fir may be taken as 1,600,000 lbs. per sq. in.; for steel 29,000,000 lbs. per sq. in. When the values of $L/\rho$ fall below the values that have been given, then the design of wood or fir struts should be based on the column curve in Fig. 181, and the design of steel struts on the column curve in Fig. 25. Center section struts will usually be included in this class of short struts. In calculating the strength of faired, hollow, or elliptical steel struts, it is customary to neglect the increase in strength afforded by the fairing. This fairing may be considered as offsetting the effect of any slight eccentricity in the strut due to the fact that it is liable to be slightly bent. In Euler's formula the strut is assumed to be perfectly straight.

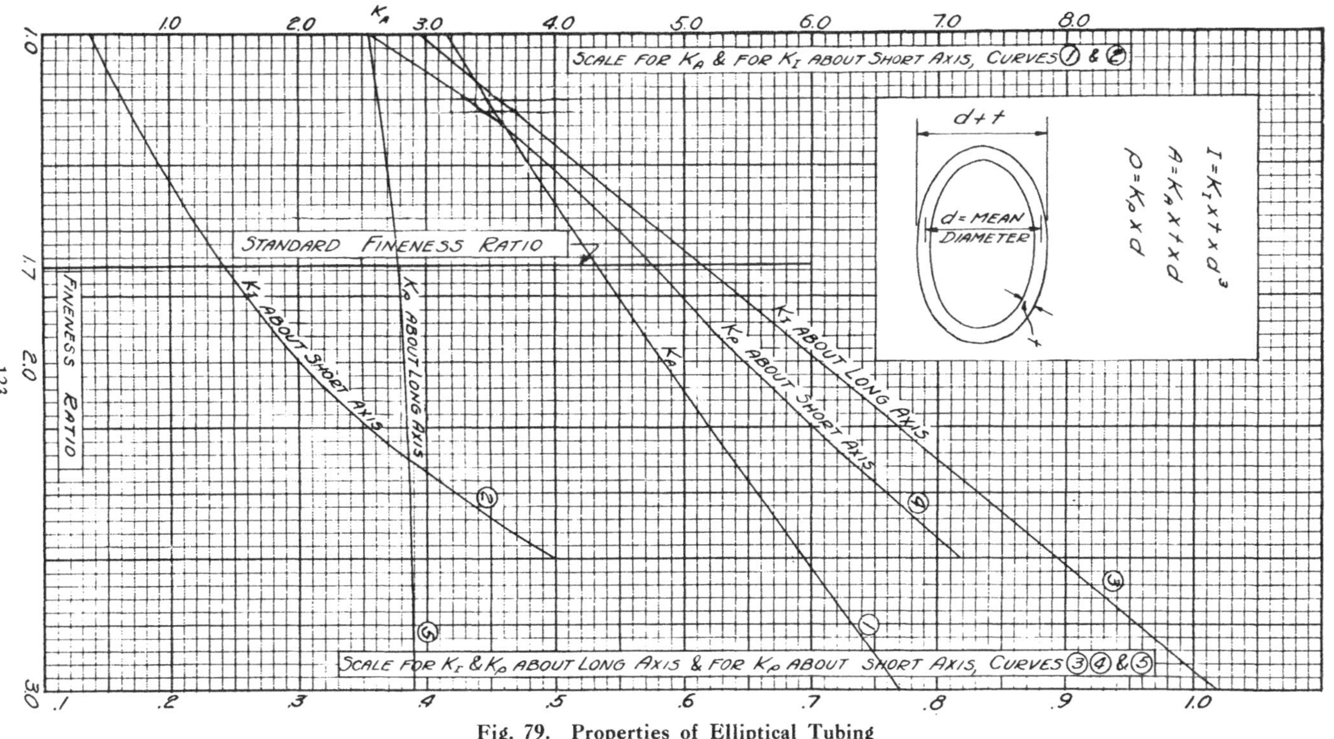


Fig. 79. Properties of Elliptical Tubing

When steel and wood are used in combination in a strut section, the strength of the composite section can be accurately determined by the formulas given in Art. 25. Should it be desirable to include the fairing in computing the strength of a faired strut, it may be done in the same way.

Instead of using directly the values obtained from the wing analysis for the loads on the struts there is a tendency to somewhat increase these stresses to allow for the secondary stresses put into the struts by the incidence wires. Incidence wires may be stressed on account of: (1) Initial tension; (2) their tendency to equalize the loads carried by the front and rear lift trusses; (3) their tendency to equalize the deflections of the upper and lower drift trusses. It is suggested that an increase of about 15 per cent be made in the original strut stresses.

Regarding the practice of laminating struts, it may be said that other things being equal, the strength of a strut is reduced by the use of laminations. Considered from a military standpoint it is probable that a laminated strut is less likely to be shattered or splintered by bullets than a solid strut. Laminating cross or spiral grained material will not improve its strength. In general, small or medium sized struts should not be laminated if material is available for making them from solid stock. On the other hand, in the case of large struts it is usually more convenient to build them in two parts. Very often large struts are routed out in the center to lighten them.

The walls of hollow wood struts should not be less than 3/8 in. thick, and in the case of very large struts 1/2 in. thick, if danger from warping is to be avoided. In this connection it may be well to suggest that in the case of hollow steel struts the use of material thinner than .035 in. is not advisable. For struts of streamline section the gage should be .049 in. or heavier for diameters greater than about 1⅛ in. For circular or elliptical struts the gage should be increased from .035 in. to .049 in. at a diameter of about 1⅜ in.

Best results are obtained from wood struts when the grain in the cross-section of the strut makes an angle between 45° and 90° with the long axis of the section. Any tendency toward warping or twisting is minimized. In a strut built up of two halves the grain should run as indicated in Fig. 80.

It is interesting to note that an Euler strut may have its section largely reduced by piercing with holes without its strength being impaired at all, providing that the direct stress on the net section is fairly well below the elastic limit of the material in compression. Both test and theory bear out this fact. For this reason the strut section shown in Fig. 80 is recommended for military airplanes subjected to severe fire. Tests have demonstrated that a bullet piercing such a strut at any angle will leave a clean hole. The plywood prevents splitting and the canvas prevents the splintering of the wood which would occur with an ordinary spruce strut.

Probably the best combination of wood and steel for a strut is shown in Fig. 81. The difficulty with such a section is in securing a proper

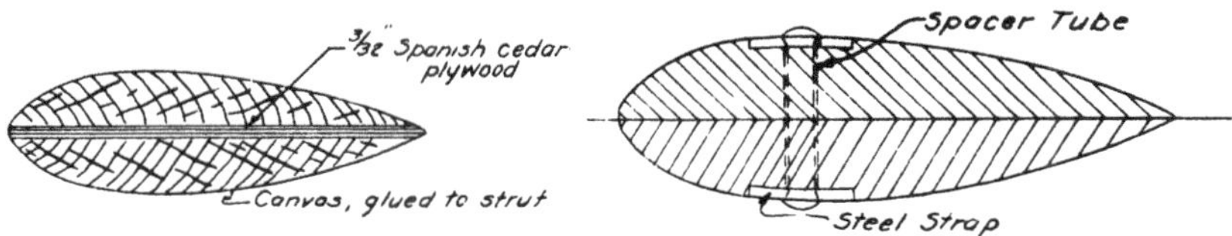


Fig. 80. Built-up Strut

Fig. 81. Combination Wood and Steel Strut

distribution of stress between the two materials. Because of the shear between the wood and steel the latter tends to slip. An effective means of preventing this action is the use of rivets, 3/16 to 1/4 in. in diameter, and spacer tubes. These require closer spacing near the ends of the strut than at the center. The steel strips should form part of the fitting at each end of the strut. Struts of this type are particularly suited to a braced monoplane cellule in which they will be in compression only occasionally.

Hollow streamline steel struts are of comparatively recent development. They have several important advantages, one of which is their suitability for production. Their equivalent weight, on all but very slow airplanes, is less than that of solid wood struts, and their reliability is much greater. Since the strength of slender struts is dependent on the modulus of elasticity, and not on the yield point or tensile strength of the material, ordinary mild steel may be used.

Elliptical tubing with fairing is seldom used for interplane struts, but more often for center section struts or for chassis struts. Balsa wood has been found satisfactory for fairing, as it is very light. So-called "portal" or "picture frame" struts built up of laminated wood are becoming well known. They present certain decided advantages, as elimination of incidence wires, and simplicity in rigging, but are considerably heavier than the usual type. For center section struts the "N" type, of either laminated wood or steel is frequently employed. Both "N" and portal struts are more expensive to manufacture than single struts. In neither type should the struts be tapered, since this adds greatly to the difficulty of production.

95. *Combined Bending*—When a strut is subjected to bending moments about both axes it may be necessary to determine what point on the strut section is most severely stressed. The curves given in Fig. 82 will give the co-ordinates of such a point in terms of the short diameter when the ratio of the two moments is known. These curves are prepared for a streamline strut of a fineness ratio of 3.5, and an ellipse of a fineness ratio of 1.7.

96. *Taper of Struts*—A curve for the standard taper of interplane struts is given in Fig. 83. It is derived from analytical considerations, and tests on struts with this taper have shown that the strength is nearly uniform throughout their length. The strength of a tapered strut may be taken as 85 per cent that of a straight strut of the same maximum

diameter and length. The ratio of the weights and volumes of tapered and untapered struts of the same maximum diameter and length is .774; the corresponding ratio for projected areas is .871.

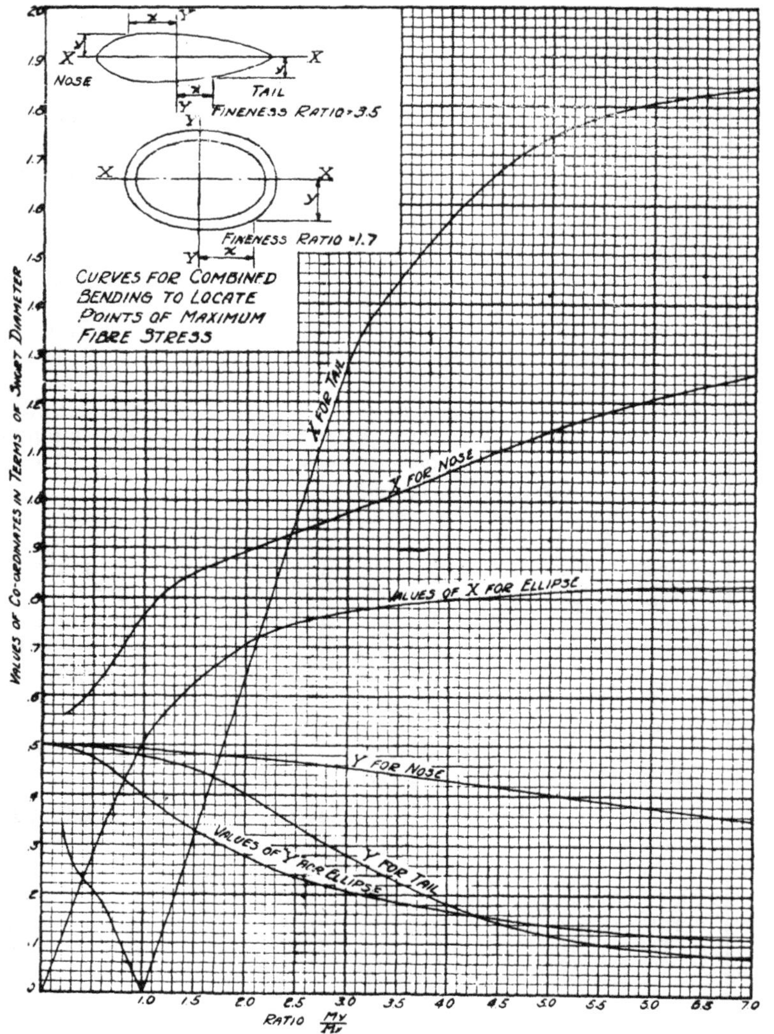


Fig. 82. Curves for Combined Bending

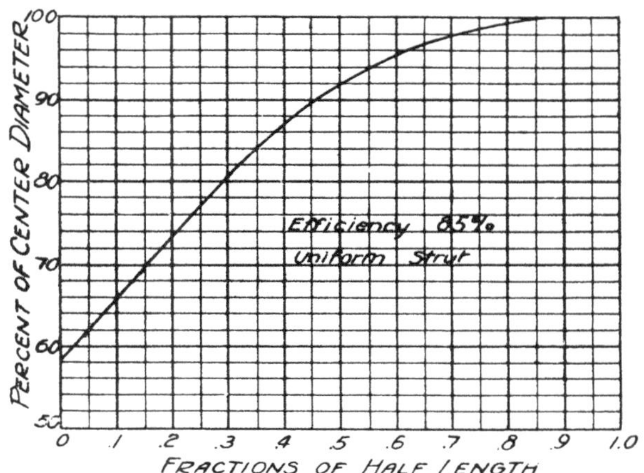


Fig. 83. Standard Taper for Struts

97. *Resistance of Struts*—The resistance of a strut with the standard streamline section and a fineness ratio between 3.5 and 4.5 is given by the formula

$$R = .000204\ LDV^2$$

where R = resistance (lbs.)
L = length of strut (ft.)
D = diameter of strut (ft.)
V = velocity (M.P.H.)

This resistance should be multiplied by the VD correction factor taken from the diagram, Fig. 84. As an example, find the resistance of a $1\frac{1}{4}$ in. diameter strut 6 ft. long, when the velocity is 90 miles per hour.

$$R = .000204 \times 6 \times \frac{1.25}{12} \times 90 \times 90 = 1.03 \text{ lbs.}$$

$$VD = 90 \times \frac{1.25}{12} = 9.4$$

The correction factor for $VD = 9.4$ is 1.03. Hence the final resistance = 1.03 × 1.03 = 1.06 lbs.

When a tapered strut is used the resistance should be computed for a straight strut with a diameter equal to the maximum diameter of the tapered one, and this resistance corrected by the ratio of the projected areas. For a standard taper this ratio of areas is .871. It is impossible to keep exactly similar sections in a tapered strut and as a result the air resistance of the strut is somewhat increased.

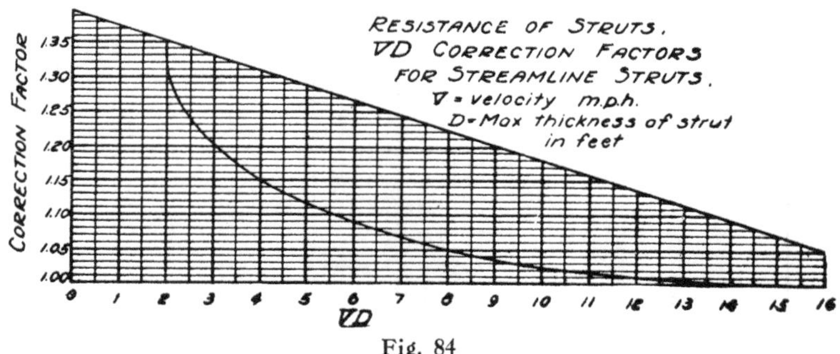


Fig. 84

98. *Fineness Ratio*—As the velocity increases the resistance is less for struts with higher ratios. The following fineness ratios are suggested for use on airplanes with various maximum speeds.

| Maximum Speed | Fineness Ratio |
|---|---|
| Over 175 M.P.H. ................ | 4.5 |
| Over 140 M.P.H. ................ | 4.0 |
| Over 100 M.P.H. ................ | 3.5 |
| Less than 100 M.P.H. ............ | 3.0 |

99. *Equivalent Weight*—When comparing two different struts it is sometimes of value to compare their equivalent weights, which are obtained by combining the resistance of a strut with its actual weight. The resistance of a strut at any velocity is reduced to an equivalent weight by multiplying the resistance of the strut in pounds by the correct value of $L/D$ (lift/drag of the airplane). A common value for this $L/D$ ratio for high speed is 3.5 to 4.5. Equivalent weight = resistance $\times$ $L/D$ + weight in pounds.

For experimental design work, the weight and resistance of a strut may be expressed in terms of some variable, such as a dimension, which is to be determined. Then, by differentiating the expression for the equivalent weight and putting the result equal to zero, the value of the unknown quantity which will give the minimum equivalent weight can be computed. For instance, suppose that a comparison is desired between wood and steel struts, or that the most efficient wall thickness for hollow struts is to be found. Both of these things can be obtained by applying the principle of equivalent weight.

100. *Center Section Struts*—These struts are most severely stressed under a reversed loading or possibly in rolling. They should be designed to carry the compressions occurring in reversed flight. In the ordinary type of center section, with four struts and diagonal incidence and cross wires, the struts should be strong enough to develop the strength of the cross wires, that is to carry a load equal to the vertical component of

the stress in the cross wires when these are stressed to their ultimate strength. The struts should be designed as pin-ended columns using the column curves of Figs. 181 and 25 for wood and steel. Faired, round or elliptical tubes are often used for center section struts.

The "N" type of strut, shown in Fig. 85, of either steel tubing or laminated wood, is quite frequently used. It is a rigid type of construction and easier to assemble than the separate struts and wires. The diagonal member should be designed to carry in compression all the drift loads coming on the center section.

Still another kind of center section construction is indicated in the line drawings, Fig. 86, which are front and side views of what may be termed the tripod type. With this construction the necessity for diagonal cross wires is removed, thus giving the pilot a clear view ahead.

Since the calculation of the stresses in such a system is somewhat difficult a solution will be made for the system shown in Fig. 86. The same general methods are used that are explained in Chapter V under the solution for the stresses in chassis struts and wires. In Table XIII are given the components of the cabane struts and wires expressed as decimal parts of the lengths of the members.

*TABLE XIII*

*COMPONENTS OF CABANE STRUTS AND WIRES*

(Expressed as decimal parts of the lengths of the members)

*Front Cabane*

| Component | Drag Wire | Lift Wire | Strut |
|---|---|---|---|
| Vertical | —.756 | —.967 | +.765 |
| Horizontal (drift) | —.497 | +.094 | —.486 |
| Horizontal (Perpen. to Prop. axis) | +.429 | +.237 | —.420 |

*Rear Cabane*

| Component | Front Strut | Lift Wire | Rear Strut |
|---|---|---|---|
| Vertical | +.588 | —.957 | +.428 |
| Horizontal (drift) | +.300 | —.107 | —.668 |
| Horizontal (Perpen. to Prop. axis) | —.748 | +.251 | —.613 |

*Note*—For vertical components, + is upward.
For horizontal, drift components, + is backward.
For horizontal components perpendicular to propeller axis, + is toward center line of airplane.

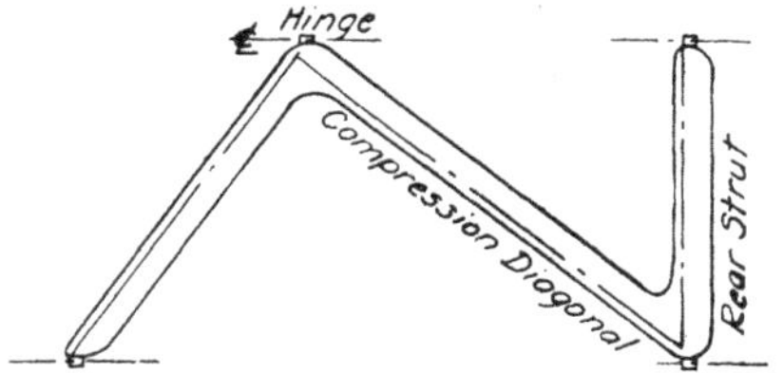


**Fig. 85. N Type of Center Section Strut**

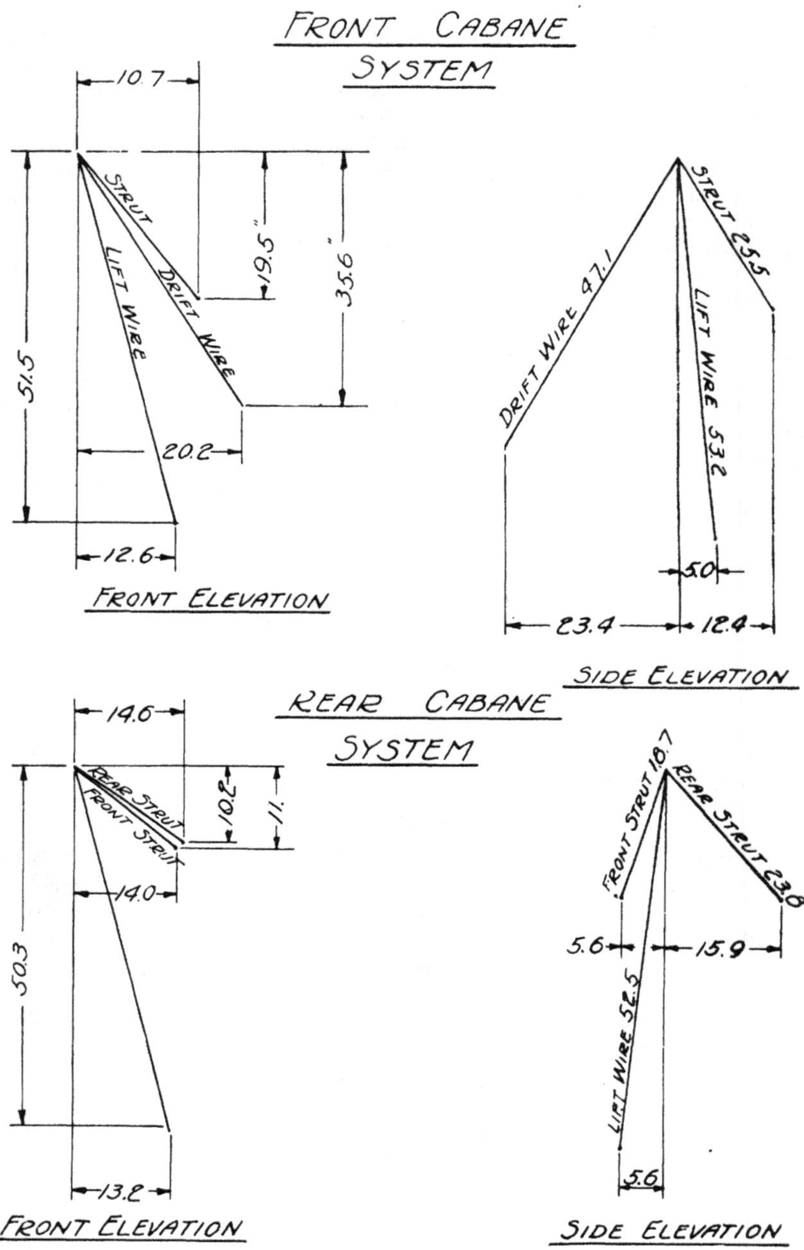


Fig. 86. Tripod System of Center Section Struts

The conditions of equilibrium are applied to the components of all the forces and loads applied at a point and resolved along three axes at right angles. The three equations so obtained are solved simultaneously. Each component of the stresses is expressed as a percentage of the total stress in the member, using the values given in Table XIII. As an example, the equations for the reversed flight condition are given below:

*Front System*

$$1285 \text{ lbs.} = \text{vertical load}$$
$$942 \text{ lbs.} = \text{drift load}$$
$$\Sigma V = -1285 + .765\, C_S = 0$$
$$\Sigma D = +\ 942 - .486\, C_S - D = 0$$
$$\Sigma H = +H - .420\, C_S = 0$$
$$C_S = 1680 \text{ lbs.} = \text{compression in front strut}$$
$$D = -126 \text{ lbs.} = \text{drift force necessary for equilibrium}$$
$$H = +705 \text{ lbs.} = \text{tension in spar necessary for equilibrium}$$

*Rear System*

$$782 \text{ lbs.} = \text{vertical load.}$$
$$942 + 126 = 1068 \text{ lbs. drift load.}$$
$$\Sigma V = -782 + .428\, C_R + .588\, C_F = 0$$
$$\Sigma D = +\ 942 + 126 - .668\, C_R + .300\, C_F = 0$$
$$\Sigma H = +H - .613\, C_R - .748\, C_F = 0$$
$$C_F = 125 \text{ lbs.} = \text{compression in front strut.}$$
$$C_R = 1656 \text{ lbs.} = \text{compression in rear strut.}$$
$$H = +1108 \text{ lbs.} = \text{tension in spar necessary for equilibrium.}$$

The cabane struts and wires should be calculated for the high and low incidence, diving, and reversed flight loading conditions. They should be further investigated for a lateral load perpendicular to the propeller axis. In the design under consideration this latter loading proved a limiting case only for the strut of the front system which, as designed, can carry a lateral load of 1100 lbs. plus a loading under the high incidence condition corresponding to a factor of 6.5.

The front system is not stable within itself because, under several conditions, the drag wire tends to be put in compression. In those cases this wire was assumed out of action, and the horizontal forces necessary for equilibrium were applied to the system. For reversed loading none of the wires in either the front or rear systems are stressed.

101. *Center Section Cross Wires and External Drift Wires*—The design of these wires, as in the case of the incidence wires, is largely a matter of judgment and experience. The center section cross wires come into action when there is an unbalanced lift on the wings, as in rolling or when a puff of air hits one wing. Table XIV gives the sizes of these two types of wires in use on various airplanes.

*TABLE XIV*

*CAPACITIES OF CENTER SECTION CROSS WIRES AND EXTERNAL DRIFT WIRES*

| No. | Machine | Type | Gross Weight | Horse-power | Center Section Cross Wires | | External Drag Wires | |
|---|---|---|---|---|---|---|---|---|
| | | | | | Type of Wire | Capacity | Type of Wire | Capacity |
| 1 | Nieuport Light Scout (Rotary). | Advanced Training | 1287 | 110 | | | Cable | 2—2100 |
| 2 | Orenco Type C ....... | Advanced Training | 1090 | 80 | Cable | 2100 | | |
| 3 | Vought VE-7... | Advanced Training | 1928 | 180 | Cable | 3200 | Cable | 2—3200 |
| 4 | JN-6H ........ | Advanced Training | | 180 | Cable | 3200 | Cable | 4—3200 |
| 5 | Curtiss JN-4D. | Primary Training | 1966 | 90 | Cable | 3200 | Cable | 4—3200 |
| 6 | Thomas-Morse MB-3 ........ | Single-seater Pursuit | 2075 | 300 | S.L. Wire | 1900 | S.L. Wire | 2—1900 & 2—3400 |
| 7 | Orenco Type D....... | Single-seater Pursuit | 2355 | 300 | Cable | 2100 | Cable | 4—3200 |
| 8 | LePere C-11... | Two-seater Fighter | 3712 | 400 | Cable | 4600 | Cable | 4—6100 |
| 9 | U.S. DH-4..... | Observation | 3780 | 400 | Cable | 4600 | Cable | 4—4600 |
| 10 | U.S. XB-1..... | Observation Fighter | 3545 / 2995 | 300 | Cable | 3200 | S.L. Wire | 4—5700 |
| 11 | U.S. D-9A..... | Day Bomber / Night Bomber | 4987 / 5000 | 400 / 400 | Cable | 6100 | Cable | 4—4600 |
| 12 | Glenn Martin Bomber ....... | Night Bomber | 10225 | 800 | Cable | 12500 | | |

102. *Incidence Wires*—These wires serve to equalize the drift on the wings and, in case of injury to one lift truss, to transfer the loads to the remaining truss. Their design is based very largely on judgment and past experience. A few typical examples are given in Table XV.

*TABLE XV*

*CAPACITIES OF INCIDENCE WIRES*

| Machine | Outer Wires | | Inner Wires | | Center Section | |
|---|---|---|---|---|---|---|
| | Type | Capacity | Type | Capacity | Type | Capacity |
| Vought VE-7 | Cable | 2100 | Cable | 2100 | Cable | 2100 |
| Thomas-Morse MB-3 | S.L. Wire | 1900 | S.L. Wire | 1900 | S.L. Wire | 1900 |
| Orenco Type D | Cable | 2100 | Cable | 2100 | Cable | 2100 |
| Orenco Type C (Rotary) | Cable | 1100 | Cable | 1100 | Cable | 1100 |
| U.S. DH-4 | Cable | 2100 | Cable | 4600 | Cable | 3200 |
| LUSAC-11 | | | | | Cable | 4600 |
| U.S. D-9A | Cable | 2100 | Cable | 4600 | Cable | 2100 |
| U.S. XB-1 | S.L. Wire | 1900 | S.L. Wire | 3400 | Cable | 3200 |
| Glenn Martin Bomber | Cable | 4600 | Cable | 4600 | Cable | 12500 |
| Curtiss JN-6H | Cable | 2100 | Cable | 2100 | Cable | 2100 |
| Nieuport Light Scout (Rotary) | | | | | Cable | 1100 |

103. *Built-up or Laminated Beams*—Of the very large number of types of built-up spars developed, the type possessing the greatest advantages for spars of ordinary depth is that shown in Fig. 87. For the flanges straight-grained, clear material should be used. But in the web either cross or spiral-grained material will give entirely satisfactory results. The principal features in favor of this spar may be summarized briefly: I. The small size of the material that can be utilized. II. The possibility of employing defective stock in the web. III. Simplicity and low production cost. IV. Reliability and strength of the construction due to the broad gluing areas and opportunity of applying heavy, direct pressures in the gluing operation. V. Minimizing of shrinkage stresses which with some types of built-up sections are very severe.

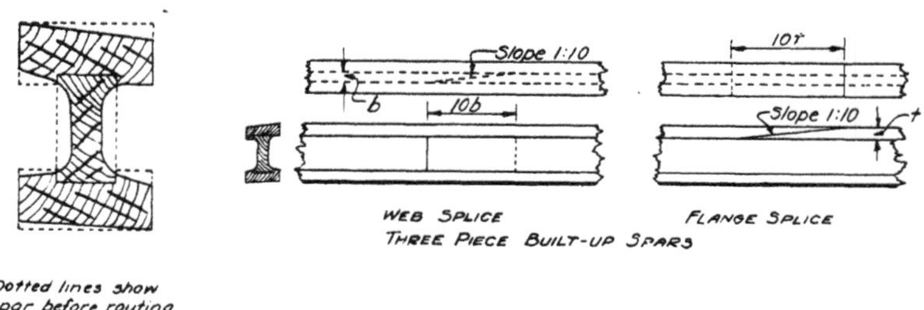


**Fig. 87. Cross-Section and Splice for Built-up Beam**

The spar described in the preceding paragraph is, of course, intended as a substitute merely for the usual "I" section, one-piece spar that is used in the so-called "thin" wing sections such as the U.S.A. 15, 16, or the R.A.F. 15. For this purpose it is much superior to any kind of laminated construction in which a number of thin laminations of wood are glued together, generally with the grain of all the plies parallel. All laminated construction is heavy because of the weight of the large amount of glue required. It is also relatively expensive to manufacture. The laminating and gluing of wood do not increase its strength, hence the modulus of rupture of a laminated section is no greater than if the same material were in one piece. For the same reason it is not possible to utilize inferior material in such a section and get the strength that would be obtained from normal material.

Recently several "intermediate" or "thick" wing sections have been developed, especially for cantilever construction. Because of the depth of these wings the customary one-piece spar section cannot be used. Up to the present, the "box" spar with solid flanges of spruce or ash and webs of plywood has been generally used, with not entirely satisfactory results. Such a member has a modulus of rupture intermediate between that for a one-piece beam, which with spruce varies from 10,300 lbs. per sq. in. to about 7,500 lbs. per sq. in. depending on the routing, and the ultimate compressive strength, which for spruce is 5,500 lbs. per sq. in. It may be taken as between 6,500 and 7,000 lbs. per sq. in. when the flanges are spruce. If the spar is deeper than 8 or 9 in. either the plywood must be fairly thick, and consequently heavy, or else small vertical stiffeners, as shown in sketch III, Fig. 88, must be used, in which case the plywood can be made thinner. Best results for box spars have been secured with the grain of the plies at an angle of 45 degs. to the horizontal and with all the plies of equal thickness. It is not advisable to have a shoulder on the sides of the flanges to protect the top of the plywood, since unless the flange is very deep the amount of gluing surface between the webs and the flanges is seriously curtailed. The plywood should, in general, run up nearly flush with the outside of the flanges. For box spars the thickness of the webs will vary from 3/32 to 3/16 in. depending on whether the plywood is of hard wood, such as birch, or of spruce and poplar, and depending also on the size of the spar.

For greater depths than 9 or 10 in. the construction shown in Sketch I, Fig. 88, is well adapted. Such a beam is light for its strength because all the material is efficiently used, much more so than in the ordinary

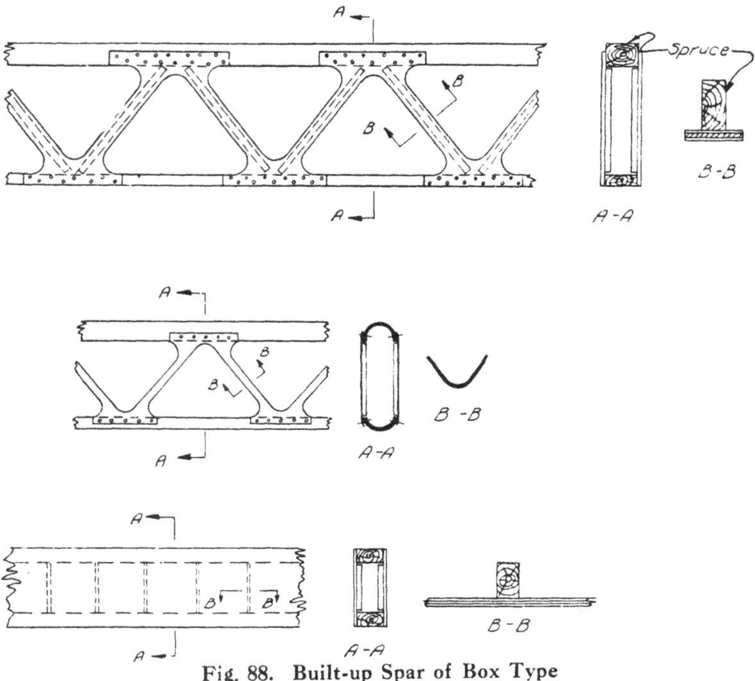


Fig. 88. Built-up Spar of Box Type

box construction. It is believed that greatest strength will be secured by having the grain of the face plies run parallel to one set of diagonals, and that of the core, parallel to the other diagonals. As this type of beam is in process of development this point has not been definitely determined. The core thickness should be 50 per cent of the total thickness of the plywood. Somewhat greater thickness of plywood can be economically used in this construction than in box beams; 1/8 to 1/4 in., depending on the character of the wood used and the spar size, is a suitable range. It will be noted that the reinforcing diagonals are of uniform section throughout their length and that the ends are square. In calculating the stresses in the members of this type of spar, the spar should be considered as a truss. The vertical component of the stress in a diagonal equals the shear at that point. The direct compression or tension in the flanges equals the bending moment at any section divided by the distance between the centers of gravity of the flanges. The diagonals can be assumed as fixed ended in one direction and as half fixed in the other. To allow for this difference in fixity the diagonal reinforcing is rectangular rather than square in section. The flanges can be taken as fixed ended columns with an unsupported length equal to the distance between panel points. The reduction in stress due to column action will usually be so small as to be negligible, and the full ultimate compressive strength of the material may be used. The strength of such a truss beam will be less variable than that of a box spar. Truss beams are particularly applicable to internally braced wings. The flange size can be readily varied according to the stress. It should also be noted that where, as in cantilever construction, the direction of the bending moment does not change, the lower flange, which is in tension, can be made smaller than the upper or compression flange. The spar must, however, be calculated for a reversed load equal to a factor of about 40 per cent that for direct load. In the case of a large airplane in which a braced cellule and an "intermediate" wing section are employed the bays could be made so large that it would be worth while to reduce the size of the upper flange near the center of the bay when the bending moment is negative, and to reduce the size of the lower flange at the strut points at which the moments are positive. The flange which is in compression must of course be increased in section where the tension flange is decreased.

What is considered the best type of steel or duralumin spar, especially for deep beams, is shown in sketch II, Fig. 88. The flanges are of heavier gage material than the webs since the web stresses are small compared with the flange stresses. Local crinkling or buckling in the compression flange is prevented by having the flange curved. Metal ribs can be easily attached to the spar by riveting through the inner edges of each flange. All the material in a truss spar of this type is effective. The remarks in the last paragraph about decreasing the size of the tension flange are also applicable to metal spars. A change in spar section can be easily effected by riveting together flanges of different gages.

104. *Splices in Spars and Longerons*—The Forest Products Laboratory of the U. S. Department of Agriculture has conducted a series of tests to determine the best method of splicing spars (Fig. 89). The results are assembled in a report entitled "Project L-228-3, Tests of Airplane Wing Beam Splices." The conclusions set forth in that report are quoted as follows:

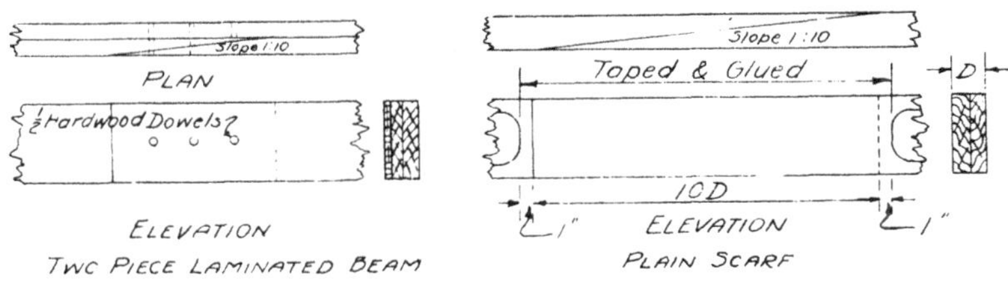


Fig. 89. Spar Splices

1. "Of the four different kinds of scarfs tested the plain sloped scarf is the best."
2. "In solid beams and parts of built-up beams in which the height is less than the width, the plane of the scarf should be perpendicular to the plane of bending; that is, the scarf should be diagonally across the side."
   "In beams, or parts of beams, in which the height is greater than the width, the plane of the scarf should be vertical; that is, the scarf should be diagonally across the top. In beams with height equal to width either type of splices may be used."
3. "The tests show that, for splices in spruce beams with high-grade glue and good gluing, a slope of 1 to 10 is sufficient to develop practically the full strength of the section."
4. "Roughening the scarf does not increase the efficiency of a splice."
5. "Dowels or bolts do not materially affect the efficiency of a well-made splice; they do not come into action to any extent until the glue fails. Now if the glue fails in the splice in a one-piece (unlaminated) beam, dowels or bolts ordinarily will not prevent an immediate collapse of the beam unless the joint is subjected to a moment less than 20 per cent of the maximum moment which the unspliced beam will withstand. In the two-piece laminated beam, however, with a splice in one lamination, dowels or bolts may keep up the strength enough to prevent collapse if the moment at the joint is not more than 65 per cent of the maximum moment which the unspliced beam will withstand. Although results of tests are slightly more favorable with bolts than with dowels, dowels are, as a rule, preferable in practice."

## Design of Fittings

Most fittings used in airplane construction depend upon bolts or pins to transfer stresses from one member to another. It is essential that the bolts have sufficient strength in shear, bearing, and also in tension, if necessary. These considerations are easily taken into account. Such fittings as clevises, lugs, eye-bolts, turnbuckles, etc., have been standardized so that the proper sizes for given loads may be chosen from tables without stress calculations. A few of the general features of the design of these fittings will be taken up to serve as guides for the design of special cases.

105. *Bolts and Pins*—Table XVI gives the cross-sectional area of several sizes of pins or bolts. In the case of bolts in tension the net area should be taken at the root of the threaded portion.

*TABLE XVI*

*AREAS OF BOLTS AND PINS*

| Diameter | Area |
|---|---|
| 1/16 in. | .0031 sq. in. |
| 3/32 in. | .0069 sq. in. |
| 1/8 in. | .0123 sq. in. |
| 3/16 in. | .0276 sq. in. |
| 1/4 in. | .0491 sq. in. |
| 5/16 in. | .0767 sq. in. |
| 3/8 in. | .1105 sq. in. |

Where pins or bolts are used the stress is often transferred through a clevis. This puts the pin in double shear, *i.e.* there are two areas resisting the shearing forces. For example, determine the size of pin required to transmit a stress of 600 lbs. in double shear, assuming that the material of the pin has an ultimate shearing strength of 50,000 lbs. per sq. in. and that a factor of safety of 4.0 is desired.

$$\text{Required area of pin} = \frac{4 \times 600}{2 \times 50{,}000} = .0240 \text{ sq .in.}$$

The nearest size pin to this has a 3/16 in. diameter and an area of .0276 sq. in.

A method of dealing with bolts which bear on wood will be derived. First assume that the fittings are not wrapped around the spar but are similar to those shown in Fig. 90.

Let P = component of pull in a direction parallel to the neutral axis of the spar.

a = distance from neutral axis to line of action of the component, *P*.

d = depth of spar.

b = diameter of bolt.

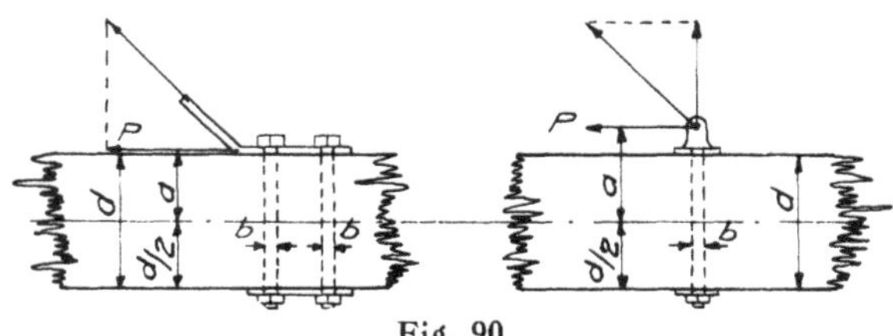

Fig. 90

The bearing pressure of the bolts against the wood is composed of a uniform pressure due to the load $P$ and a uniformly varying pressure due to the moment $Pa$. Let $f_c$ = unit pressure due to load $P$, and $f_b$ = unit pressure due to moment $Pa$. When there are two bolts, as in Fig. 90, assume the load and moment to be equally divided between the bolts.

With one bolt:

$$f_c = \frac{P}{bd} \quad \text{and} \quad f_b = \frac{6\,Pa}{bd^2}$$

$$f_c + f_b = \frac{P}{bd} + \frac{6\,Pa}{bd^2} = \frac{P}{bd}\left(1 + \frac{6a}{d}\right)$$

The sum of $f_c$ and $f_b$ shall not exceed 5500 lbs. per sq. in. for spruce.

When the bolts pass through a metal sleeve or fitting which entirely surrounds the spar there can be no exact method of solution, but the assumption may be made that the bolts carry 25 per cent of the moment in bearing on the wood, the fitting transferring the remaining 75 per cent.

The component of the pull at right angles to the spar is taken entirely by the bolts, and sufficient area of fitting under the bolt heads must be supplied in order to distribute the bearing pressure.

If the pull on a bolt, which bears on wood, has a component that makes an angle with the grain of the wood then care must be exercised in securing the wood against splitting.

106. *Lugs and Eye-bolts*—As the result of a series of tests on lugs the following procedure in their design has been suggested. Given the desired maximum strength:

I. The pin size is chosen in conjunction with the lug thickness, so that the maximum bearing stress shall not be more than twice the tensile strength of the material. The pin must be large enough to carry safely the shearing stress. Pin sizes for lugs of different capacities are given in Table XXXIII in the appendix.

II. The diameter of the head should be made sufficiently large that computed stress in direct tension on the net section through the eye shall, at the maximum desired capacity, not exceed the tensile strength of the material. It is necessary, however, to offset the center of the radius for the outer part of the head to care for extra stresses in the head due to bending. The amount of this offset for each lug capacity is given in Table 33.

III. The width of the shank should at least be sufficient to give shank area enough to stand the maximum desired load in tension.

IV. The hole in the lug should be equal to the diameter of the pin, with limits of —.000 in. and +.010 in.

Eye-bolts may be used when the angle of pull is not greater than 45° with the axis of the bolt. Beyond that limit the strength of the shank decreases very rapidly. When the eye-bolts bear directly against wood instead of a fitting they bend and crush the wood in front of them. This bending raises the apparent strength of the bolts by decreasing the angle of pull. Where the eye-bolt passes through wood enclosed in or covered by a metal fitting the ability of the fitting to distribute the stress in the bolt over a sufficient area of wood is of importance. The larger sizes are unsuited for use at critical sections or in important members, because of the amount of material which has to be removed in the bolt hole, with the resultant weakening of the member.

In lugs and eye-bolts of high strength steel it is well to offset the center of the pin hole by 1/32 of an inch from the center of the head in order to provide additional metal in front of the pin. With high strength steel the metal does not yield as readily as with ordinary steel, causing the pin to bear more nearly on a point than on an extended surface.

107. *Horizontal vs. Vertical Bolts*—It is evident that a bolt passing vertically through a spar affects the moment of inertia of the spar much more than a horizontal bolt at or near the neutral axis. Horizontal bolt holes near the top and bottom of spars are as detrimental to the strength of the spars as are the vertical holes. Due allowance for bolt holes should be made, wherever they occur, and the size of hole should be taken as 1/32 in. to 1/16 in. larger than the diameter of the bolt in order to allow for the injury to the fibres. One good method which is

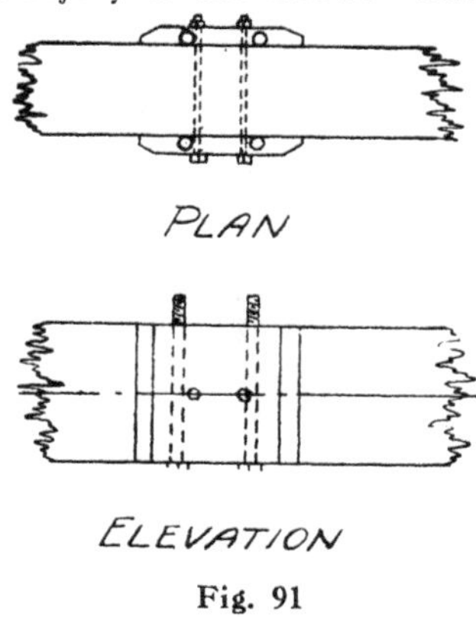


Fig. 91

employed with vertical bolts is illustrated in Fig. 91. The vertical bolts are placed on the sides of the spar and are held in position by blocks which are glued to the spar. Two horizontal bolts at the neutral axis also assist in holding the blocks to the spar. Such an arrangement as this overcomes the objection to vertical bolt holes in the spar itself.

108. *Pin Plates*—In many cases it is possible to save weight in the metal fittings by using thinner material and brazing a pin plate to the fitting at the bolt hole in order to reduce the bearing stress. An example will illustrate the method. Referring to Fig. 92, which represents a portion of a fitting carrying a pull of 840 lbs., assume that a factor of safety of 5.0 is desired, and that the steel has the following ultimate strengths:

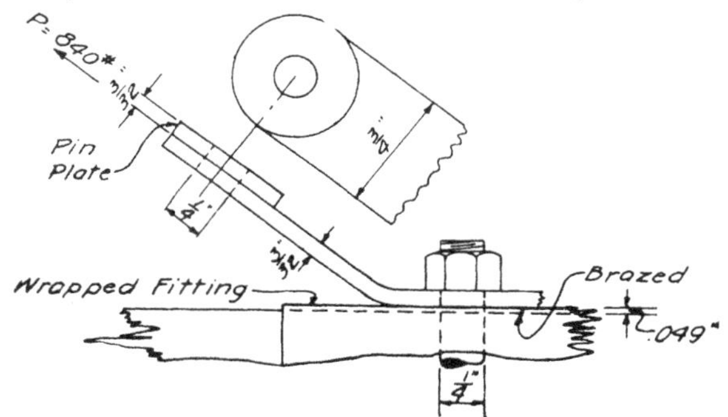


Fig. 92. Lug with Pin Plate

| | | |
|---|---|---|
| Ultimate shearing strength | = | 45,000 lbs. per sq. in. |
| Ultimate tensile strength | = | 60,000 lbs. per sq. in. |
| Ultimate bearing strength | = | 100,000 lbs. per sq. in. |

The ultimate pull on fitting $= 840 \times 5 = 4200$ lbs. per sq. in.

Tensile stress on body of fitting $= \dfrac{4200}{3/32 \times 3/4} = 59{,}700$ lbs. per sq. in.

The 1/4 in. pin or bolt is in double shear.

Shearing stress on pin $= \dfrac{4200}{2 \times .049} = 43{,}000$ lbs. per sq. in.

First assume that there is no pin plate. The bearing stress will equal $\dfrac{4200}{1/4 \times 3/32} = 179{,}000$ lbs. per sq. in. This value is much in excess of the ultimate bearing strength, although the remainder of the fitting is satisfactory. Let $t =$ required thickness to give a bearing stress of 100,000 lbs. per sq. in.; $t = \dfrac{4200}{1/4 \times 100{,}000} = .1680$ in. The thickness required is (.1680—.0938) = .0742 in. in excess of the thickness of the fitting. The pin plate should be at least .0742 in. in thickness and in this case a value of 3/32 in. will be chosen. The final bearing stress equals

$$\frac{4200}{1/4 \times 3/16} = 89600 \text{ lbs. per sq. in.}$$

Bearing on bolt: The component of the stress parallel to the beam axis is $4200 \times .70 = 2940$ lbs.

$$\text{Bearing stress on bolt} = \frac{2940}{.25\ (.0930 + .049)} = 82400 \text{ lbs. per sq. in.}$$

109. *Wing Fitting*—Fig. 93 is a wing-spar fitting developed by the Airplane Design Section. This is essentially a box fitting and illustrates very clearly how eccentricities of struts and wires may be avoided. It will be noted that the lines of action of the struts and wires intersect at a common point on the center line of the spar. The channel across the top of the spar acts as a bearing plate to transfer a portion of the strut load directly into the wood.

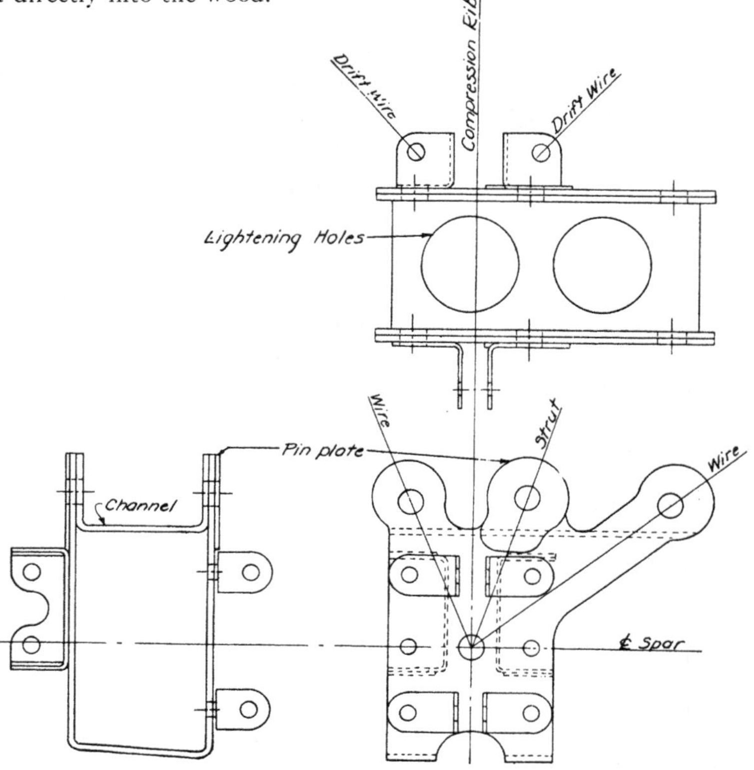


**Fig. 93. Wing Spar Fittings Without Eccentricities**

## CHAPTER V

# AIRPLANE CHASSIS

110. *General Considerations*—Until recently comparatively little had been done on the accurate stress analysis of chassis. Consequently, there has been less opportunity to corroborate by static test the results of the present methods of chassis analysis than is the case with the wing structure. Some uncertainty also exists in regard to the proper factors of safety and conditions of loading for various types of airplanes. However, the methods of analysis given in this chapter are based entirely on the principles of mechanics and it is believed will give a reasonably exact solution for the stresses produced by known external loads, and therefore form a suitable basis for design. Further experimental work may be largely directed toward obtaining a better knowledge of the loadings that will reproduce actual conditions of service, and more correct methods for computing the ultimate strength of struts subjected to combined bending and compression.

111. *Conditions of Loading*—The external loads to which a chassis may be subjected can be resolved into vertical forces, horizontal forces acting in a direction parallel to the propeller axis of the airplane, which may be termed backward or forward thrusts, and horizontal forces perpendicular to the propeller axis, which may be termed side thrusts. Each of these forces, applied initially to the wheels, is transmitted to the axle, and from the latter into the chassis structure itself, either by means of elastic cords or by direct bearing of the axle on the axle guides. It is customary to assume that the backward, forward, and side thrusts are each a certain percentage of the vertical load. For airplanes of different types and sizes these percentages should vary. In present analysis the back thrust is taken equal to .5 of the vertical load, the forward thrust .2, and the side thrust .25.

Below are given different loadings which should simulate nearly all the conditions that will be found in service. In each case the propeller axis is considered horizontal.

Case 1. Vertical load in level landing plus horizontal back thrust sufficient to make resultant of load pass through center of gravity of airplane. This component will usually be between .16 and .25 depending on the type of airplane.
Case 2. One wheel landing with vertical load.
Case 3. Case 1 plus horizontal back thrust equal to .5 of the vertical load.
Case 4. Case 2 plus horizontal back thrust equal to .5 of the vertical load.
Case 5. Case 1 plus horizontal forward thrust equal to .2 of the vertical load.
Case 6. Case 2 plus horizontal forward thrust equal to .2 of the vertical load.
Case 7. Case 1 plus side thrust equal to .25 of the vertical load.
Case 8. Case 2 plus side thrust equal to .25 of the vertical load.

Case 9. Case 1 plus both backward thrust equal to .5, and side thrust equal to .25 of the vertical load.

Case 10. Case 2 plus both backward thrust equal to .5, and side thrust, in the direction of low wing, equal to .25 of the vertical load.

To calculate the dynamic factor for each of these cases would be impracticable and also unnecessary. The type of airplane should determine not only the factor needed, but also the cases that should be calculated. For example, a training airplane is liable to be subjected to treatment both more severe and varied than would an airplane in the hands of a skilled pilot. The principle to be followed in selecting the cases to be considered is that sufficient cases should be investigated to stress severely each member of the chassis. Case 1 does not produce such stresses, but, because most chassis tests have been made with this loading, the dynamic factors so obtained have a certain comparative value. Case 4 is a limiting condition for the rear strut and usually for the diagonal wires. Case 6, which is intended to reproduce the conditions of a tail-low landing, is apt to be a limiting case for the front strut. These three cases should always be computed, and in addition it is advisable to investigate Case 7 which may prove to be the limiting case for the diagonal wires.

112. *Characteristics of the Chassis Structure*—There are several features which are important in the analysis and design of a chassis: I. The method of attachment of the spreader tubes to the struts or the guide plate: If the tubes are pinned they will carry compression only, and may be very light, while if rigidly connected they will also serve to transmit bending moment, thereby somewhat relieving the struts. It is necessary in this case to strengthen the spreader tubes. With a rigid connection, the fixity of the struts in column action is increased. II. The character of the fitting securing the chassis struts to the longerons: In some instances this is in the form of a socket receiving the strut end which is bolted through and rigidly held. A connection of this type will practically fix the upper end of the strut. Aside from largely increasing its column strength, such a connection will affect the distribution of the moments in the struts, or between the struts and spreader tubes. When this fitting is hinged, the pin may be either in the plane of the struts, that is parallel to the propeller axis, or it may be perpendicular to this direction. The first type is the more common. As the pin is parallel to the long axis of the strut the latter must be computed as a column hinged at one end at least. Moments perpendicular to the plane of the struts must become zero at the pin, but in determining the distribution of moments in this plane the struts are assumed to be fixed. On the other hand, when the pin is perpendicular to the plane of the struts, moments in this plane must reduce to zero at the pin, and the strut may be taken as fixed in determining the distribution of moments perpendicular to the plane of the struts. Although the struts are undoubtedly partially fixed at their upper ends it is hardly safe to count on

much increase in column strength from this cause. III. The bracing wires: Whether diagonal bracing wires are in the planes of both the front and rear struts, or in the plane of one set of struts only, affects the methods of analysis and also the distribution of the stresses. IV. The eccentricities of the members in a chassis: One of the marks of a good design is the absence of such eccentricities, or their reduction to an amount where they affect the stresses but slightly. Eccentricities which produce moments perpendicular to the plane of the struts are usually the most serious. Because in a level landing the struts and spreader tubes are in action, while in a one-wheel landing the struts and wires are the members that are stressed, it is possible to have an eccentricity in one case and none in another. Fig. 94 illustrates this. Slight eccentricities that cause moments in the plane of the struts are difficult to avoid and do not cause large stresses. As shown in Fig. 95 both struts and wires may be eccentric. V. The material used in the struts: Wood and steel are the materials usually employed. It is in the calculation of the allowable stresses and in solutions involving the method of least work that the nature of the material affects the analysis.

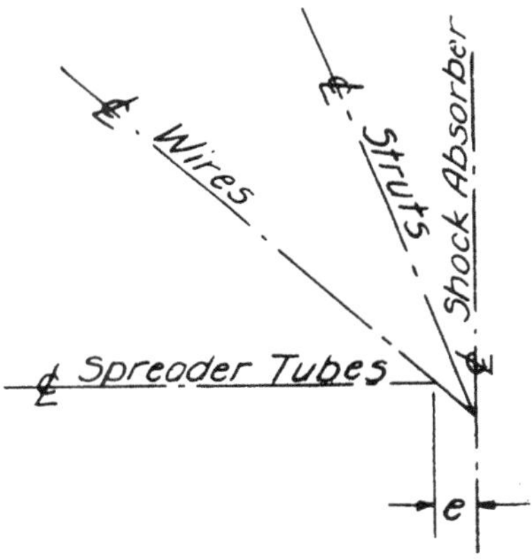


Fig. 94

113. *Concerning the Chassis Used as an Example.*

I. The following conditions of loading are investigated: cases 1, 4, 6 and 7.

II. The weight of the fully loaded airplane is 1650 lbs.

III. The design of this chassis is such that in front elevation there is no eccentricity of struts, wires or spreader tubes. In side elevation, a slight eccentricity of struts and axle is present, as well as an eccentricity of the wire. Fig. 95 gives line drawings of the chassis members, showing the eccentricities.

IV. The struts in their own plane are assumed to be fixed at the ends.

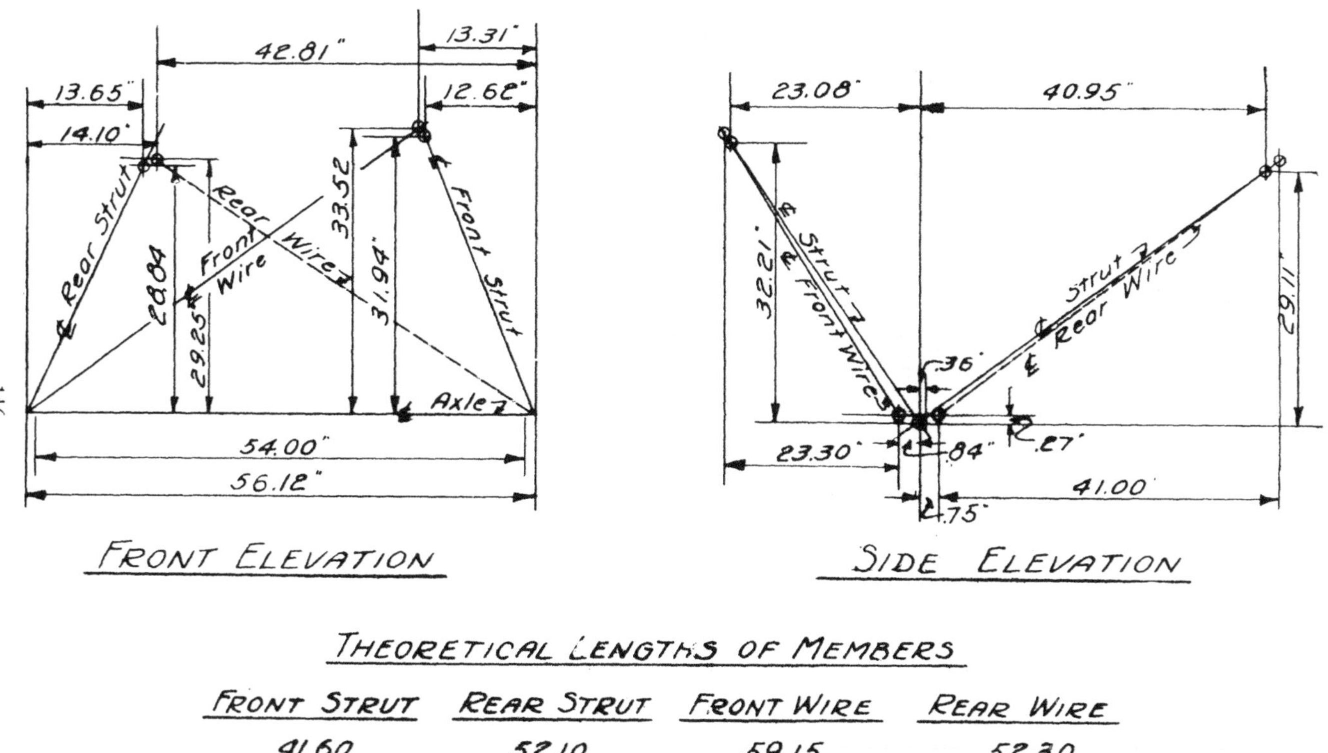


Fig. 95. Front and Side Elevation of Chassis Analyzed

V. In determining the distribution of the moments in a vertical plane perpendicular to the plane of the struts, it is assumed that the connection between the struts and the vertical guide plate and between the spreader tubes and this plate is rigid.

VI. The struts are hinged at their upper end with pins running in the direction of the propeller axis.

VII. In the actual chassis there are wires in the plane of the front struts only, but to illustrate the method of solution when wires are present in the plane of the rear struts too, the case of a one-wheel landing will be worked out for this latter condition, in addition to the solution for the original chassis.

114. *Kinds of Stresses to be Considered*—As previously suggested, there are two types of stresses: those due to direct column loads, and those caused by bending moments. The methods for obtaining the column loads and bending moments are different and will be presented separately. Bending moments are due to two causes: fixed eccentricities, and eccentricities produced by the rise of the axle in its slot.

115. *Calculation of Direct Stresses*—The first step in the analysis is to compute the theoretical length of each of the members in the chassis, and the length of each of their projections on three axes at right angles to each other, one axis vertical, the other two parallel and perpendicular to the propeller axis respectively. The length of each of the projections should be expressed as a decimal part of the length of the member. Then the values of the components, along the three axes, of the unknown stress in a member may be expressed by the corresponding decimal parts of the unknown stress. Since it is necessary to consider the direction of these components the conventions shown in Fig. 96 are followed. In Table XVII the struts are assumed to be in compression, the wires and spreader tubes in tension.

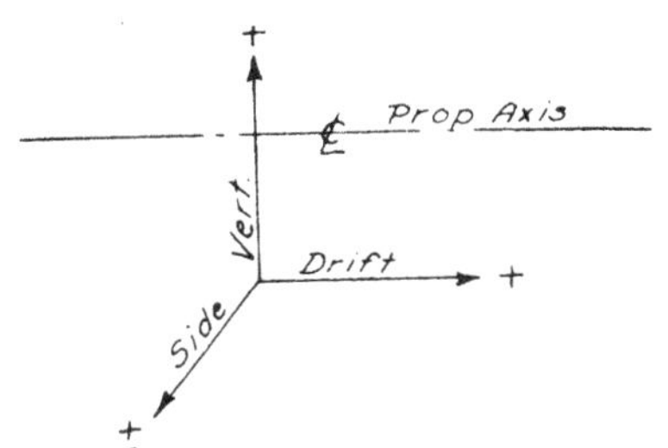


Fig. 96. Convention for Signs of Forces

## *TABLE XVII*

## *COMPONENTS OF STRESSES IN MEMBERS*

| | Vertical | Drift | Side |
|---|---|---|---|
| Rear Strut ........... | $-.558\ C_r$ | $-.786\ C_r$ | $+.265\ C_r$ |
| Front Strut ........... | $-.774\ C_f$ | $+.555\ C_f$ | $+.306\ C_f$ |
| Spreader Tubes ....... | $.000\ T_s$ | $.000\ T_s$ | $-1.000\ T_s$ |
| Front Wire ........... | $+.566\ T_f$ | $-.394\ T_f$ | $-.723\ T_f$ |
| Rear Wire (assumed).. | $+.446\ T_r$ | $+.625\ T_r$ | $-.640\ T_r$ |

$C_r$ = stress in rear strut
$C_f$ = stress in front strut
$T_s$ = stress in spreader tubes
$T_f$ = stress in front wire
$T_r$ = stress in rear wire.

*Solution by Statics*—Whenever the structure is statically determinate, that is where not more than three members carrying stress meet at a point, the principles of statics are employed. The conditions of equilibrium—namely, that the algebraic sum of the components along each of three axes at right angles, of any system of forces which is in equilibrium must equal zero—are applied to the components of the forces in question. Where but one set of wires is used the direct stresses can be obtained by statics for all types of loading. But, in the special case that is being given, where there are wires in the planes of both front and rear struts, the principle of least work must be called upon to obtain the value of the fourth unknown after the other three have been computed in terms of the fourth by the method of statics.

The external loads are considered separately, so that the stress in the different members produced by each type of load can be obtained. Then the total stress caused in any member by any combination of loads is merely the sum of the separate stresses produced by the single loads. For convenience, each external load is taken as unity. Hence, to determine the actual stress produced in a member by any given load, it is necessary only to multiply the unit stresses by the amount of the actual load.

Following are the equations of equilibrium for each different type of load considered, and the stresses in the members for a load of unity. The case of a vertical load in a one-wheel landing where two sets of wires are in action is completely worked out by the method of simultaneous equations. The full solution for the same case with but one set of wires in action is given to illustrate the use of determinants. This latter process, for statically determinate cases, is simple and somewhat more rapid than the method of simultaneous equations.

Case 1. Vertical load of unity: level landing. In this case it is assumed that the diagonal wires are not in action.

$$\Sigma V = -.558\ C_r - .774\ C_f + 1.000 = 0$$
$$\Sigma D = -.786\ C_r + .555\ C_f = 0$$
$$\Sigma S = +.265\ C_r + .306\ C_f - 1.000\ T_s = 0$$

$$C_r = +\overset{*}{.}605 \text{ compression}$$
$$C_f = +.855 \text{ compression}$$
$$T_s = +.421 \text{ tension}$$

Case 2. Vertical load of unity: one-wheel landing. It is assumed that the spreader tubes do not act in this case.

$$\Sigma V = -.558\,C_r - .774\,C_f + .566\,T_f + 1.000 = 0$$
$$\Sigma D = -.786\,C_r + .555\,C_f - .394\,T_f = 0$$
$$\Sigma S = +.265\,C_r + .306\,C_f - .723\,T_f = 0$$

$$C_f = +\overset{*}{1}.470 \text{ compression}$$
$$C_r = + .613 \text{ compression}$$
$$T_f = + .847 \text{ tension}$$

*In this part of the analysis the sign given the stress does not denote tension or compression, but merely indicates whether the direction of the stresses assumed in writing the equations was right or wrong.

Solution of Case 2 by method of determinants.

$$\begin{vmatrix} -.558 & -.774 & +.566 \\ -.786 & +.555 & -.394 \\ +.265 & +.306 & -.723 \end{vmatrix} C_r = \begin{vmatrix} -1.000 & -.774 & +.566 \\ .000 & +.555 & -.394 \\ .000 & +.306 & -.723 \end{vmatrix}$$

$$+.459 \qquad\qquad +.401 - .121 = +.280$$

$$C_r = +.610$$

$$.459\,C_f \begin{vmatrix} -.558 & -1.000 & +.566 \\ -.786 & .000 & -.394 \\ +.265 & .000 & -.723 \end{vmatrix}$$

$$+.104 + .569 = .673$$

$$C_f = +1.464$$

$$.459\,T_f = \begin{vmatrix} -.558 & -.774 & -1.000 \\ -.786 & +.555 & .000 \\ +.265 & +.306 & .000 \end{vmatrix}$$

$$+.241 + .147 = .338$$

$$T_f = +.845$$

Explanation of Determinants

$$a\,x - b\,y + c\,z = d$$
$$a'\,x + b'\,y - c'\,z = d'$$
$$-a''x + b''y + c''z = d''$$

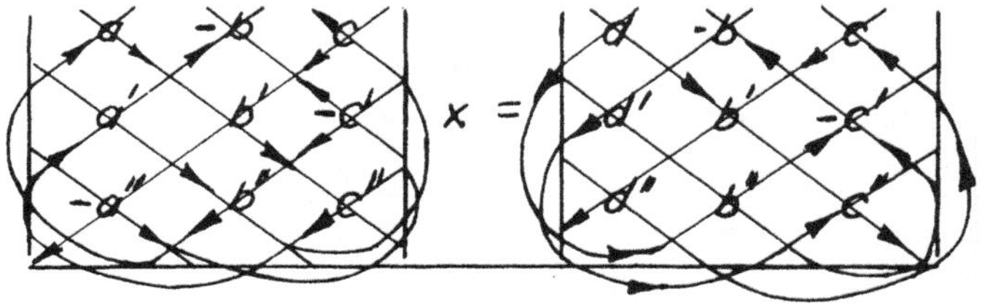

$$\begin{vmatrix} a & -b & c \\ a' & b' & -c' \\ -a'' & b'' & c'' \end{vmatrix} y = \begin{vmatrix} a & d & c \\ a' & d' & -c' \\ -a'' & d'' & c'' \end{vmatrix}$$

$$\begin{vmatrix} a & -b & c \\ a' & b' & -c' \\ -a'' & b'' & c'' \end{vmatrix} z = \begin{vmatrix} a & -b & d \\ a' & b' & d' \\ -a'' & b'' & d'' \end{vmatrix}$$

$$\left[ (ab'c'' + a'b''c - a''c'b) - (-cb'a'' - c'b''a - c''a'b) \right] X = (db'c'' + d'b''c + d''c'b) - (cb'd'' - c'b''d - c''d'b)$$

The determinant, which is the coefficient of $x$, $y$ and $z$, is invariable. In multiplying, due attention should be paid to the signs of each term.

The few brief explanations of mathematical work given in this section and elsewhere are inserted merely for the convenience of the reader to save possible reference to tables of integrals or to a handbook.

Case 3. Side thrust of unity perpendicular to axis.

$$\begin{aligned} \Sigma V &= -.558\,C_r - .774\,C_f + .566\,T_f &&= 0 \\ \Sigma D &= -.786\,C_r + .555\,C_f - .394\,T_f &&= 0 \\ \Sigma S &= +.265\,C_r + .306\,C_f - .723\,T_f + 1.000 &&= 0 \end{aligned}$$

$C_f = +1.457$ compression
$C_r = +\ .022$ compression
$T_f = +2.010$ tension

Case 4. Thrust of unity parallel to axis. In this case it is assumed that the diagonal wires are not in action.

$$\begin{aligned} \Sigma V &= -.558\,C_r - .774\,C_f &&= 0 \\ \Sigma D &= -.786\,C_r + .555\,C_f + 1.000 &&= 0 \\ \Sigma S &= +.265\,C_r + .306\,C_f - 1.000\,T_s &&= 0 \end{aligned}$$

$C_f = +.842$ compression with back thrust
$C_f = -.607$ tension with back thrust
$T_s = +.038$ tension with both back thrust and forward thrust

Case 5. Vertical load; one-wheel landing (both sets of wires) It is assumed that the spreader tubes do not act in this case.

(1) $\Sigma V = -.558\ C_r - .774\ C_f + .556\ T_f + .446\ T_r + 1.000 = 0$
(2) $\Sigma D = -.786\ C_r + .555\ C_f - .394\ T_f + .625\ T_r = 0$
(3) $\Sigma S = +.265\ C_r + .306\ C_f - .723\ T_f - .640\ T_r = 0$

Divide (1), (2) and (3) by coefficients of $C_r$

(4) $-1.000\ C_r - 1.386\ C_f + 1.014\ T_f + .799\ T_r + 1.793 = 0$
(5) $-1.000\ C_r + .706\ C_f - .502\ T_f + .795\ T_r = 0$
(6) $+1.000\ C_r + 1.154\ C_f - 2.728\ T_f - 2.414\ T_r = 0$

Subtract (4) and (5)

(7) $-2.092\ C_f + 1.516\ T_f + .004\ T_r + 1.793 = 0$

Add (5) and (6)

(8) $+1.860\ C_f - 3.230\ T_f - 1.619\ T_r = 0$

Divide (7) and (8) by coefficients of $C_f$

(9) $-1.000\ C_f + .725\ T_f + .002\ T_r + .858 = 0$
(10) $+1.000\ C_f - 1.736\ T_f - .870\ T_r = 0$

Add (9) and (10)

$$-1.011\ T_f - .868\ T_r + .858 = 0$$
$$T_f = -.858\ T_r + .848$$
$$C_f = -.620\ T_r + 1.473$$
$$C_r = +.789\ T_r + .610$$

Solution for $T_r$ by method of least work. (See Art. 116)

$$W_{total} = \frac{C_r^2 L}{2AE} + \frac{C_f^2 L}{2AE} + \frac{T_f^2 L}{2AE} + \frac{T_r^2 L}{2AE} = \frac{(+.789\ T_r + .610)^2\ 48.1}{2 \times .228 \times 29{,}000{,}000} +$$
$$\frac{(.620\ T_r + 1.473)^2\ 38.1}{2 \times .228 \times 29{,}000{,}000} + \frac{(-.858\ T_r + .848)^2\ 57.7}{2 \times .0313 \times 29{,}000{,}000} +$$
$$\frac{T_r^2 \times 63.8}{2 \times .0313 \times 29{,}000{,}000}$$

$$\frac{dW}{dT_r} = \frac{2 \times .789\ (+.789\ T_r + .610)\ 48.1}{2 \times .228\ E} +$$
$$\frac{-2 \times .620\ (-.620\ T_r + 1.473)\ 38.1}{2 \times .228\ E} +$$
$$\frac{-2 \times .858\ (-.858\ T_r + .848)\ 57.7}{2 \times .0313\ E} + \frac{2\ T_r \times 63.8}{2 \times .0313\ E} = 0$$

The lengths of members used are the actual, not theoretical lengths. Solving this equation:

$$T_r = +\ .388$$
$$T_f = +\ .516$$
$$C_f = +1.232$$
$$C_r = +\ .916$$

The assumptions made that in the different cases the diagonal wires or spreader tubes were or were not in action are justified by investigations in which no such assumptions were made. The stresses in the members that were assumed out of action were shown to be negligible.

116. *The Method of Least Work*—As in the example just given, there are certain cases for which a solution is impossible by the methods of statics, and another condition or principle must be employed. This is known as the method of least work and may be stated as follows: when a structure is acted upon by a balanced system of external forces, the stresses will be so distributed that the total internal work done in the various members due to their deformation will be a minimum. The procedure is first to express the work done in each member by the tension, compression, or bending to which that member is subjected, in terms of the unknown stress or bending moment. By taking the sum of all these expressions for the work in the individual members, an equation is obtained for the total work in the structure in terms of one or more unknown stresses. In order to determine at what value of the unknown stress or moment the total work will be a minimum, this equation must be differentiated with respect to the unknown quantity, and the resulting equation put equal to zero. A solution for the stress or moment will give its correct value. In case there is more than one unknown stress the process of differentiation must be performed as many times as there are unknown stresses, the total work being partially differentiated with respect to each unknown in turn. All the unknowns except the one in question are treated as constants during the differentiations. The equations so obtained are solved simultaneously in the usual manner.

In a member carrying direct stress only, the expression for the internal work is $W = P \times \delta/2 = P^2L/2\ AE$.

In a member subjected to bending stresses only, the expression for the internal work is, $W = \int m^2 dx/2\ EI$

P = Total end load, tension or compression, expressed in terms of the unknown stress.
L = Length of the member.
A = Cross-sectional area of the member.
E = Modulus of elasticity of the material.
$\delta$ = Total deformation of the member.
I = Moment of inertia of the section.
m = Bending moment, expressed in terms of the unknown quantity and $X$.
$W = w_1 + w_2 + w_3 + \dots$ = total internal work in structure.

The origin in each case is taken at one end of the member for which the work expression is being written.

The process of differentiation is purely mechanical and not at all difficult. With direct stresses, practically the only types of expression to be differentiated are in the form $W = aP^2$ or $W = a(b—cP)^2$. The differentials of these expressions are $dW/dP = 2aP$ and $dW/dP = —2ac\ (b—cP)$, respectively. The quantities $a$, $b$ and $c$ are known constants. In the second type, if the sign is + both signs in the differential are +.

This method of work is far more accurate than any arbitrary way of proportioning stresses, since it is based only on mathematics and on the properties of the members, and contains no assumptions. Once understood, its application is very simple.

For further explanation of the method, reference can be made to Fuller and Johnson's "Applied Mechanics," Vol. II, p. 242, and to Spofford's "Theory of Structures," pp. 359-367.

117. *Distribution of Moments*—In discussing the distribution of the different moments between the members of a chassis, the case of the chassis which has been chosen as an example is first taken up, and then an explanation is given of the difference in methods which are necessary with a different type of construction.

*Moments in a Vertical Plane Perpendicular to the Plane of the Struts*—Moments in this plane are caused by eccentricities due to the fact that the intersection of the struts and spreader tubes, or of the struts and wires, does not fall on the line of action of the rubber cord in the shock absorber, or to the rise of the axle when side load is present, which causes the collar on the axle to bear against the vertical guide plate. The manner in which the total moment carried by the front and rear struts is apportioned between them is computed by the method of least work. As the struts are hinged at the fuselage, with pins running parallel to the long axis of the struts, the moment at this point is zero. It increases uniformly to its full value at the lower end of the strut.

Let the total moment going into the front and rear struts on one side be unity.

$m_f$ = proportion of total moment in front strut.
$m_r$ = $(1.00—m_f)$ = proportion of total moment in rear strut.
$m_x$ = moment at any point in the struts.
$r_f$ = reaction at upper end of front strut.
$r_r$ = reaction at upper end of rear strut.
$L_f$ = length of front strut.
$L_r$ = length of rear strut.
$I_f$ = moment of inertia of front strut.
$I_r$ = moment of inertia of rear strut.

Case 1

$$r_f = \frac{m_f}{L_f}\ ;\ m_x = \frac{m_f \cdot x}{L_f} \qquad r_r = \frac{m_r}{L_r}\ ;\ m_x = \frac{m_r \cdot x}{L_r}$$

$$w_f = \int_0^{L_f} \frac{m_f^2\, x^2\, dx}{2\, L_f^2\, EI_f} \qquad w_r = \int_0^{L_r} \frac{m_r^2\, x^2\, dx}{2\, L_r^2\, EI_r}$$

$$W = w_f + w_r = \frac{m_f^2}{2\, L_f^2\, EI_f}\int_0^{L_f} x^2\, dx + \frac{m_r^2}{2\, L_r^2\, EI_r}\int_0^{L_r} x^2\, dx$$

$$= \frac{m_f^2\, x^3}{3 \cdot 2\, L_f^2\, EI_f}\Bigg]_0^{L_f} + \frac{m_r^2}{3 \cdot 2\, L_r^2\, EI_r}\Bigg]_0^{L_r}$$

$$= \frac{m_f^2\, L_f^3}{6\, L_f^2\, EI_f} + \frac{(1.00 - m_f)^2\, L_r^3}{6\, L_r^2\, EI_r}$$

by differentiating

$$\frac{dW}{d\, m_f} = \frac{2\, m_f\, L_f}{6\, EI_f} - \frac{2\,(1.00 - m_f)\, L_r}{6\, EI_r} = 0$$

$$m_f = \frac{L_r}{EI_r} \Bigg/ \left[\frac{L_f}{EI_f} + \frac{L_r}{EI_r}\right] \qquad \text{Let } \frac{L_f}{EI_f} = K_1 \text{ and } \frac{L_r}{EI_r} = K_2$$

$$= L_r\, I_f \,/\, (L_f\, I_r + L_r\, I_f)$$

$$= K_2 \,/\, (K_1 + K_2)$$

Case 2

In this case the upper ends of the struts are considered to be fixed either by a rigid socket or by a pin perpendicular to the long axis of the struts. The moment throughout the strut is constant.

$$W = w_f + w_r = \int_0^{L_f} \frac{m_f^2\, dx}{2\, EI_f} + \int_0^{L_r} \frac{m_r^2\, dx}{2\, EI_r}$$

$$= \frac{m_f^2\, x}{2\, EI_f}\Bigg]_0^{L_f} + \frac{m_r^2\, x}{2\, EI_r}\Bigg]_0^{L_r}$$

$$= \frac{m_f L_f}{2 EI_f} + \frac{(1.00-m_f)^2 L_r}{2 EI_r}$$

$$\frac{dW}{dm_f} = \frac{2 m_f L_f}{2 EI_f} - \frac{2 (1.00-m_f) L_r}{2 EI_r} = 0$$

$$m_f = \frac{L_r}{EI_r} \Big/ \left(\frac{L_f}{EI_f} + \frac{L_r}{EI_r}\right) = \frac{K_2}{(K_1 + K_2)}$$

These two solutions show that the distribution of the total moment between the struts is independent of the condition of the upper ends of the struts. It is also clear that by following the same mathematical reasoning the same results would be reached with regard to the distribution of a moment in the plane of the struts. It may, therefore, be stated that for a moment, either in the plane of the struts or perpendicular to it, the distribution between the struts is inversely proportional to their length and directly proportional to their moment of inertia.

Owing to the rigid connection of the struts and spreader tubes to the guide plate, any moment in a vertical plane perpendicular to the plane of the struts cannot be concentrated in the struts on either side, but is transmitted from one set to the other through the spreader tubes. Here again the principle of least work will be resorted to. Assume an external moment of unity applied at $B$, Fig. 97.

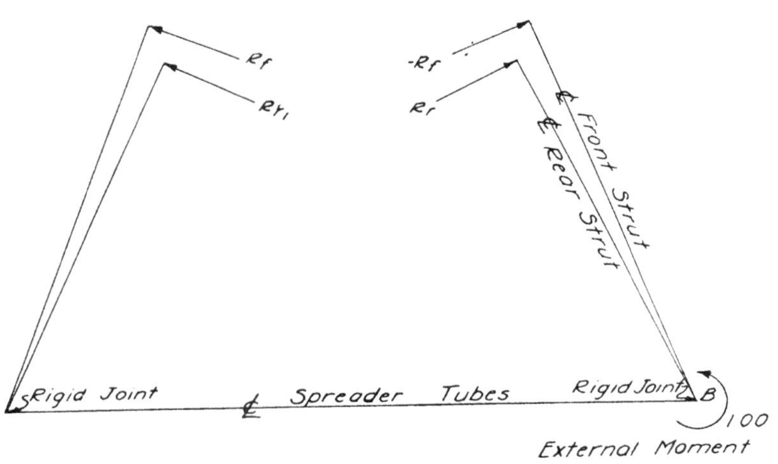


Fig. 97

Sum of maximum moments in struts on right side

$m_s = m_f + m_r$

$m_c =$ Moment in spreader tubes

$= 1.00 - m_s = m_{f_1} + m_{r_1}$

$=$ Sum of maximum moments in the struts on the left side.

As the front and rear struts are the same size, the moment taken by each is in inverse proportion to its length.

Maximum Values

$$\begin{cases} m_f = \dfrac{m_s\,48.1}{86.3} = .558\,m_s \\[2ex] m_r = \dfrac{m_s\,38.1}{86.3} = .442\,m_s \\[2ex] m_{f_1} = .558\,(1.00 - m_s) \\[1ex] m_{r_1} = .442\,(1.00 - m_s) \end{cases}$$

$$r_f = \frac{.558\,m_s}{38.1} = .01464\,m_s$$

$$r_r = \frac{.442\,m_s}{48.1} = .00916\,m_s$$

$$r_{f_1} = .01464\,(1.00 - m_s)$$

$$r_{r_1} = .00916\,(1.00 - m_s)$$

The upper end of each strut, and the left ends of the spreader tubes are taken as the origins. The moment in the struts at a distance "x" from the origin equals the reaction at that point times "x".

$W = w_f + w_r + w_s + w_{f_1} + w_{r_1} =$ total work in system.

$$W = \int_0^{38.1} \frac{(.01464\,m_s)^2\,x^2\,dx}{2\,EI_f} + \int_0^{48.1} \frac{(.00916\,m_s)^2\,x^2\,dx}{2\,EI_r} +$$

$$\int_0^{54.0} \frac{(1 - m_s)^2\,dx}{2\,EI_s} + \int_0^{38.1} \frac{.01464^2\,(1 - m_s)^2\,x^2\,dx}{2\,EI_{f_1}} +$$

$$\int_0^{48.1} \frac{.00916^2\,(1-m_s)^2\,x^2\,dx}{2\,EI_{r_1}}$$

$$W = \left.\frac{(.01464\,m_s)^2\,x^3}{6\,EI_f}\right]_0^{38.1} + \left.\frac{(.00916\,m_s)^2\,x^3}{6\,EI_r}\right]_0^{48.1} +$$

$$\left.\frac{(1-m_s)^2\,x}{2\,EI_s}\right]_0^{54.0} + \left.\frac{(0.1464)^2\,(1-m_s)^2\,x^3}{6\,EI_{f_1}}\right]_0^{38.1} +$$

$$\left.\frac{(.00916)^2\,(1-m_s)^2\,x^3}{6\,EI_{r_1}}\right]_0^{48.1}$$

$$\frac{dW}{d\,m_s} = \frac{2 \times .01464^2\,m_s\,38.1^3}{2\,E \times .0375} + \frac{2 \times .00916^2\,m_s\,48.1^3}{6\,E \times .0375}$$

$$- \frac{2\,(1-m_s)\,54.0}{2\,E \times .0210 \times 2} - \frac{2 \times .01464^2\,(1-m_s)\,38.1^3}{6\,E \times .0375}$$

$$- \frac{2 \times .00916^2\,(1-m_s)\,48.1^3}{6\,E \times .0375} = 0$$

Since the members are all of the same material $E$ can be cancelled from each expression.

$$\frac{dW}{dm_s} = +105.4\,m_s + 83.0m_s - 1285 + 1285m_s - 105.4 + 105.4\,m_s$$

| | |
|---|---|
| $m_s = .885$ | $m_c = .115$ |
| $m_f = .494$ | $m_{f_1} = .064$ |
| $m_r = .391$ | $m_{r_1} = .051$ |

In a chassis where, in addition to rigid connections at the guide plate, the struts are fixed at their upper ends also, the general procedure is the same as in the example just given except that the moment in the struts is constant throughout their length. The difference in the form of the integral is shown in Case 1 and Case 2 in which the distribution of the moment between struts is determined.

If the spreader tubes are hinged, the entire moment is taken by the struts on one side. If in this case the struts are fixed at their upper ends, the moment in them is constant.

The processes of integration which have been used in the last few examples are as simple as the differentiations previously explained.

It should be clearly brought out that there are two distinct operations performed in the examples given. First, an integration is made to determine the value of the work done in each member and in the system, and second, the work thus found is differentiated with respect to the unknown moment. During the integration the unknown moment is treated as a constant while the "$x$" factor is integrated. Once the integration has been made and the limits substituted, "$x$" drops out and in the differentiation the unknown moment becomes the variable. The example which follows illustrates the simple, mechanical operation of integration and substitution of limits for the cases that ordinarily arise.

$$\int_b^a c\,x^n\,dx = \left[\frac{c\,x^{n+1}}{n+1}\right]_b^a = \frac{c\,(a^{n+1} - b^{n+1})}{n+1}$$

## Moments in the Plane of the Struts

118. *Resolution of Moments*—Moments in this plane are caused by fixed eccentricities of the struts or wires with respect to the center-line of the axle slot, or by the variable eccentricity of backward or forward thrusts due to the rise of the axle. The distribution of moments in this plane between the struts has already been explained. However, as the external moments are usually not in the plane of the struts, but in a vertical plane parallel to the propeller axis, an explanation must be given of the method used in resolving these moments into the proper plane. Moments, like forces, can be represented by vectors in magnitude and direction, and can be resolved or composed graphically the same as forces. In this work the moments are laid off on moment axes so that the vectors representing the moments are always perpendicular to the plane in which the moments act. Reference to Fig. 98 illustrates the problem. Line *ab* is the vector of the external moment of unity acting in a vertical plane. Line *ac* is the vector of the moment in the plane of the struts, while *bc* represents the moment going into the spreader tubes. This latter moment causes compression in one tube and a tension in the other equal to the moment divided by the spacing of the tubes.

It has been assumed that the spreader tubes do not assist in taking moments that are in the plane of the struts. Some moment is without doubt carried by the spreader tubes in torsion, but it is so small that it may be neglected.

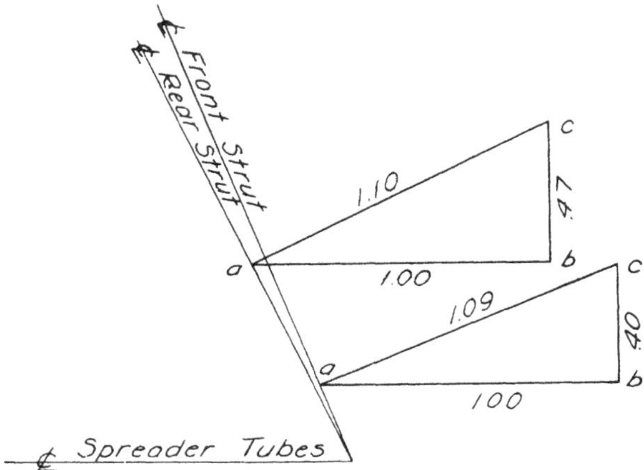


**Fig. 98. Resolution of Moments**

119. *Actual Stresses and Moments in Members.*

I. Vertical load on one wheel in level landing = 825 lb.
II. Backward thrust on one wheel in level landing = 413 lb.
III. Forward thrust on one wheel in level landing = 165 lb.
IV. Side thrust on one wheel in level landing = 206 lb.
V. Total moment due to strut eccentricity in side view, level landing = .36 × 825 = —297 in. lb.
VI. Total moment due to horizontal eccentricity of wire in side view = 33.5/59.2 × .847 × .84 × 1650 = +665 in. lb.
VII. Total moment due to vertical eccentricity of wire in side view = 23.3/59.2 × .847 × .27 × 1650 = —148 in. lb.

For a one-wheel landing the loads in I, II, III, IV and V must be doubled. The moments caused by the wire eccentricity are calculated by finding the horizontal and vertical components of the stress in the wire due to a load of unity and multiplying by the corresponding eccentricities. For a one-wheel landing the moments given by V, VI and VII may be combined, giving a total moment of —77 in. lbs. This moment is in a vertical plane and must be resolved into the plane of the struts. The eccentric moments produced by the thrusts acting through the rise of the axle are computed by multiplying the thrust in question by the rise in the axle. The latter factor may be determined closely, as follows: Assume that the shock absorber will allow the axle to reach the top of its slot when a certain dynamic factor, such as 5.0 or 6.0 is applied. This factor should be somewhat less than is calculated for new cord to allow for deterioration. From this relation can be calculated the axle rise for each dynamic load. Then by the method of trial the dynamic factor which the struts will hold is found, and the axle rise computed for this factor.

In Table XIX is given a summary of the stresses and loads on the different members of the chassis for the four important conditions of loading.

Detailed calculations will be made for cases, first in which there are moments only in the plane of the struts, and then in which moments are present both in the plane of the struts and perpendicular to this plane.

*Rear Strut, Case 4* (one wheel landing with .5 back thrust).

Calculation of moment due to rise of axle.

Total rise of axle for a load factor of 6.0 in level landing = 5.12 in.
Rise of axle for a load factor of 1.0 in one wheel landing = 1.71 in.
Back thrust = 825 lbs. for a load factor of 1.0 in one wheel landing.

Assume the factor of safety to be 2.1. The maximum eccentricity of back thrust then equals $2.1 \times 1.71 = 3.59$ in. Total eccentric moment due to axle rise $= 825 \times 3.59 = +2960$ in. lbs. The portion of this moment going into the rear strut $= 2960 \times .442 = +1310$ in lbs. (See p. 156.)

To resolve this moment, which is in a vertical plane, into a moment in the plane of the struts, it must be multiplied by the factor 1.10 (See Fig. 98) $+ 1310 \times 1.10 = +1440$ in. lbs.

This moment must then be combined with that due to the strut and wire eccentricities, so that the final moment $= +1440 - 37 = +1403$ in. lbs.

$$f_b = \frac{1403}{.0912} = 15{,}400$$

$$f_c = \frac{(1010 + 695)}{.228} = 7{,}490$$

$$f_t = 22{,}870 \text{ lbs. per sq. in.}$$

Calculation of allowable unit stress.

$P/A = 36{,}000 - 0.37\,(85)^2 = 33{,}300$ lbs. per sq. in. in straight compression.

Ultimate allowable stress in bending = 55,000 lbs. per sq. in.

The formula used in obtaining the ultimate allowable stress in combined bending and compression is the same as that given in Art. 174 for the stresses in spars.

$$F_a = \frac{f_b}{f_b + f_c}(F - C) + C$$

$$F_a = \frac{15{,}400}{15{,}400 + 7470}(55{,}000 - 33{,}300) + 33{,}300 = 47{,}900 \text{ lbs. per sq. in.}$$

$$F.S. = \frac{47{,}900}{22{,}870} = 2.1$$

*TABLE XVIII*

| | Front Strut | Rear Strut | Front Wire | Spreader Tubes |
|---|---|---|---|---|
| **Vertical Load (level landing)** | | | | |
| Direct Stress | *— 706 lb. | — 499 lb. | 0 lb. | +347 lb. |
| Moment in plane of struts due to strut eccentricity | — 180 in. lb. | — 145 in. lb. | | |
| **Vertical Load (one-wheel landing)** | | | | |
| Direct Stress | —2425 lb. | —1010 lb. | +1397 lb. | |
| Moment in plane of struts due to strut and wire eccentricity | — 47 in. lb. | — 37 in. lb. | | |
| **Back Thrust of .5** | | | | |
| Level landing | + 251 lb. | — 348 lb. | 0 lb. | + 16 lb. |
| One-wheel landing | + 501 lb. | — 695 lb. | | + 31 lb. |
| **Forward Thrust of .2** | | | | |
| Level landing | — 100 lb. | + 139 lb. | 0 lb. | + 6 lb. |
| One-wheel landing | — 201 lb. | + 278 lb. | | + 12 lb. |
| **Side Thrust of .25** | | | | |
| | — 601 lb. | — 9 lb. | + 828 lb. | —413 lb. |
| **Back Thrust of .25 (for Case I)** | | | | |
| Level landing | + 125 lb. | — 174 lb. | 0 lb. | + 8 lb. |

*In this table the signs denote tension or compression, or positive or negative moment.

This factor of 2.1 applies to the vertical component of the load, for convenience taken as the weight of the airplane. (To obtain the true factor of safety based on the assumption that the resultant load, obtained by combining the vertical and horizontal loads, equals the weight of the airplane, the factor 2.1 must be multiplied by 1.115 which equals $\sqrt{.50^2 + 1.00^2}$, F.S. = 2.1 x 1.115 = 2.33). A similar factor must be applied to the other factors of safety which are given in Table XIX.

At their lower ends the struts are rigidly connected to a vertical steel plate. At the upper ends they are connected to the longerons by pins in the plane of the long axis of the strut. In computing the effect of column action caused by the direct stress, a fixity factor of 3 was used in the parabolic column formula of Art. 24 when the bending moments on the struts were in the plane of the struts. This value of the fixity coefficient, rather than the value of 4, was adopted for the reason that although the struts were assumed to be fixed in their own plane, it is practically impossible to obtain theoretical fixity. When the main bending moments were perpendicular to the plane of the struts, the fixity factor was reduced to 2 on account of the direction of the hinge pins at their upper ends.

The solution for a case in which there are external moments about both axes will now be considered. The side thrust, acting through an eccentricity equal to the rise of the axle, produces a moment which is treated in the same manner as that caused by the back thrust in Case 4.

*Rear Strut, Case 7* (Level landing with .25 side thrust)

Calculation of moments perpendicular to plane of the struts caused by rise of the axle.

Rise of axle for load factor of 1.0 = .853 in.

Side thrust = 413 lbs.

Assume that the factor of safety is 4.3. The maximum eccentricity of the side thrust then equals 4.3 × .853 = 3.67 in.

Total eccentric moment due to axle rise = 413×3.67 = 1515 in. lbs.

The portion of this moment going into the rear strut = 1515× .391 = 592 in. lbs. (See p. 157.)

Location of most stressed fiber.

$m_1$ = —145 in. lbs. = moment in the plane of the struts.

$m_2$ = 592 in. lbs.

$m_1/m_2$ = .245 from Fig. 82  y = .495 × 1.125 = .556 in.

x = .150 × 1.125 = .169 in.

$$f_{b_1} = \frac{145 \times .169}{.0872} = 280$$

$$f_{b_2} = \frac{592 \times .556}{.0375} = 8750$$

$$f_c = \frac{(499 + 9)}{.228} = 2230$$
$$f_t = 11{,}260 \text{ lbs. per sq. in.}$$

Calculation of allowable ultimate stress.

$P/A = 36{,}000 - .56\ (128)^2 = 26{,}800$ lbs. per sq. in. in straight compression.

Ultimate stress in bending = 55,000 lbs. per sq. in.

$$f_a = \frac{9{,}030}{11{,}260}(55{,}000 - 26{,}800) + 26{,}800 = 49{,}400 \text{ lbs. per sq. in.}$$

$$F.S. = \frac{49{,}400}{11{,}260} = 4.4$$

Calculation of Spreader Tubes.

Case 7.

Length = 54.0 in.; $L/\rho = 163$

Assume pinned ends

$$P/A = \frac{9.86 \times 30{,}000{,}000}{163^2} = 11{,}100 \text{ lbs. per sq. in.}$$

Compression in the two tubes = 413 lbs.

Assume that the axle rises to the top of the slot.

Portion of moment due to axle rise going into spreader tubes = $413 \times 5.12 \times .115 = 243$ in. lbs. (See p. 157.)

$$f_b = \frac{243}{.084} = 2890$$

$$f_c = \frac{413}{.384} = 1080$$
$$f_t = 3970$$

$$f_a = \frac{2890}{3970}(55{,}000 - 11{,}100) + 11{,}100 = 43{,}100$$

$$F.S. = \frac{43{,}100}{3970} = 10.8$$

## TABLE XIX

### STRESSES AND FACTORS OF SAFETY

| Member | *Case 1* | | | | | | | Case 4 | | | | | | | |
|---|---|---|---|---|---|---|---|---|---|---|---|---|---|---|---|
| | Direct Stress | $M_1$ | $f_d$ | $f_b$ | $f_t$ | $F_a$ | F.S. | Direct Stress | $M_1$ | $M_2$ | $f_d$ | $f_b$ | $f_t$ | $F_a$ | F.S. |
| Front Strut ..... | — 561 | +463 | — 2550 | 5070 | 7620 | 47300 | 6.2 | —1924 | +1585 | | — 8430 | 17400 | 25830 | 47500 | 1.8 |
| Rear Strut ..... | — 673 | +368 | — 2950 | 4030 | 6980 | 45800 | 6.6 | —1705 | +1403 | | — 7470 | 15400 | 22870 | 47900 | 2.1 |
| Front Wire ..... | 0 | | | | | | | +1397 | | | +37100 | | 37100 | 152000 | 4.1 |
| Spreader Tubes.. | + 347 | | + 903 | | 903 | 55000 | 61. | + 31 | | | | | | | |
| | *Case 6* | | | | | | | *Case 7* | | | | | | | |
| Front Strut ..... | —2626 | —770 | —11520 | 8450 | 19970 | 41790 | 2.1 | —1307 | — 180 | 556 | — 5730 | 8610 | 14340 | 45040 | 3.2 |
| Rear Strut ..... | — 732 | —997 | — 3210 | 10920 | 14130 | 50100 | 3.5 | — 508 | — 145 | 592 | — 2230 | 9030 | 11260 | 49400 | 4.4 |
| Front Wire ..... | +1397 | | +37100 | | 37100 | 152000 | 4.1 | + 828 | | | +22000 | | 22000 | 152000 | 6.9 |
| Spreader Tubes.. | + 12 | | | | | | | — 413 | | 243 | — 1080 | 2890 | 3970 | 43100 | 10.8 |

Both struts are $1\frac{1}{8}$ in. O.D., elliptical tubing, .049 in. gage and 1.7 fineness ratio. Their properties are as follows:

Area = .228
Moment of Inertia = .0375 about long axis
Moment of Inertia = .0872 about short axis
Section modulus = .0666 about long axis
Section modulus = .0912 about short axis
Radius of gyration = .406 about long axis
Front Wire 1—5700 lb. streamline wire; area = .0376
Spreader tubes: 2—1 in. O.D. .065 in. gage round tubes
Total area = .384; total I = .042; $\rho$ = .331; total I/y = .084
$m_1$ is moment in plane of struts.
$m_2$ is moment perpendicular to plane of struts.
$L/\rho$ for front strut is 103 about long axis
67 about short axis
$L/\rho$ for rear strut is 128 about long axis
85 about short axis
Parabolic column formula for 55,000 lbs. per sq. in. steel is

$$P/A = 36{,}000 - .56\,(L/\rho)^2 \text{ for } C = 2$$

$$P/A = 36{,}000 - .37\,(L/\rho)^2 \text{ for } C = 3$$

In computing this table it is assumed that the shock absorber has a service factor of 6.0.

It will be observed from a study of Table XIX that the greatest stresses are in general produced by bending moments and not by direct compression. It will also be noted that the moments caused by back thrusts acting through the rise of the axle are of opposite sign to the moments resulting from the eccentricity of the struts. The effect of this is especially important in Case 1. Therefore, it is suggested that an eccentricity of the struts in side view may be advantageous in neutralizing the moments produced by back thrusts. In Case 6, however, the forward thrust produces a moment which is in the same direction as that caused by the strut eccentricity. It is evident that in determining the eccentricity to give minimum stresses, both Cases 4 and 6 must be considered.

It is further suggested that the front strut, which usually is much more severely stressed than the rear strut, be made of a heavier section than the rear so that the factors for both struts will be nearly equal.

120. *Hinged Axle Chassis*—As a number of chassis are designed with the axle hinged at two points instead of being continuous, some of the features of such a design will be discussed. In what is probably the best type, the elastic cord is wrapped over the axle on each side of the struts and under the struts; the spreader tubes are secured to the struts by a pinned connection; the axle is hinged to a cross member between the spreader tubes, and at this point the latter are supported by a wire taken off the small cross tube. Fig 99 shows these features diagrammatically. With this design no bending moment is put into the spreader tubes, nor any moments in the struts except those due to strut eccen-

tricities. The load on the struts is increased by the amount of the stress in the wire which takes the reaction from the axle at its hinged end. However, this affects only slightly the design of the struts, and the chassis as a whole will be much lighter for the same strength than the other types of hinged axle chassis. Referring to Fig. 99 the vertical component of the load transmitted by the elastic cord to the struts equals $W \cdot a/b$, and the reaction at $c$ that is carried by the support wire equals $W \frac{(a-b)}{b}$. The fact that the load which the shock absorber has to carry is increased by the amount of this reaction is one of the most disadvantageous features of all hinged axle designs.

If the wire at $c$ is omitted, the spreader tubes must carry in bending the axle reaction at that point. This adds largely to their weight. To have them rigidly connected to the struts would partly relieve the bending moment on the tubes. But, as the moment would go directly into the struts and have to be carried by them in bending about their long axes, such a rigid connection would not be at all advisable.

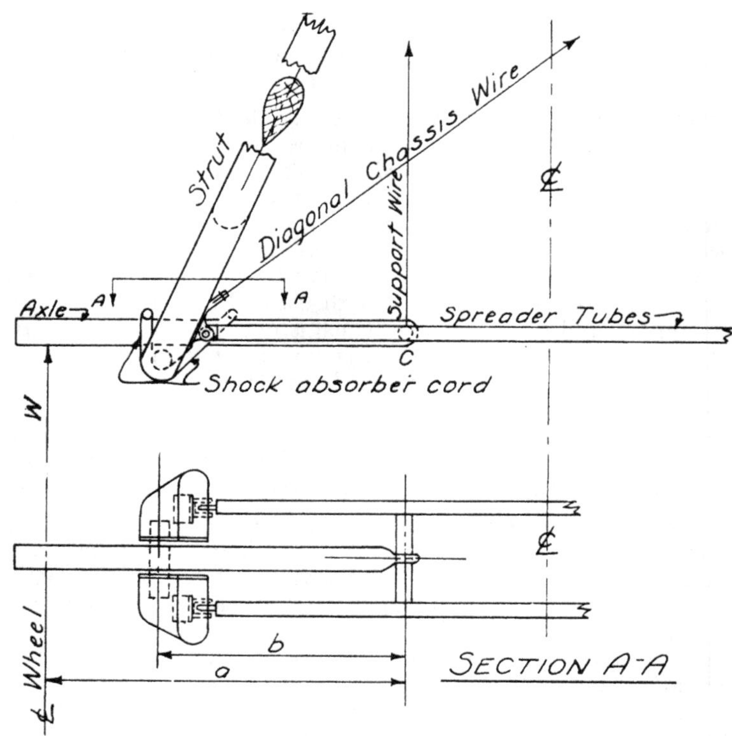


Fig. 99. Chassis with Hinged Axles

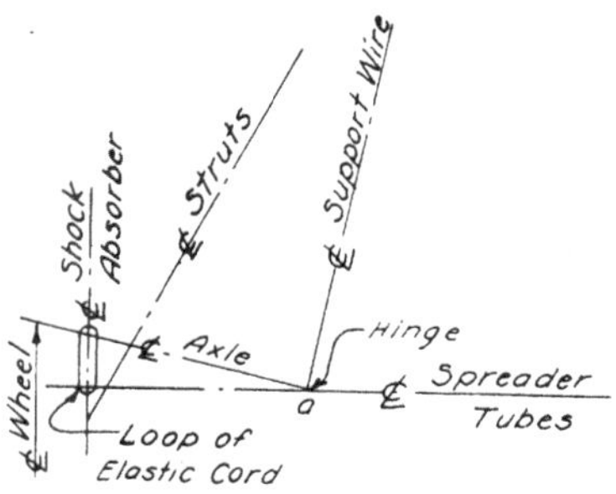


Fig. 100

Fig. 100 is a line sketch illustrating a design in which the elastic cord is attached to the spreader tubes, which pass through the struts, rather than directly to the struts. Since the spreader tubes are subjected to a considerable bending moment their diameter and weight have to be increased. There is also danger that some of the moment from the tubes will be transmitted into the struts. The wire at *a* to take the axle reaction may be omitted, though only by largely increasing the size of the tubes. One advantage with this type of chassis is the general cleanness of design and the good streamlining that is possible. The distance between the centerlines of the wheel and shock absorber can be made less than in the more usual type, thereby decreasing the bending moment on the axle and, hence, its size. The remarks made in Art 122, on the design of axles, relative to the loss in strength caused by heating for the purpose of brazing, as to the quality of steel used, and to the piercing of a member by holes for a pin, etc., are applicable to spreader tubes in hinged axle chassis, which are heavily stressed in bending. In some tests on this type of chassis failure has first occurred in the spreader tubes because these principles were not adhered to.

Chassis with hinged axles have been somewhat of a novelty, but designers are in general more inclined to use the one-piece axle. The difference between the two types, in so far as their riding qualities are concerned, is negligible, while the design using the one-piece axle is simpler and generally stronger for the same weight.

121. *Problems in Chassis Design*—General cleanness of design should be the aim in chassis construction. This and other desirable characteristics may be obtained by meeting the following requirements:

I. Adequate but not excessive strength.

II. Minimum weight.

III. Low structural air resistance.

IV. Simplicity of construction.

V. Easy riding qualities.

The various problems in chassis design are concerned with and affect these factors.

In connection with the first point, the question of suitable factors of safety will be discussed elsewhere. It should be stated here, however, that the strength of the struts and wires should be as uniform as possible. In general, the axle, and, in the case of hinged axles, the spreader tubes, should be no stronger than the weaker of the two struts. It is permissible to have the factor on the shock absorber no higher than on the axle. As the whole object of a chassis is to break the fall of the main portion of the airplane, there should be a progressive failure of the various component parts: first, the shock absorber, then the axle, and lastly the struts. The struts, too, should fail before the load becomes heavy enough to crush in the fuselage or longerons. However, but a small difference in factor is required to effect this. In many chassis tests the shock absorber has failed at a factor of 3.0, the axle at 4.5, and the struts at 8.0. Such a wide variation in strength indicates poor design.

Minimum weight is secured, partly by seeing that no members have excessive strength, and partly by the elimination or reduction of eccentricities that cause large stresses. It may be noted again that eccentricities producing moments in a plane perpendicular to the plane of the struts are much more serious than eccentricities producing moments in the plane of the struts. The following method is suggested for reducing the eccentricity produced by the rise of the axle, which, when a back thrust is present, causes a moment in the plane of the struts. As is shown in Fig. 101, the axle should be dropped vertically below the intersection of the struts and a line through the centers of the spreader tubes. The axle fairing should be lowered till it rests on top of the spreader tubes, and the axle till it touches the bottom of the fairing. When the axle is forced up in the slot, the eccentricity of the back thrust is reduced by the amount the axle is dropped. This feature will also lessen the height of the unsupported part of the vertical guide plate, and, hence, make it more rigid in resisting side thrust loads when the axle is forced nearly to the top of the slot.

The careful streamlining of shock absorbers by aluminum fairing, and of the axle either by this or by suitable wood fairing, the use of the standard strut section described in Art. 91 for the struts whether they are all wood or are faired steel tubing, the elimination of wires in the plane of the rear struts where practicable and the streamlining of those used, care in cleaning up fittings and minor details, and the attachment of the wheel fairing disks to the tire at its maximum diameter instead of to the rim of the wheel should largely reduce the structural resistance of a chassis.

In designing, an effort should always be made to eliminate complicated fittings and those which present difficulties in manufacture. As a rule, simplicity and strength in construction go together to give dependability and long service. The article dealing with shock absorbers

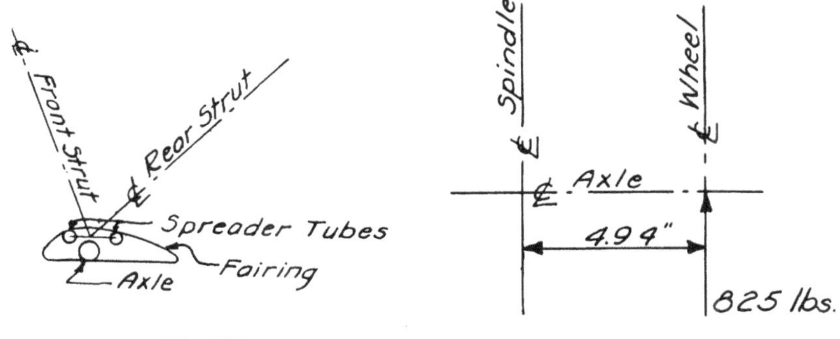


Fig. 101 Fig. 102

discusses the resiliency of elastic cord and its effect on the easy riding qualities of a chassis.

122. *Design of the Axle*—Axles are calculated for the vertical load in level landing, and are designed for a factor equal to or slightly less than the factor for the worse stressed struts under the same condition. In a one-wheel landing the factor is cut in half, and with a back thrust it is still further reduced, perhaps even more than the factor for the struts. However, if an axle is overstressed it merely bends and no damage is caused to the airplane as a whole. Furthermore, such failure can be readily detected and the axle easily replaced. Under no circumstances should any part of an axle be heated as high as a low red heat after it has once been given its final heat-treatment, or its strength will be decreased to about 1/3 its proper value. Also, soft steel or plain high carbon steel should never be used in axle design. Alloy steel of an ultimate tensile strength of 200,000 lbs. per sq. in., conforming to U.S.A. Signal Corps Specification No. 10,229, can be obtained and depended upon. According to this specification, all axles, to pass inspection, must be subjected to a bending stress of 180,000 lbs. per sq. in. without injury.

Fig. 102 gives the necessary data for the design.

The bending moment on the axle in level landing

$$= 825 \text{ lb.} \times 4.94 \text{ in.} = 4075 \text{ in. lb.}$$

$$\text{O.D. of axle} = 1\tfrac{1}{2} \text{ in.; Wall} = 3/32 \text{ in.; } I/y = .137$$

$$f_b = \frac{4075}{.137} = 29{,}700 \text{ lb. per sq. in.}$$

$$\text{F.S.} = \frac{200{,}000}{29{,}700} = 6.7$$

123. *Design of the Shock Absorber*—There are two general types of shock absorbers; one in which the cord is wound directly over the axle and under the struts or spreader tubes, and one in which the cord is wound on spindles, the lower spindle being attached to the struts or spreader tubes and the upper one to the axle. The former type is no

longer used in careful, refined design for the following reasons: I—Owing to the way in which the cord is wound it is impossible to determine with accuracy its elongation for any rise of the axle, and therefore to calculate the depth of slot required. II—As the cord is wrapped in several layers the chafing and cutting of the under layers by the upper when the cord is under strain causes rapid deterioration. III—For the same reason, the distribution of load between the cords is unequal. Some cords are likely to take enough load to break the fabric. IV—With this type of shock absorber streamlining of the cord is very difficult. The later designs of spindle shock absorbers permit the cord to be wound on the spindles when they are removed from the chassis. As the cord is wrapped in a single layer chafing and unequal load distribution are a minimum. The elongation of the cord is easily computed. Because of the compactness and small section of a spindle shock absorber, it is readily streamlined and offers but little resistance. Fig. 103 shows the absorber used on the chassis which is taken as the example in this analysis.

The function of a shock absorber is to lessen the shock of landing by absorbing the kinetic energy of the airplane which is falling with a certain vertical velocity. The energy thus taken up by each strand of cord equals the elongation of the cord times the average tension in the cord during this elongation. For example, assume that the airplane used in this chapter, whose weight is 1650 lbs., falls freely through a distance of 10 in. The kinetic energy of the airplane at the instant of contact $= W =$ 16,500 in. lbs. It is assumed that 1/2 in. cord whose average

Fig. 103

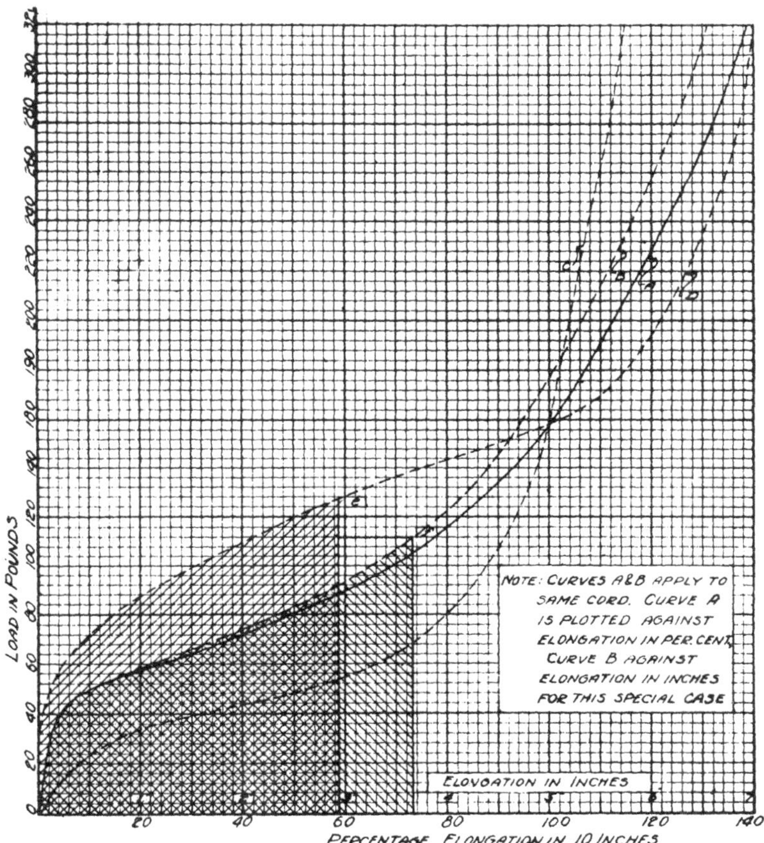


Fig. 104. Typical Load Elongation Curve for ½-in. Rubber Cord

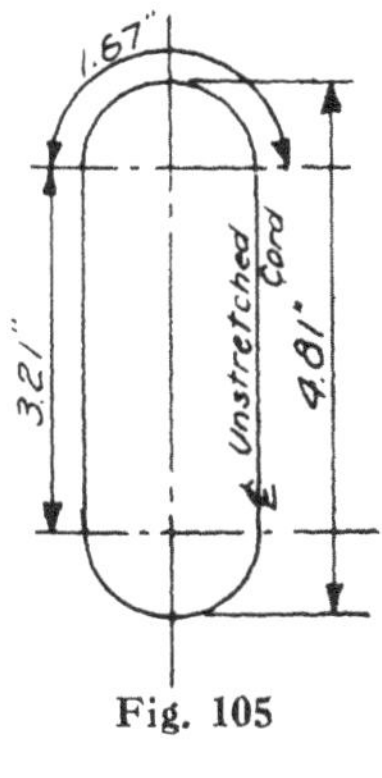


Fig. 105

load-elongation curve is shown in Fig. 104 is used on the airplane.

Below are given the necessary data for the shock absorber.

*One Unit*

Length of 4 partial loops = 14.1 in.
Length of 1 complete loop = 2(1.67+3.21) = 9.76 in. (See Fig. 105)
Number of complete loops = 13
Total length of unstretched cord
$= 13 \times 9.76 + 14$ = 141.1 in.
Total possible rise of axle = $5\frac{1}{8}$ in.
Total elongation of cord $= 5.12 \times 30$ = 153.6 in.
Percentage maximum elongation $= \frac{153.6}{141.1}$ = 109 per cent

In the two shock absorbers on the airplane there are $2 \times 30 = 60$ strands of cord. The energy to be taken up by each strand $= \frac{16{,}500}{60} =$ 275 in. lbs. The distance through which the tension in each strand acts per 1 per cent elongation $= \frac{153.6}{30 \times 109} = .0470$ in. Referring to curve *B*, Fig. 104, which is obtained by replotting curve *A*, using for ordinates actual distances instead of percentage elongations, the area under the curve to the left of any ordinate, when multiplied by the scales of force and distance, equals the work done on the cord in stretching the given distance, or the amount of the energy absorbed. When the stretch equals 3.65 inches, corresponding to an elongation of 76 per cent, the area under the curve equals 6.87 square inches.

Energy $= 6.87 \times 40 \times 1 = 275$ in. lbs. The totatl energy which can be absorbed by each strand before the axle rises to the top of its slot at an elongation of 109 per cent equals 557 in. lbs. Therefore, the maximum capacity of both shock absorbers $= 557 \times 60 = 33{,}500$ in. lbs.

This study of the resiliency of elastic cord shows the importance of the character of the load-elongation curve. Cords which may require the same load to produce a 100 per cent elongation will give very different results in service. The cord of which *C* is the load-elongation curve, is so weak and has such a small resiliency that it is practically useless. On the other hand, the cord represented by curve *D* is too stiff to give easy riding qualities. The stress in the cord for a given resiliency is much greater than with the more normal cord represented by Curve *A*. The areas indicated in Fig. 104 under the two curves are the same, 275 in. lbs., but the tension in the cord at *e* is more than 20 per cent greater than in the cord at *f*. It is clear, therefore, that to absorb a given amount of energy with cord *D* will put a greater stress in the struts with the average landing. On the contrary, when the landing is so severe as to cause an elongation greater than about 95 per cent the struts will be less stressed with cord *D*.

The usual basis for determining the number of strands of cord and the necessary rise of the axle, where the cord to be used conforms to a standard specification, is that the total load required to force the axle to the top of the slot (against the resistance of the cord) divided by the weight of the airplane shall equal, or nearly equal, the factor of safety of the worse stressed struts in a level landing.

From curve *A* the load required to elongate the cord 109 per cent is 186 lbs.

$$\text{F.S.} = \frac{186 \times 60}{1650} = 6.75$$

Owing to the latitude in regard to the stiffness of cord that is necessary in the specification for elastic cord, cord which would meet the lower limit set by the specification would give a factor of safety of 6.25, which is satisfactory.

The requirements of good design limit the range of maximum elongation. This should be between 100 and 120 per cent. A value of 110 per cent will ordinarily give the best results, for if the maximum elongation is much less the design is not economical, since, owing to the steepness of the load-elongation curve for elongations greater than 100 per cent, a slight increase in elongation gives a large increase in load carried, while if the elongation is much more the action of the cord is too stiff, and the cord is liable to deteriorate rapidly. There are two ways to secure any desired elongation: one is to use a large amount of rubber and have the axle slot very deep; the other is to decrease to a minimum the distance between the spindles of the shock absorber. The latter method gives a small, compact design, but, because of the small amount of rubber used, it lacks resiliency or shock absorbing power and is not so effective in reducing the shocks of bad landings.

A moot point in connection with shock absorbers is the advisability or necessity of putting a large initial tension in the cord. Whether it is necessary to employ such a tension depends both on the quality of the cord and the design of the shock absorber. If a cord is weak, especially at low elongations, and has the general characteristics shown by curve *C*, Fig. 104, it will stretch so much, unless given a large initial tension, that the axle will rise an inch or so even under the dead load of the airplane. Or if the design does not provide a sufficient number of strands of cord to support the dead load without causing a large tension in each strand and consequently a large elongation, the axle will rise an excessive amount unless an initial tension is used. One important fact should be brought out regarding the effect of initial tension in reducing the resiliency of the shock absorber. If, for instance, the cord represented by curve *D* is wound with a tension of 50 pounds it will be initially elongated 50 per cent, and all the capacity for absorbing energy represented by the area under curve *D* to the left of the ordinate at 50 per cent elongation will be lost. A similar loss will, of course, occur when initial tension must be used because of an insufficient number of strands, even though the cord be of good quality. Another point to be mentioned is that unless a spring balance is used when the cord is being wound, or unless the elongation of the cord for the desired tension is known and the cord cut to the proper length before winding, there is no way of knowing what initial tension is actually used. The cord may be stressed twice as much as was intended. In spite of these facts it may be advisable to use an initial tension of 30 to 50 pounds where the design, owing to certain limitations, could not provide an adequate number of strands of cord. A distinction should be made between an insufficient number of strands of cord and an insufficient amount of cord. In the first case the amount of cord may be ample, but inefficiently distributed, the spindles probably being too far apart.

In general, a shock absorber should be originally designed, first for a good quality cord that meets the government specifications, and second for a small initial tension of only 10—15 lbs. or just enough to wrap the cord snugly. This will lengthen the life of the cord because, not being continually under a high stress, it will not chafe badly. If, also, enough cord is used so that the cord is very seldom stretched nearly up to the breaking of the fabric it should give good service.

*Government Specifications.*

| | Load at 100 per cent elongation |
|---|---|
| 3/8 in. cord .................. | 65— 90 lbs. |
| 1/2 in. cord .................. | 145—180 lbs. |
| 5/8 in. cord .................. | 240—275 lbs. |

124. *Strength Factors for Chassis*—The factors given in column 3 are applicable only to the struts. They represent the requirements of the present standard static test for chassis. In this test the load is equally divided between the wheels, and the testing jig is so arranged that the resultant load makes an angle with the normal to the propeller axis equal to the angle corresponding to $a$ on page 176 for the airplane in question. For most designs this is equivalent to considering the propeller axis horizontal, and combining a vertical load with a back thrust, equal to from .2 to .25 of the vertical load, so that their resultant equals the weight of the airplane.

The minimum factor for loadings (4) and (6), given in Art. 111, shall be not less than 35 per cent of the required factor for the static test, which is specified in Table XX. For the fully loaded day and night bombers the minimum factor for loadings (4) and (6) shall be not less than 40 per cent of the factor in Table XX.

When the above strength factors for the struts are computed analytically, the stresses upon which such factors are based must include stresses due to the various eccentric moments, which have been taken up in detail in the present chapter. The methods used to calculate the

124. *Factors of Safety.*

*TABLE XX*

*STRENGTH FACTORS FOR CHASSIS AND VALUES OF HORIZONTAL THRUSTS*

| Type of Airplane | | Static Load Factors | | | | Horizontal Thrust | | |
|---|---|---|---|---|---|---|---|---|
| No. | Class | Struts | Axle | | Shock Absorber | Back-ward | For-ward | Side |
| 1 | 2 | 3 | 4 | 5 | 6 | 7 | 8 | 9 |
| | Primary Training | 7.5 | 6.0 | 7.0 | 6.5 | .5 | .2 | .25 |
| 14 and 15 | All-through Training | 7.0 | 5.5 | 6.5 | 6.0 | .45 | .2 | .25 |
| | Advanced Training | 7.0 | 5.5 | 6.5 | 6.0 | .45 | .2 | .25 |
| 1, 3 and 4 | One-Seater Pursuit for Day Work | 6.0 | 5.0 | 5.5 | 5.0 | .4 | .2 | (.18) |
| 2 | One-Seater Pursuit for Night Work | 7.0 | 5.5 | 6.5 | 6.0 | .4 | .2 | .25 |
| 5, 6, 7 and 10 | Two-Seater Pursuit or Observation for Day Work | 5.0 | 4.0 | 4.5 | 4.5 | .35 | .2 | (.18) |
| 8 | Two-Seater Pursuit or Observation for Night Work | 6.0 | 5.0 | 5.5 | 5.0 | .35 | .2 | (.20) |
| 9, 11 and 12 | Fully Loaded Day Bombers | 4.5 | 4.0 | 4.25 | 4.0 | .3 | .2 | (.15) |
| 13 | Fully Loaded Night Bombers, Long Distance | 4.0 | 3.5 | 3.75 | 3.5 | .25 | .2 | (.10) |

stresses should closely follow those which have been previously discussed in this chapter.

The axle will be required to sustain, without permanent deformation, a load corresponding to the factor specified in column 4. What may be considered failure of the axle shall not occur at a factor less than the factor specified in column 5. In the case of hinged axle chassis these same factors shall apply to the spreader tubes where they carry major stresses.

The axle shall not be forced to the top of its slot by a load less than that corresponding to the factor given in column 6. Furthermore, the outside fabric of the elastic cord shall not be injured by this load.

125. *Relation of Chassis to Fuselage*—The height of a chassis is determined, first, by the clearance of the propeller with respect to the ground when the propeller axis is horizontal, and second, by the angles, $\phi$ and $\theta$. In measuring these angles the propeller axis is considered horizontal. $\phi$ is the angle formed by the ground-line and a line through the tip of the tail skid and tangent to the lower part of the wheel; $\theta$ is the angle between the ground-line and a line through a bottom wing tip and the lower part of the wheel on the same side as the wing tip; this angle shows in front elevation. The minimum value of the propeller clearance may be taken as 10 in. The longitudinal location of the chassis is fixed by the angle $\alpha$ between the vertical through the point of tangency of the wheels and the ground, and a line through this point and the center of gravity of the airplane. The usual range of values of $\phi$, $\theta$ and $\alpha$ for a number of types of American and foreign airplanes is given below.

*TABLE XXI*

| | Single Seater | Two Seater | Heavy Bombers |
|---|---|---|---|
| $\phi$ ........ | 13°—14° | 11°—12° | 10°—11° |
| $\theta$ ........ | 19°—22° | 13°—17° | 9°—12° |
| $\alpha$ ........ | 15°—18° | 10°—13° | 9°—12° |
| T/S ..... | .17 —.21 | .12 —.15 | .16 —.20 |

In each class, the value of $\alpha$ depends on the minimum speed of the airplane. It is greater for high landing speeds. Values of $T/S$ or the ratio of the tread to wing span are useful in determining the tread. The slopes of the chassis struts are determined largely by the location of suitable bulkheads, to which the upper ends of the struts may be attached. If possible, the front struts should support the engine directly.

# CHAPTER VI

# CONTROL SYSTEMS

126. *Introduction*—The application of the principles of structural analysis to the design of control systems is not so general as in the case of wing structures. As a consequence, failure of controls during a sand test on control surfaces has frequently occurred, sometimes at a very low factor. So designers, knowing of such failures, have often added unnecessary weight to the controls to insure adequate strength.

Both control surfaces and systems may be designed to withstand air loads which are determined either by what is considered the maximum force it is possible for a pilot to exert, or by definite unit loads fixed by specification for each type of machine. The basis for specifying these latter loads will be discussed in Chapter VII. It is evident that designs based on such loads will be more uniform and better balanced.

In the present chapter, an investigation will be made of the control system for a single-seater pursuit airplane. The functioning of this system is shown diagrammatically in Fig. 106. The design, though a reasonable one, might be considerably improved. It is impossible, of course, to proportion each tube and bolt so that all parts are stressed up to the limit. However, the weakest part of a control system should be somewhat stronger than the minimum factor required for the control surfaces. A margin of about 10 to 15 per cent should be allowed. Several special cases will be considered to illustrate certain features not occurring in the controls of this pursuit airplane.

## *STRESS ANALYSIS FOR CONTROLS OF PURSUIT AIRPLANE*

127. *Calculation of Moments*—A torque system of control, shown in Fig. 106, is used for the ailerons; and for the elevators, instead of the usual two masts, there is only one attached to the elevator tubes inside the fuselage. The curved rudder bar, Fig. 114, is a trifle unusual, but aside from these features the controls conform to common practice.

The first step in the design is the calculation of the moments produced by the loads on the elevators, ailerons and rudder. For a single-seater pursuit the present Government specifications require an ultimate load on the horizontal tail surfaces which shall average 35 lbs. per sq. ft.; on the vertical tail surfaces an average load of 30 lbs. per sq. ft.; and on the ailerons the same loading as on the elevators. The distribution of load between the fixed and movable surfaces, both horizontal and vertical, is shown in Fig. 107. Owing to the rounding of the ends of the surfaces, the center of gravity of the load on each must be calculated. This is done by dividing the surface into a sufficient number of strips. As the loading for the movable surfaces is triangular, the center of gravity of the load on each strip will be 1/3 the length of the strip from the leading edge of the surface.

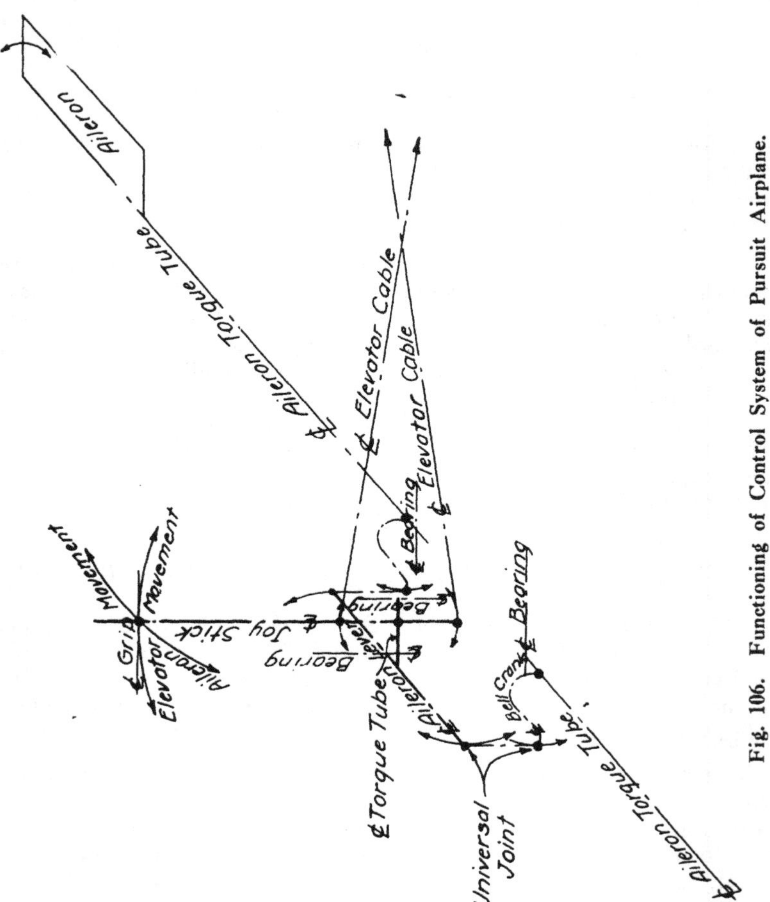


Fig. 106. Functioning of Control System of Pursuit Airplane.

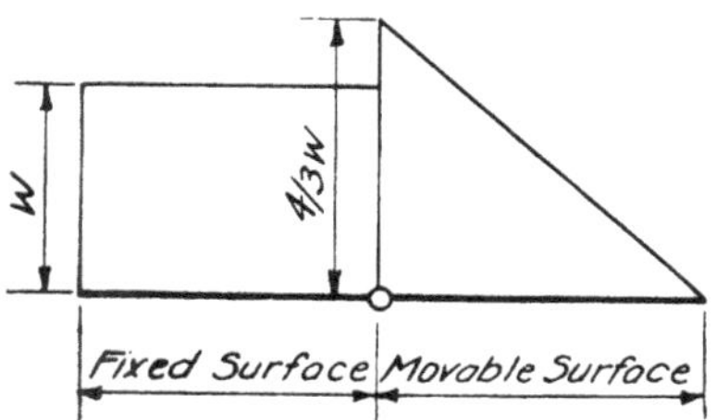


**Fig. 107. Load Distribution Between the Fixed and Movable Surfaces**

Calculation of Unit Loads.

I. Elevator.

Area of elevators = 16.7 sq. ft.
Area of stabilizer = 12.5 sq. ft.
Let "*w*" equal unit load on stabilizer.

$$12.5\,w + \frac{4}{3} \times \frac{w}{2} \times 16.7 = (12.5+16.7)\,35 = 1020 \text{ lbs.}$$

w = 43.3 lbs. per sq. ft.

2/3 w = 29.0 lbs. per sq. ft.

= Average elevator and aileron load.

II. Rudder.

Area of rudder = 8.6 sq. ft.
Area of fin = 3.8 sq. ft.
Let *w* equal unit load on fin.
$3.8\,w + 4/3 \times .5\,w \times 8.6 = (8.6+3.8)\,30 = 372$ lbs.
w = 39.0 lbs. per sq. ft.
2/3 w = 26.0 lbs. per sq. ft.
= Average rudder load.

*Calculation of Center of Gravity of Elevator Load.*

$$x = \frac{(23.5\times30.2)\,7.83+(3.75\times23.15)\,7.72+(3.55\times21.8)\,7.27+(3.4\times18.9)\,6.3+(1.9\times14.2)\,4.72+(4.46\times23.15)\,7.72+(3.7\times21.3)\,7.10+(42)\,6.0}{1200}$$

$$x = \frac{8993}{1200} = 7.5 \text{ in.} = \text{arm of load}$$

*Calculation of Moments.*

I. Elevators.

Total moment = Area × average load × mean arm
= 2 × 8.33 × 29.0 × 7.5
= 3620 in. lbs.

II. Rudder

Moment = Area × average load × mean arm
= 8.63 × 26.0 × 9.45
= 2120 in. lbs.

III. Ailerons

Total moment = Area × average load × mean arm
= 2 × 12.5 × 29.0 × 7.12
= 5160 in. lbs.

128. *Investigation of Control Stick for Elevator Loads*—Refer to Fig. 108.

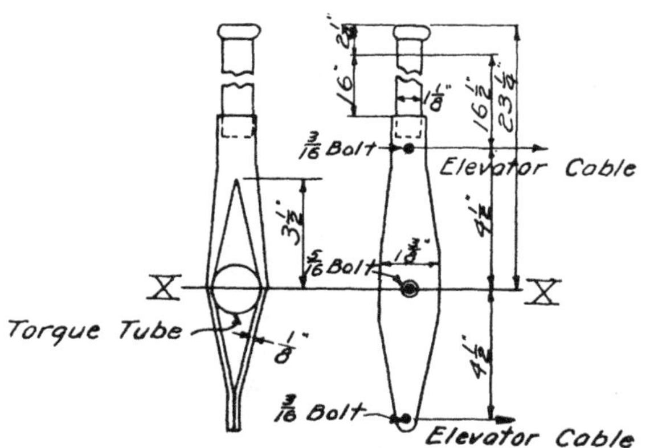


Fig. 108. Control Stick

The moment on the control stick from the elevator or aileron loads, or on the rudder bar from the rudder loads is not necessarily the same as the moment of these loads about the hinges of the movable surfaces, but is dependent on the relation between the height of the control masts and the length of the lever arms on the control stick or rudder bar. In the design under consideration the height of the control masts on the elevator is 4½ ins.

Moment of elevator load = 3620 in. lbs.

$$\text{Reaction at grip} = \frac{3620}{21} = 173 \text{ lbs.}$$

Maximum moment in single tube = 173 × 16 = 2770 in. lbs.
Tube size = 1⅛ in. O.D., 3/32 in. wall.
Section modulus = .0724

$$f_b = \frac{2770}{.0724} = 38{,}300 \text{ lbs. per sq. in.}$$

$$\text{Coefficient} = \frac{55{,}000}{38{,}300} = 1.43 \quad \text{(satisfactory)}$$

A wall thickness of 1/16 in. gives a maximum bending stress of 52,800 lbs. per sq. in., which does not provide sufficient margin when 55,000 lbs. per sq. in. steel is employed. A higher strength alloy steel may be used and the gage reduced, provided the tube is not annealed by brazing so as to destroy the effect of the heat treatment.

$$\text{Pull on elevator cable} = \frac{3620}{4.5} = 805 \text{ lbs.}$$

Elevator cable is 1/8 in. 7 × 19 extra flexible, with 2000 lbs. capacity.

$$\text{Coefficient} = \frac{2000}{805} = 2.48 \quad \text{(satisfactory)}$$

Shear on 3/16 in. nickel steel pin holding elevator cable.

Area of 3/16 in. pin = .0276 sq. in.

$$f_s = \frac{805}{2 \times .0276} = 14{,}600 \text{ lbs. per sq. in.} \quad \text{(satisfactory)}$$

Bearing on 3/16 in. pin.

$$f_c = \frac{805}{.1875 \times .25} = 17{,}200 \text{ lbs. per sq. in.} \quad \text{(satisfactory)}$$

Moment on control stick at section *X—X*, Fig. 108,
= 805 × 4.5 = 3620 in. lbs.

$$\text{Net I of section} = \frac{1.375^3 \times .25}{12} - \frac{.312 \times .25}{12} = .0535$$

$$f_b = \frac{3620 \times .688}{.0535} = 46{,}600 \text{ lbs. per sq. in.}$$

$$\text{Coefficient} = \frac{55{,}000}{46{,}600} = 1.18 \qquad \text{(satisfactory)}$$

Shear on 5/16 in. bolt in control stick

Load $= (805 + 173) = 978$ lbs.

Area $= 2 \times .0766$ sq. in.

$$f_s = \frac{978}{2 \times .0766} = 6380 \text{ lbs. per sq. in.} \qquad \text{(satisfactory)}$$

Bearing of 5/16 in. bolt

Bearing area $= 2 \times .312 \times .125 = .078$

$$f_c = \frac{978}{.078} = 12{,}500 \text{ lbs. per sq. in.} \qquad \text{(satisfactory)}$$

The short torque tube to which the control stick is attached is held in place by a 3/16 in. bolt passing through a collar on the tube.

Shearing stress on this bolt

$$f_s = \frac{978}{2 \times .0276} = 17{,}700 \text{ lbs. per sq. in.} \qquad \text{(satisfactory)}$$

Bearing stress on this bolt

$$f_c = \frac{978}{2 \times .187 \times .0625} = 41{,}700 \text{ lbs. per sq. in. (satisfactory)}$$

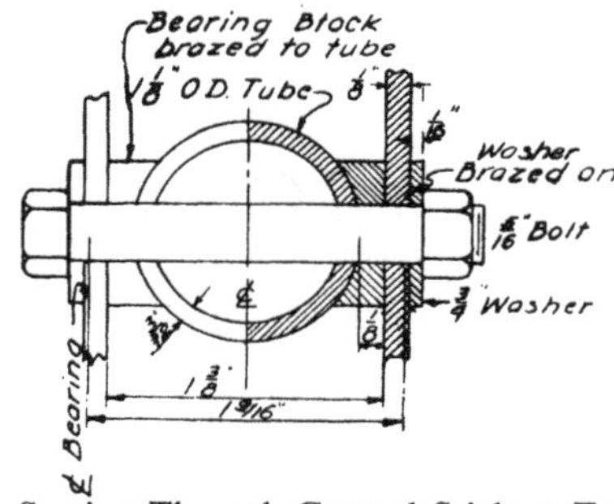


**Fig. 109. Section Through Control Stick at Torque Tube**

129. *Investigation of Control Stick for Aileron Loads*—Reference should be made to Fig. 109, which shows a section through the control stick and torque tube. The total aileron moment is 5160 in. lbs.

Shear on 5/16 in. bolt in transmitting this moment

$$\text{Shear} = \frac{5160}{1.375} = 3750 \text{ lbs.}$$

$$f_s = \frac{3750}{.0766} = 48{,}900 \text{ lbs. per sq. in.} \qquad \text{(satisfactory)}$$

Bearing of 5/16 in. bolt on control stick

$$\text{Compression} = \frac{5160}{1.562} = 3300 \text{ lbs.}$$

$$f_c = \frac{3300}{.312 \times .1875} = 56{,}500 \text{ lbs. per sq. in. (satisfactory)}$$

This value is somewhat high, but bearing stresses may be greater than the ultimate strength of the material without harm.

Compressive stress in sides of control stick

Compression = 3440 lbs.

Compressive stress on net area through 5/16 in. bolt hole=

$$f_c = \frac{3400}{(.125 \times 1.062 + .0625 \times .437)} = \frac{3440}{.160}$$
$$= 21{,}500 \text{ lbs. per sq. in.} \qquad \text{(satisfactory)}$$

As the sides of the control stick are unsupported for a length of 3½ ins., there will be some column action and consequently a reduction in the allowable ultimate stress. Since a careful determination of the correct average value of the radius of gyration of this column would be very difficult, the extreme assumption will be made that the column section is a flat plate instead of a varying circular arc.

$$f_c = \frac{3440}{.172} = 20{,}000 \text{ lbs. per sq. in.}$$

$$\text{Coefficient} = \frac{31{,}700}{20{,}000} = 1.58 \qquad \text{(satisfactory)}$$

$$\text{Area} = .125 \times 1.375 = .172$$

$$\rho = \frac{h}{\sqrt{12}} = \frac{.125}{3.46} = .0361$$

$$L/\rho = 97$$

$$f_c = 55{,}000 - 240L/\rho = 31{,}700 \text{ lbs. per sq. in.}$$

130. *Investigation of Torque Tube to Which Control Stick is Secured*

Torsional shear stress

Aileron moment = 5160 in. lbs.

Polar I of $1\frac{1}{8} \times \frac{3}{32}$ in. tube $= 2 \times .0407 = .0814$

$$f_s = \frac{5160 \times .562}{.0814} = 35{,}700 \text{ lbs. per sq. in.}$$

See Fig. 153 for Ultimate Torsional Shear Stress

$$\text{Coefficient} = \frac{44{,}000}{35{,}700} = 1.23 \quad \text{(satisfactory)}$$

Bearing of 5/16 in. bolt through joy stick on tube. From Fig. 109 it will be observed that a bearing block is brazed to the torque tube.

Distance of center of bearing area from center of tube =
$(.469 + .5 \times .219) = .578$ in.

$$\text{Force} = \frac{5160}{2 \times .578} = 4470 \text{ lbs.}$$

$$f_c = \frac{4470}{.312 \times .219} = 65{,}400 \text{ lbs. per sq. in.}$$

This value in bearing is high, but failure will not occur.

131. *Investigation of Aileron Lever*—Refer to Figs. 110 and 111.

$$\text{Vertical force on outer end of arm} - \frac{5160}{2 \times 6.5} = 397 \text{ lbs.}$$

Since the rods connecting the ends of the arm with the bell cranks make an angle of 16 degs. with the vertical, a horizontal component is present which produces a moment about the long axis of any cross-section of the arm.

| | |
|---|---|
| Vertical component | $= 397$ lbs. |
| Horizontal component | $= 397 \times \tan 16° = 114$ lbs. |
| Moment about horizontal axis | $= 114 \times 6.5 = 740$ in. lbs. |
| Moment about vertical axis | $= 397 \times 6.5 = 2580$ in. lbs. |

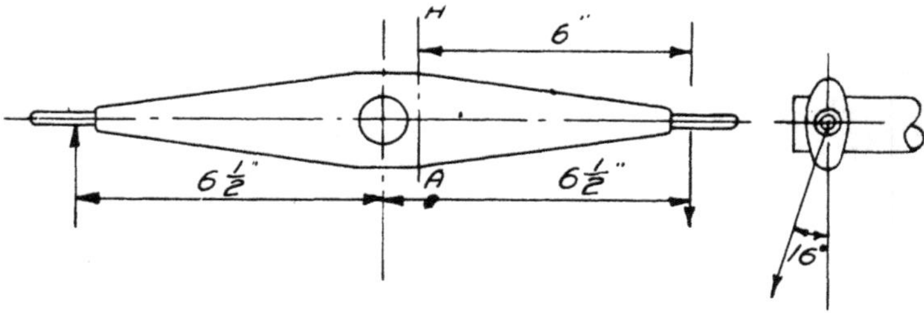


**Fig. 110. Aileron Lever on Torque Tube**

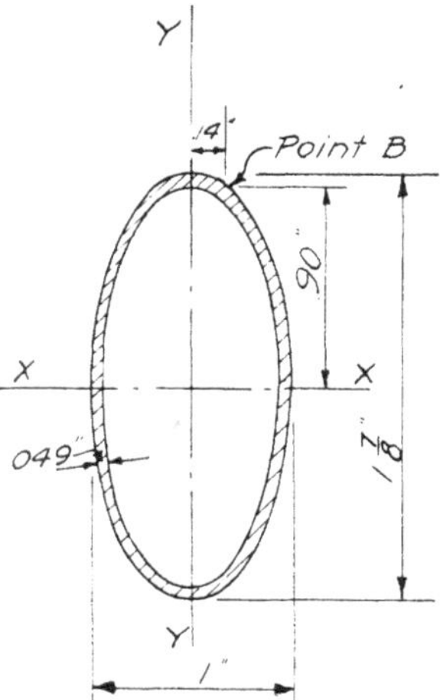


**Fig. 111. Section A-A Through Aileron Lever**

By the method of trial it is found that "*B*" is the most stressed point on the cross-section "*A—A*", Fig. 111.

$$f_{b_1} = \frac{2580 \times .90}{.0741} = 31{,}300$$

$$f_{b_2} = \frac{740 \times .14}{.0281} = \underline{3{,}700}$$

Resultant stress $= 35{,}000$ lbs. per sq. in.

$$\text{Coefficient} = \frac{55{,}000}{35{,}000} = 1.57 \quad \text{(satisfactory)}$$

Stress on weld between arm and torque tube
Minimum area of weld $= \pi \times 1.125 \times 2 \times .049 = .346$ sq. in.

$$\text{Total stress on weld} = \frac{5160}{.5625} = 9160 \text{ lbs.}$$

$$f_s = \frac{9160}{.346} = 26{,}500 \text{ lbs. per sq. in.}$$

This is a dangerously high stress on a welded section due to the uncertainty of the quality of the weld.

132. *Investigation of Bell Crank Lever*—This lever is curved, as shown in Fig. 112, in order to clear some obstruction.

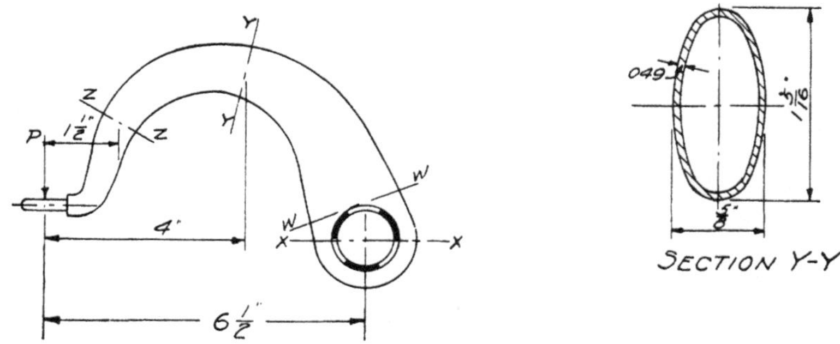


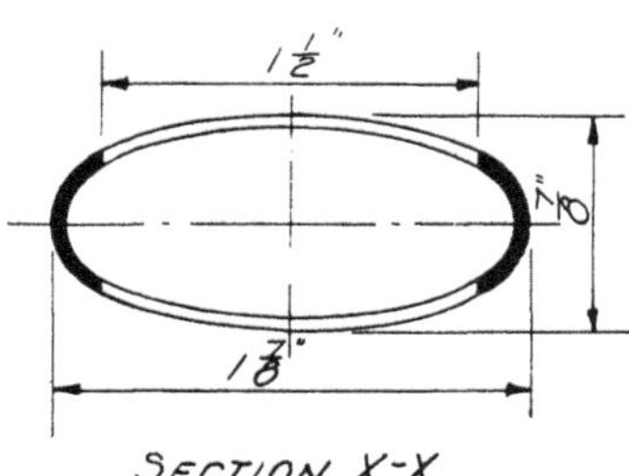


**Fig. 112. Bell Crank Lever**

Stress on weld between lever and torque tube.

Minimum area of weld $= \pi \times 1.375 \times 2 \times .049 = .423$ in.

$$\text{Moment} = 5160/2 = 2580 \text{ lbs.}$$

$$\text{Total stress} = \frac{2580}{.688} = 3750 \text{ lbs.}$$

$$f_s = \frac{3750}{.423} = 8880 \text{ lbs. per sq. in.} \qquad \text{(satisfactory)}$$

Stress on section *X—X*., Fig. 112.

It will be assumed that at this section only half the moment is carried by the crank, the rest having been taken up by the half of the weld above the section.

$$\text{Moment} = 2580/2 = 1290 \text{ in. lbs.}$$

Net I $= .0430$ on section *X—X*

$$f_b = \frac{1290 \times .937}{.0430} = 28{,}100 \text{ lbs. per sq. in.}$$

$$\text{Coefficient} = \frac{55{,}000}{28{,}100} = 1.96 \qquad \text{(satisfactory)}$$

Stress on section *Y—Y*, Fig. 112.

I = .0237 M = 397 × 4.0 = 1590 in. lbs
Z = .0361 P cos 75° = 103 lbs.
A = .146

$$f_b = \frac{1590}{.0361} = 44{,}000 \text{ lbs. per sq. in.}$$

$$P/A = \frac{103}{.146} = \frac{700}{44{,}700} \text{ lbs. per sq. in.}$$

$$\text{Coefficient} = \frac{55{,}000}{44{,}700} = 1.23 \qquad \text{(satisfactory)}$$

Although calculations for the stresses on sections *W—W* and Z—Z are not given, they should be carried through to check the design.

133. *Investigation of Teeth in Collar on Aileron Torque Tube*—Fig. 113 shows the collar and tube.

Depth of teeth = .25 in.
Area of one tooth = .25×.125 = .0312 sq. in.

$$\text{Force on each tooth} = \frac{2580}{3 \times .6875} = 1250 \text{ lbs.}$$

$$\text{Bearing stress} = f_c = \frac{1250}{.0312} = 40{,}000 \text{ lbs. per sq. in. (satisfactory)}$$

$$\text{Section area of one tooth} = \frac{.6875 \times 2 \times 3.14 \times .125}{6} = .090 \text{ sq. in.}$$

$$\text{Shearing stress} = f_s = \frac{1250}{.090} = 13{,}900 \text{ lbs. per sq. in. (satisfactory)}$$

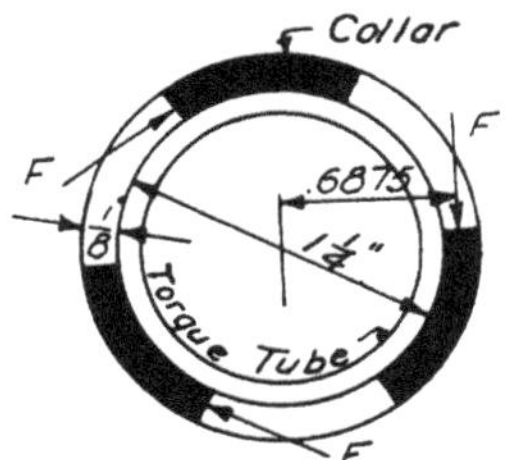


**Fig. 113. Teeth on Aileron Torque Tube Collar**

134. *Investigation of Aileron Torque Tube*—Length of the torque tube from the end in the fuselage to the center of the ailerons = 120 ins.

Tube is $1\frac{1}{4} \times \frac{3}{32}$ in.

Polar I = .1146

$$\text{Torsional shear} = f_s = \frac{2580 \times .625}{.1146} = 14{,}100 \text{ lbs. per sq. in.}$$ (satisfactory)

$$\text{Angle of twist} = \frac{57.2 \times 2580 \times 120}{12{,}000{,}000 \times .1146} = 12.9^\circ$$

$$\text{Extra movement of center of hand grip} = \frac{12.9 \times 21}{57.2} = 4.7 \text{ in.}$$

135. *Investigation of Rudder Bar*—Owing to the offset in this rudder bar, see Fig. 114, two kinds of stress are produced, bending and torsional.

Moment of rudder load = 2120 in. lbs.

$$\text{Tension in cable} = \frac{2120}{4.5} = 472 \text{ lbs.}$$

Cable is 1/8 in. 7 × 19 extra flexible of 2000 lbs. capacity.

$$\text{Coefficient} = \frac{2000}{472} = 4.23$$

$$\text{Pressure on pedal} = \frac{2120}{10.5} = 202 \text{ lbs.}$$

Stress on section *A—A* at which cable is attached.

Bending moment = 202 × 6 = 1212 in. lbs.
Torsional moment = 202 × 3 = 606 in. lbs.
Tube is 1 × 1/16 in.
I = .0203
Polar I = .0406

$$f_b = \frac{1212 \times .5}{.0203} = 29{,}900 \text{ lbs. per sq. in.}$$

$$f_s = \frac{606 \times .5}{.0406} = 7{,}450 \text{ lbs. per sq. in.}$$

Maximum combined stresses.

$$\text{Bending} = f_b = \frac{29{,}900}{2} + \sqrt{(7450)^2 + \left(\frac{29{,}900}{2}\right)^2}$$

$$= 31{,}650 \text{ lbs. per sq. in.}$$ (satisfactory)

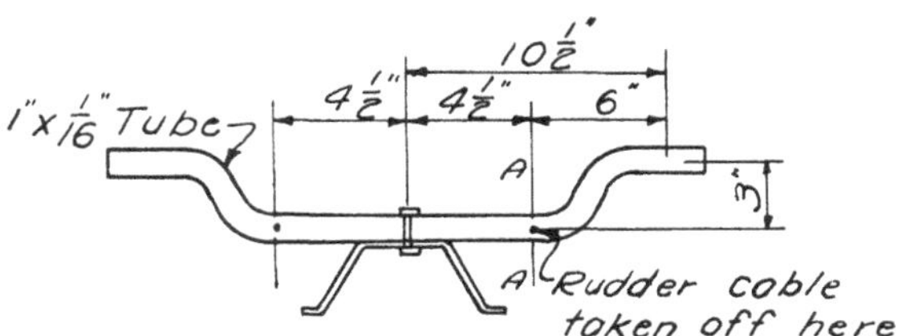


Fig. 114. Rudder Bar

$$\text{Shearing} = f_s = \sqrt{(7450)^2 + \frac{(29{,}900)^2}{(\;2\;)^2}}$$

$$= 16{,}700 \text{ lbs. per sq. in.} \qquad \text{(satisfactory)}$$

136. *Investigation of Mast on Elevator Tubes*—Moment of elevator load = 3620 in. lbs. Refer to Fig. 115.

$$\text{Pull on elevator cable} = \frac{3620}{4.5} = 805 \text{ lbs.}$$

Stress on weld between collar on elevator tube and mast.

$$\text{Total stress on weld} = \frac{3620}{.875} = 4130 \text{ lbs.}$$

$$\text{Unit stress on weld} = \frac{4130}{2 \times .0625 \times 3.14 \times 1.75}$$

$$= 6030 \text{ lbs. per sq. in. (satisfactory)}$$

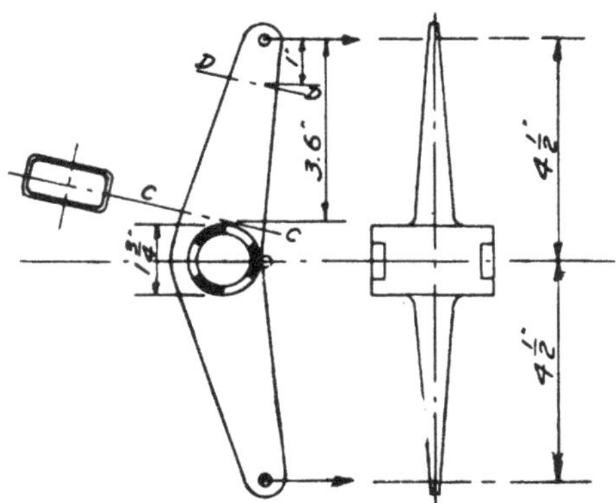


Fig. 115. Elevator Control Mast

Stress on section *D—D.*

$$m = 1.0 \times 805 = 805 \text{ in. lbs.}$$

$$I = \frac{1.20^3 \times .40 - 1.075^3 \times .275}{12} = .0291$$

$$f_b = \frac{805 \times .6}{.0291} = 16{,}600 \text{ lbs. per sq. in.} \qquad \text{(satisfactory)}$$

Stress on section *C—C.*

$$m = 3.6 \times 805 = 2900 \text{ in. lbs.}$$

$$I = \frac{1.0 \times 2.0^3 - .875 \times 1.875^3}{12} = .185$$

$$f_b = \frac{2900 \times 1.0}{.185} = 15{,}700 \text{ lbs. per sq. in.} \qquad \text{(satisfactory)}$$

137. *Attachment of Masts to Tubes*—The method of welding used in this design to secure control levers and masts to tubes is not at all satisfactory. The welding is liable to cause distortion of the tube and

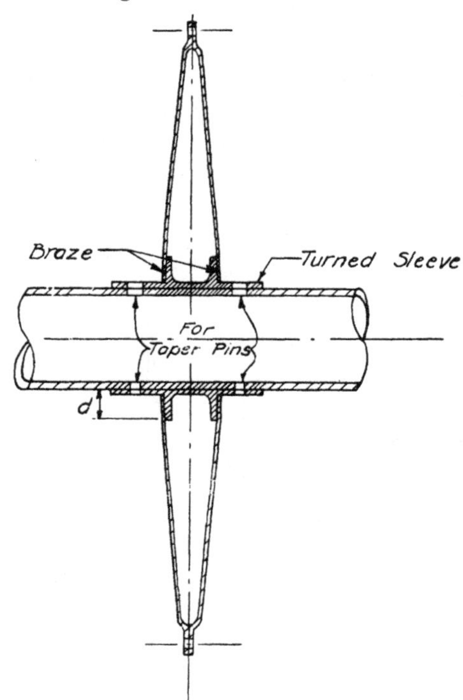


**Fig 116. Method of Securing Large Masts to Torque Tubes**

lever as well as completely annealing them at the weld. If the gage of the metal is light, a considerable part of the steel may be burned. A much better type of connection is shown in Fig. 116. The sleeve can be cheaply and easily turned. The distance *"d"* is determined by the brazing area required to carry the stress. When the brazing is properly done, 25,000 lbs. per sq. in. is a safe ultimate stress. Sometimes the sleeve can be brazed to the tube, making a unit construction; or it may have to be removable, in which case the sleeve would be suitably pinned to the tube. If brazing is employed, the sleeve should be cut off flush with the outside of the mast.

The type of sleeve shown assumes that the mast is stamped in two halves which are welded or, preferably, brazed together. Masts or levers of this general character are used for exposed work, or on large airplanes where the stresses are great and where the overall length of the lever is over 15 in.

For masts or levers to be used inside, and of a length less than 15 in., the type of mast and sleeve illustrated in Fig. 117 is well adapted. Such construction is used on the control system shown in Fig. 118. It should

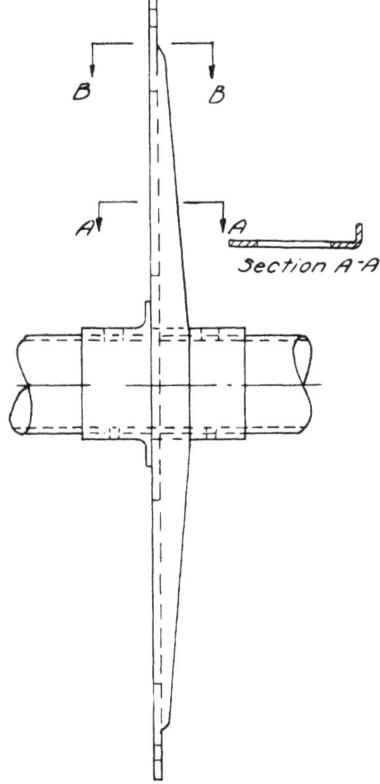


**Fig. 117. Small Internal Mast and Its Attachment to Torque Tube**

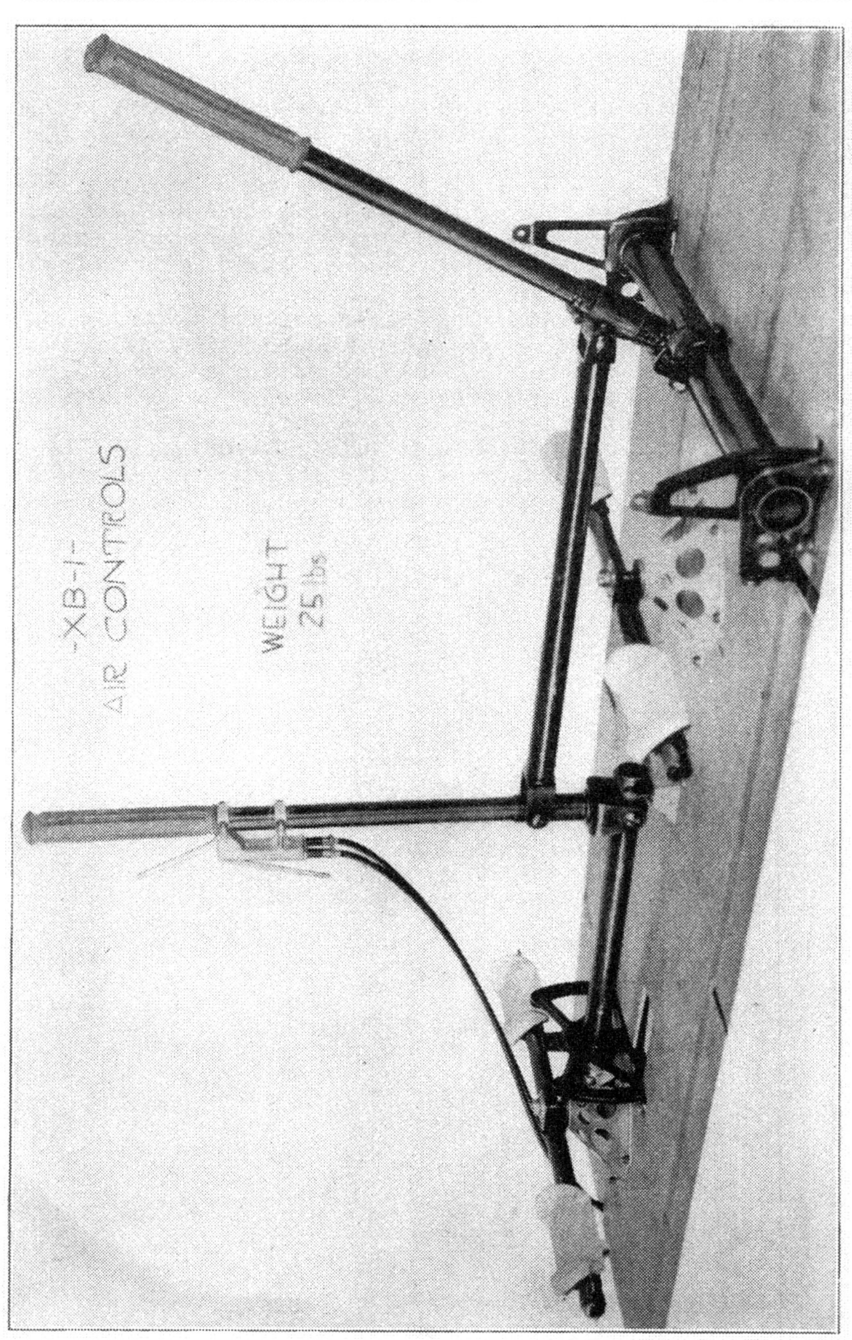


Fig. 118. Dual Control System on U.S. XBI-A

be noted that a flange is put only on the compression side of the mast. One point of weakness in this mast, and often in other types, is through section *B—B*. A slight lateral component in the stress in the control cable will cause the top of the mast to buckle sideways.

## *DUAL STICK CONTROL*

138. *General Comments*—Fig. 118 illustrates the control system of a two-seater pursuit airplane. One of the few questionable features of this design for all dual control work is the torque tube, connecting the front and rear sticks, by which the aileron torque is transmitted when the control is from the rear seat. Unless the workmanship is good, so that the clearances can be small, there will be a certain amount of play. Also, as shown in the example which follows, an angle of lag is present equal to the angle of torsion in the tube between the sticks. In the case of this design it amounts to 7 degs. with the full sand load on the four ailerons, which is not an objectionable amount.

The light, duraluminum base for the rudder bar is an efficient design. The manner in which it is secured to the floor by six widely spaced bolts tends to make the base rigid even with a light flooring. The steel strap holding the rudder bar is riveted to the duraluminum base.

The end thrust bearing for the front aileron torque tube is combined with the base of the forward rudder bar. This gives a firm bearing and eliminates the necessity for an extra base.

The mounting of the elevator levers on the torque tube at the base of the rear stick gives compactness and simplicity. These levers are attached to the tube by flanged collars of the type shown in Fig. 117, to which they are brazed. Such a method of connection is to be recommended.

139. *Investigation of Torque Tube Between Control Sticks*—When the control is through the rear control stick, the aileron torque must be transmitted by the horizontal tube connecting the control sticks. As the analysis of this tube presents some new features, it will be given in detail.

*Moment of Aileron Load.*

$$\begin{aligned}\text{Total moment} &= \text{Area} \times \text{average load} \times \text{mean arm}\\ &= 51.8 \times 24 \times 7.82 = 9720 \text{ in. lbs.}\\ \text{On each aileron} &= 2430 \text{ in. lbs.}\end{aligned}$$

Since the aileron masts and the aileron lever arms are the same length, the moment carried by the aileron torque tube is the same as the moment of the aileron loads.

Tube is $1\frac{3}{8} \times \frac{3}{32}$ in.
Polar I = .1557

$$\text{Torsional shear} = f_s = \frac{9720 \times .6875}{.1557} = 42{,}900 \text{ lbs. per sq. in. (satisfactory)}$$

Length of tube = 23 ins.

$$\text{Angle of torsion} = \frac{9720 \times 23 \times 57.2}{12{,}000{,}000 \times .1557} = 6.85°$$

In Fig. 119 *DE* represents the position of the front stick, and *BC* that of the rear stick. *ABC*, or the angle of lag, equals the torsional angle; refer also to Fig. 120. Since point *C* moves, with reference to

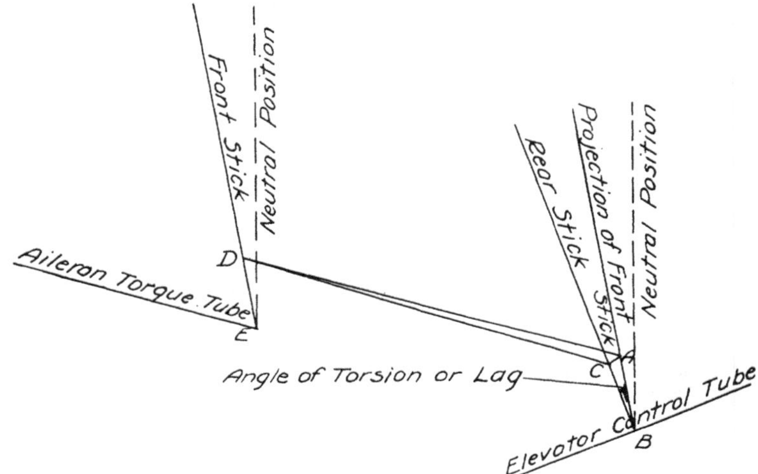


**Fig. 119. Angle of Lag Between Front and Rear Control Sticks**

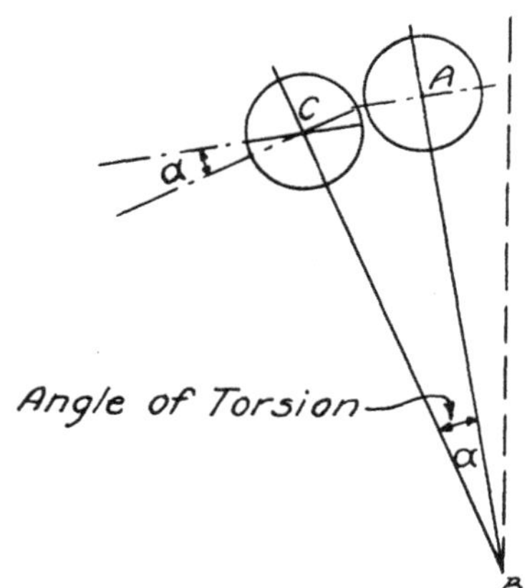


**Fig. 120. Relation Between Angle of Lag and Torsion**

point *A*, over the arc *AC*, the tube connecting points *D* and *C* must be deflected a distance, *AC*. The force applied at *C* that is required to do this produces bending in member *DC*.

Calculation of Deflection of C Due to Bending.

$$BC = 4.0 \text{ ins.} \qquad AC = \frac{6.85 \times 4}{57.2} = .478 \text{ in.} = \text{Deflection of C}$$

Calculation of Force Required to Produce Deflection.

$$.478 = \frac{F\,L^3}{3\,EI} \quad \text{or} \quad F = \frac{3 \times 29{,}000{,}000 \times .0778 \times .478}{23.0^3} = 266 \text{ lbs.}$$

Calculation of Bending Stress.

Moment $= 266 \times 21.5 = 5720$ in. lbs.

$$f_b = \frac{5720 \times .687}{.0778} = 50{,}500 \text{ lbs. per sq. in.}$$

Calculation of Combined Stresses.

Maximum bending

$$f_b = \frac{50{,}500}{2} + \sqrt{42{,}900^2 + \frac{50{,}500^2}{2^2}} = 75{,}100 \text{ lbs. per sq. in.}$$

Maximum shear

$$f_s = \sqrt{42{,}900^2 + \left(\frac{50{,}500}{2}\right)^2} = 49{,}800 \text{ lbs. per sq. in.}$$

Both of these stresses are above the allowable ultimate and hence the strength of the tube may be increased by changing to a wall thickness of 1/8 in. However, since this rear control for a pursuit airplane is really an emergency control, the present design is strong enough for that type of airplane.

140. *Torque Tube Carrying Elevator Masts and Rear Stick.*

Moment of Elevator Load.

Total moment = Area × average load × mean arm
= 23.0 × 24.0 × 7.65 = 4220 in. lbs.

The elevator masts and the elevator lever arms on the control stick are the same length, so that the moment on the stick equals the moment of the elevator loads.

Calculation of Reaction on Grip of Control Stick.

Length of stick to centerline of grip = 24⅛ in.

$$F = \frac{4220}{24.12} = 175 \text{ lbs.}$$

Tension in Elevator Cables.

$$T = \frac{4220}{2 \times 5.0} = 422 \text{ lbs.}$$

Fig. 121 shows this tube with its applied loads.

Maximum bending moment $= 509.5 \times 12 - 422 \times 11$
$= 1470$ in. lbs.

Maximum torsional moment $= 2110$ in. lbs.

Tube is 1¾ × .049 in.  Polar I = .190

$$f_b = \frac{1470 \times .875}{.095} = 13{,}550 \text{ lbs. per sq. in.}$$

$$f_s = \frac{2110 \times .875}{.190} = 9{,}720 \text{ lbs. per sq. in.}$$

Combined Stresses.

$$\text{Maximum } f_b = \frac{13{,}550}{2} + \sqrt{9720^2 + \left[\frac{13{,}500}{2}\right]^2}$$

$$= 18{,}630 \text{ lbs. per sq. in.}$$

$$\text{Maximum } f_s = \sqrt{9720^2 + \left[\frac{13{,}500}{2}\right]^2}$$

$$= 11{,}850 \text{ lbs. per sq .in.}$$

As far as these stresses are concerned, the tube could be decreased to perhaps 1½ in. diameter.

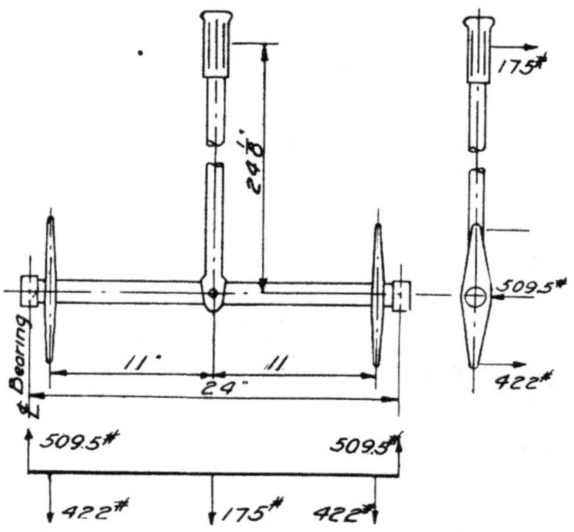


**Fig. 121. Forces Acting on Rear Stick and Transverse Torque Tube**

141. *Pursuit Airplane Control System*—Fig. 122 illustrates diagrammatically the functioning of a control system which has certain good points. For the design in question, the aileron and elevator levers are to the rear of the seat where there is ample room instead of in a crowded cockpit. As shown in Fig. 123, the aileron and elevator levers are combined in a compact manner. The former is securely riveted to the torque tube, and the latter, pinned through the torque tube, rotates within the aileron lever in a plane passing through the axis of the tube. The weight of the entire system, excluding the torque tube in the wings, is 8 lbs.

142. *Relation of Air Controls to Pilot*—It is very difficult to establish fixed rules to govern the layout of a standard cockpit because of the different types of fuselages found in military airplanes. Yet the dimensions directly affecting the pilot's comfort can be standardized. In Figs. 125 and 126 are shown three views of a cockpit in which the main dimensions are indicated by symbols. Table XXII gives average values for these various dimensions, which are based on nine two-seater and eight single-seater airplanes of American, British, French and German designs. It is recommended that, in general, these values be closely adhered to. The angular movements of the rudder bar and control stick, and the lengths of the control arms on the elevator and rudder are more dependent on the power ratio and linkages desired than on the pilot's reach and height. For this reason the values given in the table for these angles and dimensions should be regarded merely as reasonable values. They will vary with the type of airplane.

It is very necessary that there should be an adjustment which will allow for variation in a pilot's height. The adjustable footpad on the rudder bar shown in Fig. 124 is the simplest and most effective method of adjustment. The dimension $w$ given in Table XXII is with the footpad in mid-position. In the case of a very tall pilot, the use of a small cushion in the front of the seat adds greatly to his comfort by supporting his thigh under the knee. Such a cushion would conform to the curve of the seat in plan view and be triangular in section.

The seat is designed in accordance with the specifications of the Parachute Branch, of the Equipment Section of the Engineering Division. The maximum dimensions of any parachute pack for which provision must be made are 26 in. × 14 in. × 6 in. In most cases the pack will be smaller. The 6 in. given as the maximum thickness determines the location of the bulkhead just to the rear of the pilot. It is essential that the head-rest should not impede a pilot in climbing out of the cockpit with a parachute pack on his back.

143. *Controls for Large Airplanes*—For multi-engined bombers and commercial airplanes the simple stick control is no longer practicable. Some form of wheel control for the ailerons is commonly resorted to. A recent design of this type embodying many excellent features is illustrated in Fig. 127. The transmission used is the sprocket-chain combination which is mechanically highly efficient, and also simple. Some designs employ a worm and gear transmission. With this type, however, the loss of power from friction is larger than with the sprocket and chain,

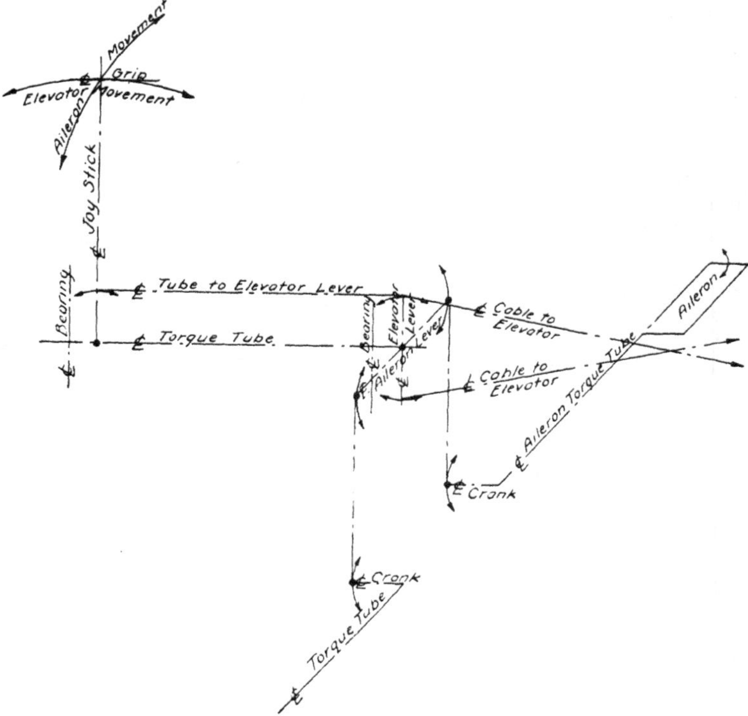


Fig. 122. Functioning of Control System on VCP-1

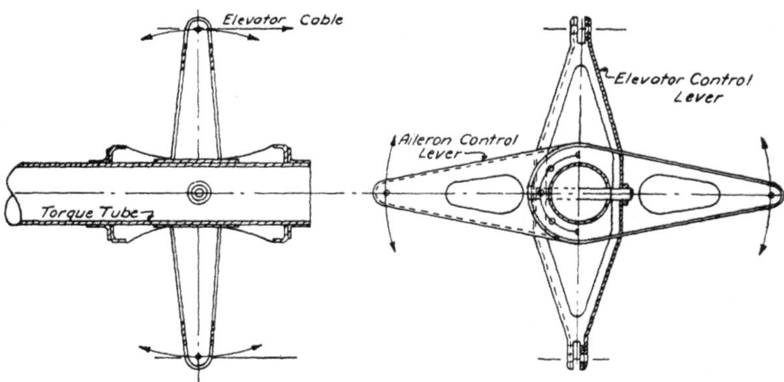


Fig. 123. Aileron and Elevator Lever Unit

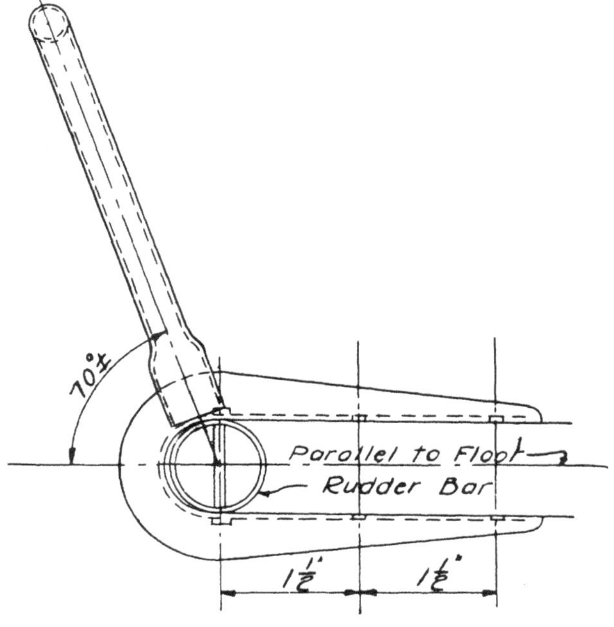


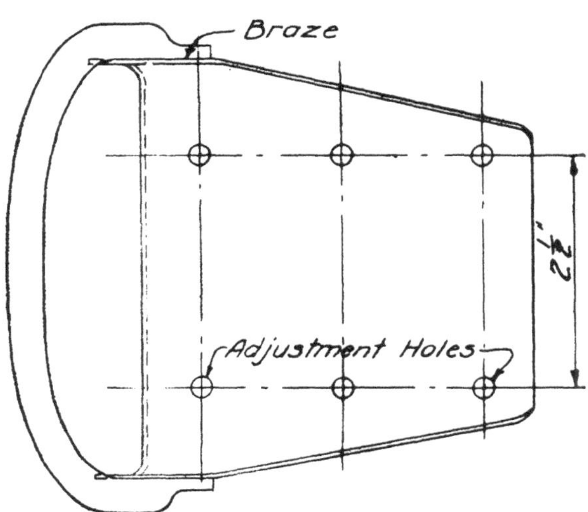


Fig. 124. Adjustable Foot Pad for Rudder Bar

# *TABLE XXII*

| | Depth from Cowling to Bottom of Seat. | Depth of Seat. | Height of Back of Seat. | Distance from Vert. ℄ thru Rudder Bar to Back of Seat. | Distance from Vert ℄ of Rudder Bar to Fire-wall of Bulkhead. | Height of Rudder Bar above Floor. | Height of Seat above Horizontal Line thru Rudder Bar. | Length of Rudder Bar | Distance from Rudder Bar Pivot to Rudder Cable Attachment | Lever Arm at Rudder | Distance from Top of Stick to Pivot Point. | Dist from Stick Pivot to Elevator Attachm't. | Lever Arm of Elevator | Distance from Bottom of Seat to Floor | Width of Cock-Pit Opening. |
|---|---|---|---|---|---|---|---|---|---|---|---|---|---|---|---|
| SYMBOL | A | B | C | D | E | F | G | H | I | J | K | L | M | N | O |
| AVERAGE FOR 2-PLACE PLANES | 26 | 16.5 | 12.5 | 38 | 11 | 3.75 | 11.5 | 25.5 | 11 | 5.25 | 27 | 5.0 | 5.0 | 15.0 | 26 |
| AVERAGE FOR SINGLE SEATERS | 28 | 16.5 | 19.5 | 36 | 17 | 3.75 | 6.0 | 23.0 | 4.5 | 3.75 | 23 | 5.0 | 4.0 | 7.5 | 23 |

| | Width of Seat | Distance from Top of Cowling to Floor. | Length of Cock-Pit Opening. | Dist from Top of Cock-Pit to Bottom of Back of Seat. | Dist from Back of Seat to ℄ of Stick | Dist from ℄ of Stick to Vert ℄ of Rudder Bar. | Position of Head Rest with Reference to Top of Back of Seat. | Clearance between Top of Stick and Edge of Instrument Board | Maximum Angle of Movement Forward from Neutral of Stick. | Maximum Angle of Movement of Stick Back from Neutral. | Maximum Angle of Side Movement of Stick. | Maximum Angle of Movement of Rudder Bar | Angle of Incline of Back of Seat | Angle of Inclination of Seat Bottom. | |
|---|---|---|---|---|---|---|---|---|---|---|---|---|---|---|---|
| SYMBOL | P | Q | R | U | V | W | X | Y | $\theta$ | $\theta_1$ | $\theta_2$ | $\theta_3$ | $\theta_4$ | $\theta_5$ | |
| AVERAGE FOR 2-PLACE PLANES | 19 | 39.5 | 23 | 24.5 | 20.5 | 17.5 | 1.75 | 1.5 | 25° | 25° | 25° | 15° | 9° | 3° | |
| AVERAGE FOR SINGLE SEATERS | 20 | 32.0 | 21 | 24.5 | 20. | 16. | 1.75 | 1.5 | 25° | 25° | 30° | 22½° | 9° | 3° | |

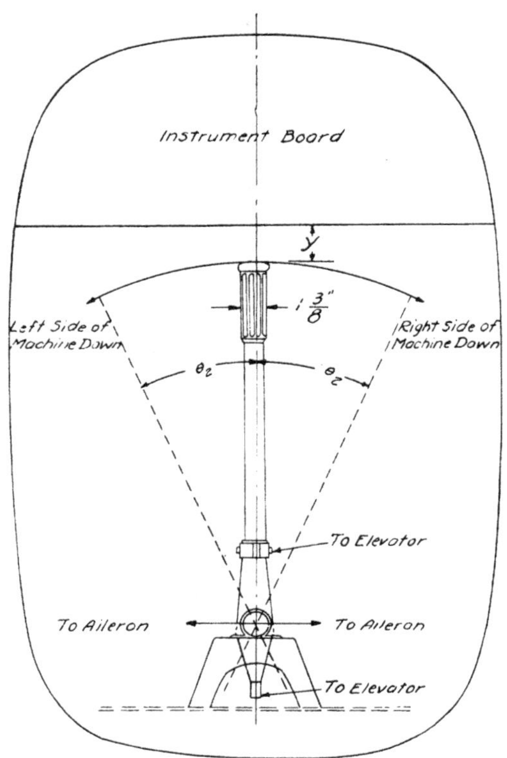


END VIEW LOOKING FORWARD OF PILOT'S SEAT

**Fig. 125**

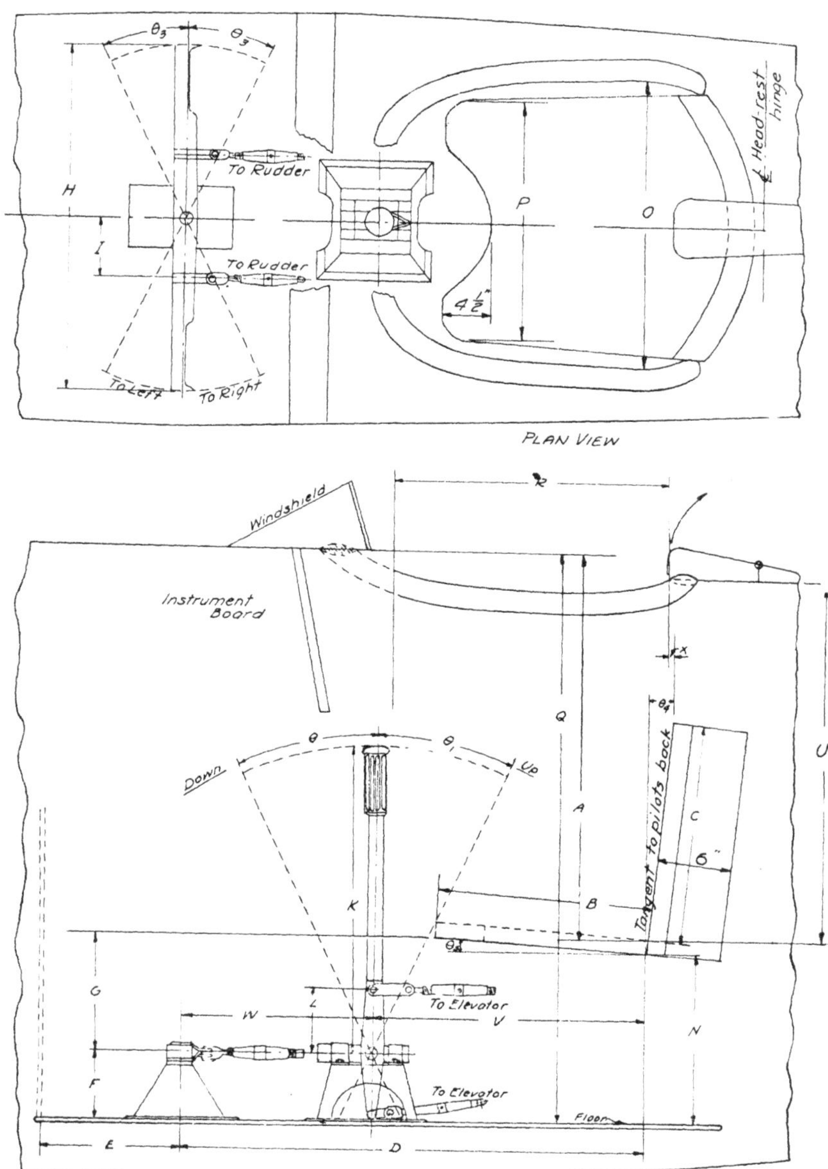


Fig. 126. Side Elevation and Plan View of Standard Cockpit

which is a most important consideration when the power required to operate the control must be reduced to a minimum. An improvement in the design illustrated would be the use of ball bearings on the sprocket and on the two pulleys at the base of the steering column. This would be necessary with the largest types of airplanes. The question of the power ratio is an important one. In the old "dep" control this ratio was about 4 to 1. For heavy airplanes a greater ratio is required, or the controls become too sensitive and require too much power to operate. The present design has a ratio of 9 to 1, which is ample and insures ease of control. A 200-pound pull on the rim of the wheel gives a cable tension of 1800 pounds, which is applied to a 12¾ inch aileron mast. To give a maximum movement of the ailerons of 25 degs. either side of the neutral position requires, with the design shown, 1¼ turns of the wheel. The entire weight of this control, including pulleys, guards and cables, is less than fifteen pounds.

Where it is desirable, the horizontal tube at the base of the steering column can be extended to the sides of the fuselage and have walking beams mounted on it for the elevator cables. This arrangement makes the control a unit.

The fact that a worm and gear control is practically irreversible, is one feature much in its favor. For example, if one engine stops in a twin-engine airplane the unbalanced torque is partly counter-balanced by the ailerons. With a chain and sprocket control a constant force must be exerted on the wheel, while in the case of the worm and gear type the ailerons, once having been forced to a certain position, will tend to remain there, thus relieving the pilot of continued strain.

144. *Ultimate Allowable Stresses*—Most of the material used in a control system will be either quarter hard carbon steel of a tensile strength of 55,000 lbs. per sq. in. or 3½ per cent nickel steel, usually cold rolled, of a tensile strength of 100,000 lbs. per sq. in. The low carbon material is covered by U.S.A. Signal Corps Specifications 10,201 and 10,225, the first for sheet stock, the second for tubing. For alloy steel either specification 10,029 or 10,045 may be used. All bolts are made of this nickel steel, and also certain other members which carry high stresses and which would otherwise be unduly heavy. However, if members are made of high strength material it is not permissible to braze them, because their strength would be very largely reduced by this operation.

The ultimate strength of all steels in shear may be assumed to be very closely 2/3 of its ultimate tensile strength, unless the steel is heat treated. This means that the maximum allowable shearing stress on bolts may be taken as 65,000 lbs. per sq. in. It should be remembered, however, that, on account of necessary clearances, bolts will always be subjected to bending stresses in addition to pure shearing stresses. For small bolts especially it is not advisable to allow more than 45,000 lbs. per sq. in. shear.

Torsional stresses may be considered the same as shearing stresses. The usual formula for torsional stress, $f_s = M \cdot r/J$, will not give the

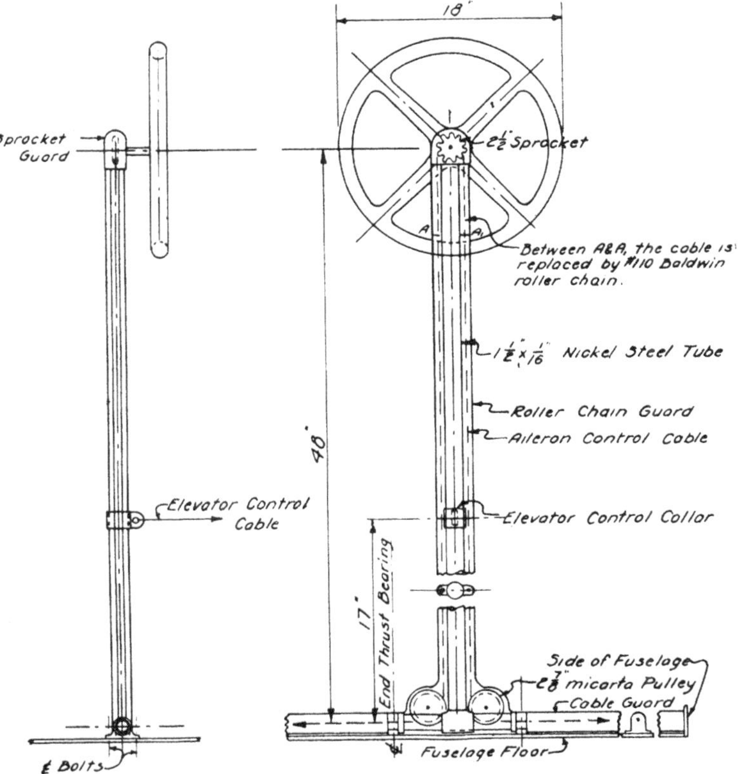


Fig. 127. **Wheel Control for Large Airplanes**

true shearing stress when failure in torsion occurs. Upton has developed a formula, $f_s = .75\ M\ r/J$, which gives closely the true value of the torsional stress at failure in solid or thick walled tubes.

The actual ultimate torsional stress so calculated may be taken as equal to 2/3 the ultimate tensile strength. In thin walled tubes secondary stresses are produced by local crumpling of the metal so that the allowable ultimate stresses are lower than those given above. Fig. 153 is a curve for this reduction in ultimate torsional shear stress plotted against the ratio of diameter to wall thickness. This curve is based on a few experimental values and it is believed will give safe and reasonable results, but it is subject to change when more data is available.

Definite recommendations in regard to ultimate bearing strengths can not be made. Data on a large number of lug tests indicate that ultimate bearing strengths may be taken as nearly twice ultimate tensile strengths for any given steel.

## CHAPTER VII

# CONTROL SURFACES

145. *General*—For the most part an exact analysis of control surfaces is impossible. The present chapter indicates such methods of analysis as are practicable, but its chief aim is to suggest certain features of design that tests on control surfaces have shown to add to the strength of the surfaces. Perhaps the best means of estimating the strength and weight of any given design, and of determining the size of the main members is a tabulation of similar data for a number of surfaces that have been tested. Such a tabulation is given in Table XXIII. With few exceptions the surfaces tabulated are representative of recent good design. Four or five types of construction are each represented by several examples. Some of the designs, especially in steel, are somewhat of an experimental nature. General problems in design will be taken up first, then a discussion of the important features of various designs which are illustrated by photographs and sections, followed by an explanation of methods of analysis.

146. *Control Masts*—Probably the most important single detail of a design is the mast and the method of attaching it to the surface. The necessity of placing a mast in line with its control cable is clearly illustrated by Fig. 129, in which the angle between the cable and the line

**Fig. 129. USD 9-A Control Mast**

of the mast is 9 degs., giving a side component equal to .16 of the pull in the cable. Particularly for the type of mast shown, a core of sheet steel with wood fairing, a lateral force causes very severe stresses, owing to the small moment of inertia of the section about its long axis. If a mast is secured to a rib, or is braced by an external stay wire, these also should be in line with the control cable. Of all types of masts, the best for exposed work is the hollow, streamline mast of sheet steel, welded along the trailing edge, and spread at the base. The mast shown in Fig. 149 is an example of this type. A plywood mast with a steel core, which is built into a strong box rib that transmits a large amount of torque, is illustrated in Fig. 130. Where the height of the mast is small this design is entirely satisfactory. If a bracing wire is present, running from the mast to the trailing edge or to a stringer rib, the base of the mast can be shortened because the torque that the mast carries is much reduced.

The method of securing a mast to a tube by welding to the main spar and to a tubular rib, as shown in Fig. 131, can be used when a circular tube is employed for the leading edge of the movable surface. Since most of such tubes have a very high ratio of diameter to wall thickness it is always advisable to reinforce the tube, at a point where a control mast is attached, by a 3 in. length of 1/16 in. tubing, fitting inside, and spot welded or brazed to the main tube. With the type of construction in Fig. 129, a steel box, shown in detail in Fig. 132, is a standard means of securing a mast to a torque tube. The strip forming the top and bottom of the box passes around the tube and is welded to it, and also the sides of the box are flanged out and welded to the tube. A rib may or may not be fastened to the box.

147. *Bracing*—The maximum speed of an airplane is often the determining factor in deciding whether or not tail surfaces should be externally braced. With velocities over 150 M.P.H. the resistance of exposed struts and wires becomes so large that it is more economical to eliminate them, and to strengthen the spars of the surfaces so that they can carry the load as cantilever beams. Internally braced surfaces will thus be confined almost wholly to high speed pursuit airplanes which have surfaces of a relatively short span and of a roughly semi-circular or elliptical shape. As will be noted in describing the different types of surfaces, the surface with a fairly thick section and with wooden spars is best adapted to internal bracing. As the possibility of doing away with external bracing depends not only on the strength and rigidity of the control surfaces themselves, but also on the character of the support afforded by the fuselage, this character must be considered. A fuselage that is narrow at the end is unsuited to an internally braced empennage.

Type 1. Fig. 130 is a general view of this tail unit, showing the internal construction of the horizontal surfaces, and the character of the fuselage. Reference to Table XXIII shows that although these surfaces are unbraced their unit weight is reasonably low. The efficient design which is the cause of this low weight makes a careful study of the details worth while.

# TABLE XXIII

| | | | ELEVATOR | | | | | | | | STABILIZER | | | | | | | | RUDDER | | | | | | | | |
|---|---|---|---|---|---|---|---|---|---|---|---|---|---|---|---|---|---|---|---|---|---|---|---|---|---|---|---|
| 1 | 2 | 3 | 4 | 5 | 6 | 7 | 8 | 9 | 10 | 11 | 12 | 13 | 14 | 15 | 16 | 17 | 18 | 19 | 20 | 21 | 22 | 23 | 24 | 25 | 26 | 27 | 28 |
| Airplane | Number | Material | Total Area | Total Weight Covered | Weight Per Sq. Ft. | Ultimate Load Per Sq. Ft. | $S_e$* | $B_e$* | Surface Braced | Mast Braced | Material | Total Area | Total Weight Covered | Weight Per Sq. Ft. | Ultimate Load Per Sq. Ft. | $S_s$* | $B_s$* | Surface Braced | Material | Total Area | Total Weight Covered | Weight Per Sq. Ft. | Ultimate Load Per Sq. Ft. | $S_r$* | $B_r$* | Surface Braced | Mast Braced |
| U. S. D-9A | 1 | W | 23.6 | 21.6 | .915 | 20. | 99 | 24 | Yes | Yes | W | 38.9 | 39.4 | 1.01 | 20. | 80 | 37 | Yes | W | 13.5 | 14.6 | 1.08 | 40. | 67 | 31 | Yes | Yes |
| Lepere C-11 | 2 | W | 33.4 | 28.0 | .84 | 20. | 78 | 32 | Yes | No | W | 17.0 | 14.0 | .825 | 20.0 | 49 | 33 | Yes | W | 13.1 | 10.0 | .765 | 30.0 | 63 | 30 | No | No |
| Pomilio BVL-12 | 3 | W | 24.8 | 28.0 | 1.13 | 57.5 | 72 | 29 | Yes | No | W | 34.4 | 26.3 | .766 | 57.5 | 54 | 62 | Yes | W | 14.0 | 12.25 | .875 | 32.5 | 59 | 37 | Yes | No |
| Pomilio FVL-8 | 4 | W | 11.6 | 14.3 | 1.34 | 108. | 55 | 17 | Yes | No | W | 10.1 | 11.1 | 1.09 | 108. | 42 | 26 | Yes | W | 6.55 | 6.5 | .99 | 58.5 | 51 | 24 | Yes | No |
| Thomas-Morse MB-3 | 5 | W | 17.0 | 17.75 | 1.04 | 52.5 | 53 | 21 | No | No | W | 10.25 | 14.0 | 1.37 | 79 | 34 | 27 | No | W | 7.25 | 8.25 | 1.14 | 60. | 45 | 20 | No | No |
| Ordnance Type C | 6 | W | 13.3 | 8.9 | .66 | 37.5 | 49 | 24 | Yes | No | W | 12.2 | 8.0 | .655 | 37.5 | 45 | 21 | Yes | W | 5.4 | 3.2 | .592 | 45.0 | 38 | 24 | No | No |
| Verville U. S. VCP-1 | 7 | S & W | 13.0 | 13.3 | 1.02 | 40.0 | 61 | 19 | No | No | W | 18.0 | 21.2 | 1.17 | 60.0 | 60 | 31 | No | W | 8.0 | 6.55 | .82 | 32.0 | 49 | 28 | No | No |
| Ordnance Type D | 8 | W | 17.0 | 14.75 | .87 | 55. | 51 | 21 | No | No | W | 15.0 | 14.5 | .97 | 55. | 48 | 30 | No | S & W | 7.9 | 10.0 | 1.27 | 75. | 37 | 26 | No | No |
| Curtiss SE-5 | 9 | W | 15.2 | 10.7 | .702 | 32.5 | 68 | 20 | Yes | Yes | W | 14.6 | 16.0 | 1.09 | 32.5 | 61 | 20 | Yes | S & W | 5.4 | 5.2 | .95 | 45.0 | 49 | 19 | Yes | No |
| Victor Training | 10 | W | 9.0 | 7.5 | .833 | 32.5 | 47 | 19 | Yes | No | W | 11.4 | 8.0 | .70 | 32.5 | 42 | 24 | Yes | S & W | 6.65 | 4.75 | .64 | 40.0 | 40 | 29 | Yes | No |
| Vought VE-7 | 11 | S & W | 17.2 | 14.3 | .88 | 40. | 61 | 24 | Yes | Yes | S & W | 19.2 | 21.3 | 1.11 | 40. | 59 | 24 | Yes | S & W | 8.6 | 7.6 | .88 | 42.5 | 47 | 29 | Yes | Yes |
| Bristol U. S. XB-1A | 12 | S | 23.0 | 13.5 | .585 | 26.5 | 81 | 24 | Yes | No | S | 22.5 | 20.4 | .905 | | | | Yes | S | 7.2 | 5.3 | .735 | | | | Yes | No |
| Bristol U. S. XB-1A | 13 | S | 23.0 | 13.5 | 585 | 34.7 | 81 | 24 | Yes | No | | | | | | | | | | | | | | | | | |
| Aileron U. S. XB-1A | 14 | S | 13.1 | 8.8 | .672 | 30. | 86 | 24 | No | No | | | | | | | | | | | | | | | | | |
| U. S. A. C-2 | 15 | S | 26.5 | 21.0 | .795 | 42.0 | 83 | 24 | Yes | No | S | 34.0 | 32.8 | .96 | 30.5 | 67 | 30 | Yes | S | 12.2 | 10.5 | .86 | 44.8 | | | Yes | No |

## TABLE XXIII (Continued)

| | ELEVATOR | | | | | | | | | | STABILIZER | | | | | | | | RUDDER | | | | | | | | |
|---|---|---|---|---|---|---|---|---|---|---|---|---|---|---|---|---|---|---|---|---|---|---|---|---|---|---|---|
| 1 | 2 | 3 | 4 | 5 | 6 | 7 | 8 | 9 | 10 | 11 | 12 | 13 | 14 | 15 | 16 | 17 | 18 | 19 | 20 | 21 | 22 | 23 | 24 | 25 | 26 | 27 | 28 |
| Airplane | Number | Material | Total Area | Total Weight Covered | Weight Per Sq. Ft. | Ultimate Load Per Sq. Ft. | $S_e$ | $B_e$ | Surface Braced | Mast Braced | Material | Total Area | Total Weight Covered | Weight Per Sq. Ft. | Ultimate Load Per Sq. Ft. | $S_s$ | $B_s$ | Surface Braced | Material | Total Area | Total Weight Covered | Weight Per Sq. Ft. | Ultimate Load Per Sq. Ft. | $S_r$ | $B_r$ | Surface Braced | Mast Braced |
| Standard #1 E-1 | 16 | S | 12.7 | 13.5 | 1.06 | 45.0 | 48 | 24 | Yes | Yes | S & W | 12.1 | 10.5 | .87 | 45.0 | 45 | 21 | Yes | S | 6.9 | 5.5 | .80 | 60.0 | 46 | 30 | Yes | Yes |
| Standard #2 E-1 | 17 | S | 14.2 | 14.0 | .99 | 51.9 | 62 | 19 | Yes | Yes | S & W | 11.2 | 15.0 | 1.34 | 43.8 | 54 | 24 | Yes | S | 6.4 | 5.4 | .83 | 27.5 | 37 | 29 | Yes | Yes |
| Standard #3 E-1 | 18 | S | 7.7 | 9.0 | 1.17 | 51 | 40 | 20 | No | Yes | S | 5.55 | 6.1 | 1.09 | 51.0 | 40 | 19 | Yes | S | 6.6 | 2.8 | .428 | 25. | 37 | 29 | Yes | Yes |
| Kawneer JN-4 & JN-6 | 19 | S | 11.2 | 9.75 | .87 | 58. | 62 | 34 | Yes | Yes | S | 11.0 | 10.75 | .975 | 75.0 | 49 | 36 | Yes | S | 12.3 | 10.6 | .865 | 44. | 63 | 37 | Yes | Yes |
| Curtiss JN-6 HB | 20 | | | | | | | | | | | | | | | | | | S | 11.0 | 10.5 | .96 | 25. | 53 | 35 | Yes | No |
| JN-6 HB | 21 | | | | | | | | | | | | | | | | | | S & W | 11.0 | 9.75 | .865 | 30. | 53 | 35 | Yes | Yes |
| Curtiss-Kirkham 18 | 22 | S & W | 13.0 | 13.8 | 1.06 | 40.0 | 65 | 19 | No | No | W | 14.3 | 15.8 | 1.10 | 60.0 | 68 | 27 | No | S & W | 8.65 | 8.1 | .934 | 30.0 | 46 | 32 | No | No |
| Fokker D-7 | 23 | S | 18.4 | 11.5 | .625 | 46.0 | 60 | 20 | Yes | No | S | 20.4 | 15.2 | .745 | 69 | 45 | 51 | Yes | S | 6.8 | 6.5 | .955 | 55. | 51 | 21 | Yes | No |
| U.S. G.A.X. | 24 | W | 45.8 | 43. | .94 | 41.0 | 96 | 35 | Yes | No | W | 55.3 | 55. | .99 | 54 | 81 | 78 | Yes | W | 36.4 | 41. | 1.13 | 42 | 94 | 51 | Yes | No |

* See Fig. 128

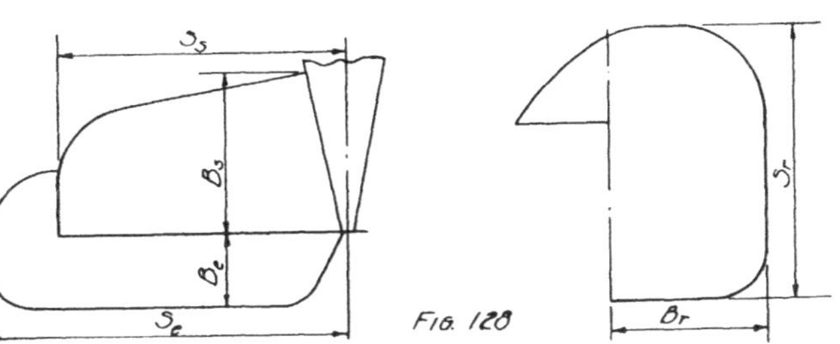

Fig. 128

Fig. 130. Empennage of Orenco Type D

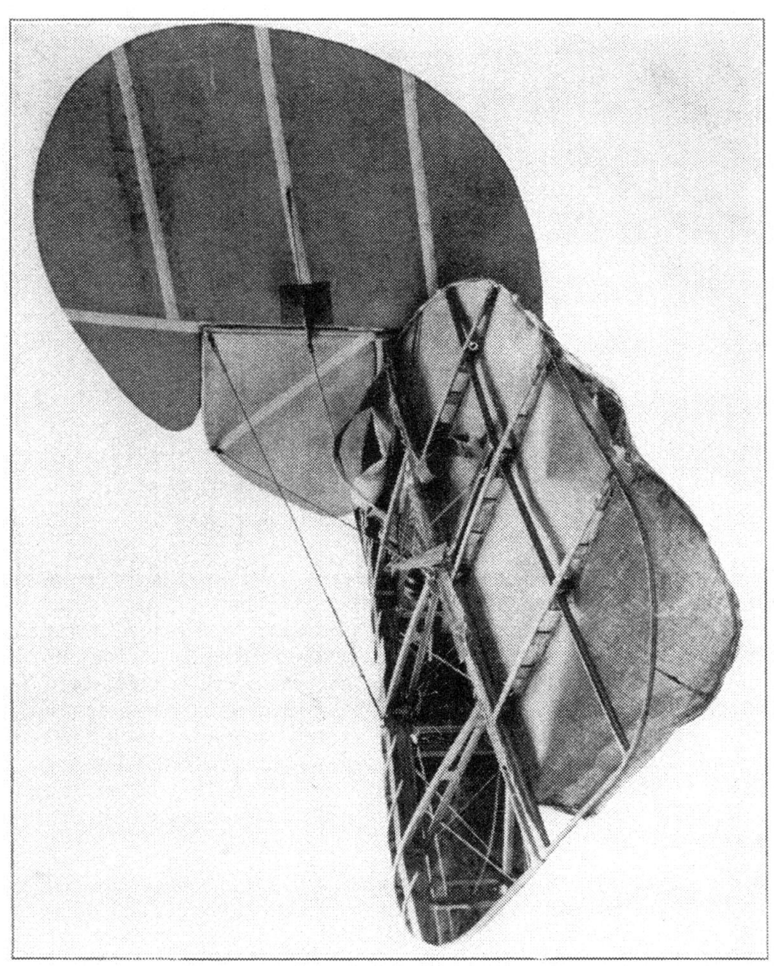

Fig. 131. Empennage of Vought VE-7 Training Airplane

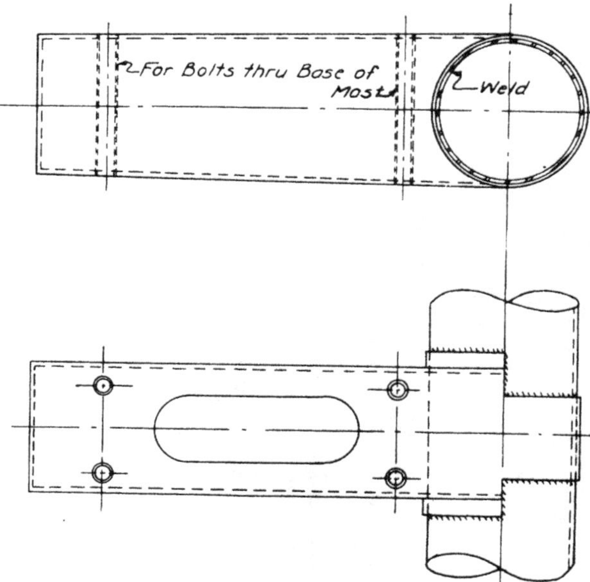


**Fig. 132. Steel Box for Supporting Control Mast**

The heavy mast ribs carry a considerable portion of the torque into an auxiliary spar. Owing to the fact that the elevator is a single unit this spar is continuous and is, therefore, peculiarly well fitted to transmit loads to the ribs. This condition will not occur with a double elevator unless each mast is located nearly at the center of its elevator. The auxiliary spar performs two functions; it reduces the torque carried by the main spar, and it supports the ribs, as well as bracing them laterally. A section of this member is shown in Fig. 133. Since the moment resisted by this secondary spar decreases from the center, the width of its flange could be reduced and circular lightening holes cut in the web toward the ends.

A section of the mast rib is shown in Fig. 134. It will be noticed that this rib is of a box section out to the cantilever spar. The balance of the ribs are of light construction with 7/16 × 1/8 in. spruce capstrips and 3/32 in. 3-ply webs. The manner in which the cap strips are attached to the main spar is worth attention. The large surface provided for gluing and nailing insures a rigid connection. Fig. 135 is a section of the heavy, stabilizer box rib. The short stabilizer ribs are similar to the light elevator ribs, while the transverse rib is of sturdier construction with 1/8 in. web and 1/2 × 1/8 in. cap strips. All plywood, both on the spars and ribs, is 3-ply spruce. The leading and trailing edges of

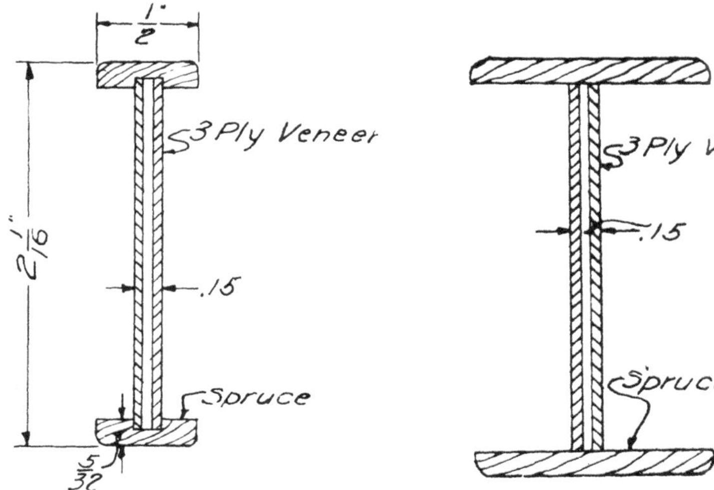


Fig. 133. Auxiliary Spar in Orenco Elevator

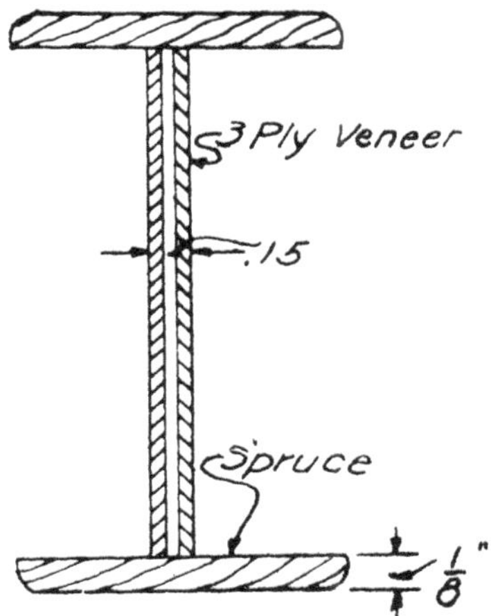


Fig. 134. Section of Mast Rib in Orenco Elevator

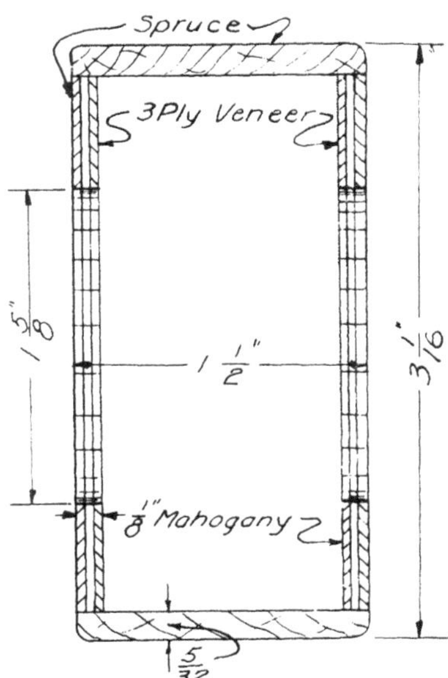


Fig. 135. Section of Box Rib in Orenco Stabilizer Next to Fuselage

stabilizer and elevator, respectively, are 7/16 in. O.D. .035 in. gage aluminum tubing, fastened to the ribs by means of a copper strap wrapped around the tube and nailed to the rib.

The front spar of the stabilizer is spruce of rectangular section $2^{11}/_{16}$ x ¾ in. For economy this spar should decrease in width from the body out. In Fig. 136 is shown a section of the rear stabilizer spar, and in Fig. 137, the forward elevator spar. The box construction of the latter has been followed in most of the best recent designs. The wide spar is strong in torsion and provides a broad base to which elevator masts can be secured. At hinge and mast points the routing is omitted.

The construction of the rudder is made clear in Fig. 138. The somewhat excessive weight of the rudder is due to its excess strength. Aside from the two light wood ribs, the rudder is all steel. The torque tube is 2 in. O.D. and .065 in. gage; all other tubing is 7/16 in. O.D. and .035 in. gage. These gages can be reduced to at least .042 in. and .028 in., respectively, and in the case of the 7/16 in. tubing and the metal ribs perhaps to .022 in., changes which would bring the weight of the rudder down to about 1.00 lb. per sq. ft.

The Orenco Type D fuselage, tapering down to a horizontal edge in the rear, is well suited to internally braced horizontal surfaces because of the broad, firm support afforded the spars. Conditions are not quite as favorable for the rudder, which had to be located rather far forward on the fuselage in order to get sufficient depth in the latter to give adequate bracing. The chief effect of this is to place the rudder area too high above the thrust line. Taken as a whole, however, the design of the empennage is excellent.

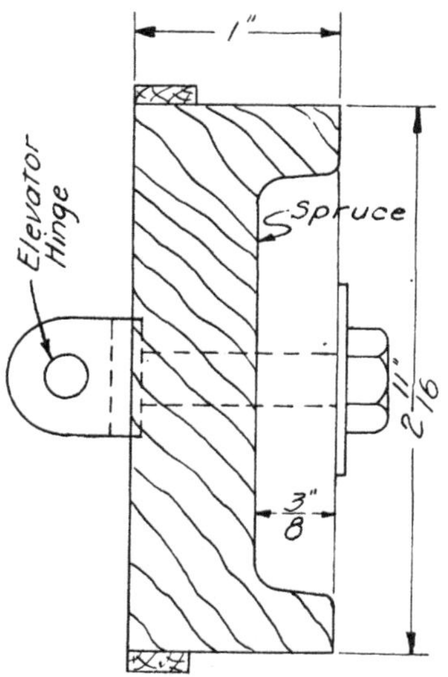


**Fig. 136. Rear Spar in Orenco Stabilizer**

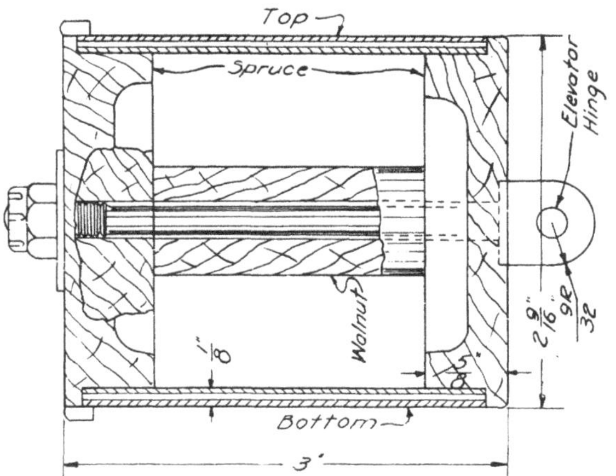


Fig. 137. Leading Edge of Orenco Elevator

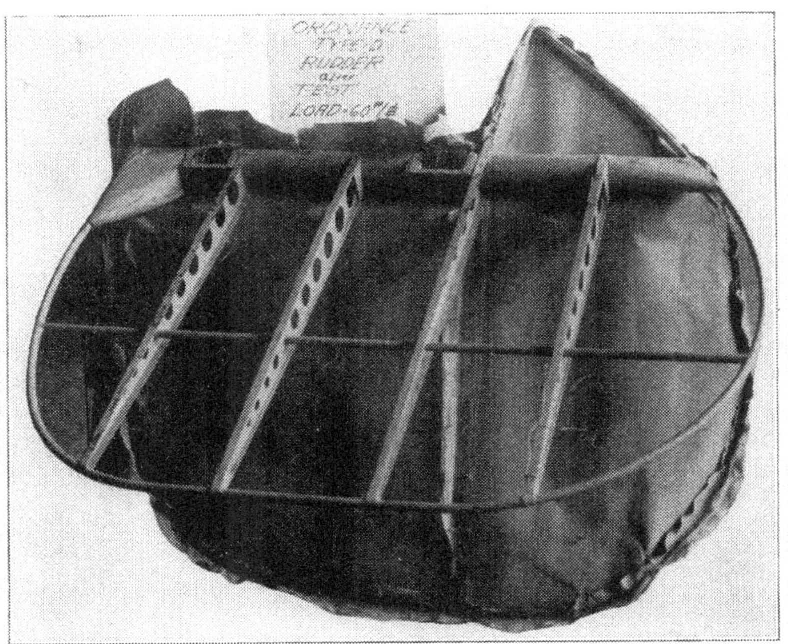


Fig. 138. Internal Construction of Orenco Rudder

Type 2. The tail surfaces of the Curtiss Kirkham 18-T, shown in Fig. 139, are of the same general type as those just discussed, but as they present considerable differences in detail they will be taken up here. The original manner in which, with a fuselage of rather narrow elliptical section, a sufficiently rigid support is obtained for the unbraced surfaces should be noted. The fuselage is cut off short and the blunt end streamlined by false fairing. As will be seen, all controls are internal so that this design practically eliminates structural resistance.

Both fixed tail surfaces are entirely covered with 1/16 in., 3-ply Spanish cedar plywood, covered with linen, doped and painted. Movable surfaces have ribs of pressed steel, similar to those shown in Fig. 138, welded or brazed to a steel torque tube. The leading edge of the fin and stabilizer is a small spruce member of triangular section, and the rudder and elevator trailing edges are of .022 in. gage steel bent to a U section. The ribs in the fixed surfaces are spaced 9 in., and are of the type shown in Fig. 140. The ribs in the movable surfaces are laterally braced by light wood spacers of rectangular section 3/8 x 5/8 in. One is used in the elevator and two in the rudder.

The construction of the rudder is evident from Fig. 141. One feature deserves attention, the close spacing of the ribs, 6 in. This is made necessary by the unusually great width of the rudder which causes the moment on each rib near the torque tube to be large. As a general rule, the ribs in control surfaces are spaced too widely. Since the ribs are such

**Fig. 139. Empennage of Curtiss Kirkham 18-T Triplane**

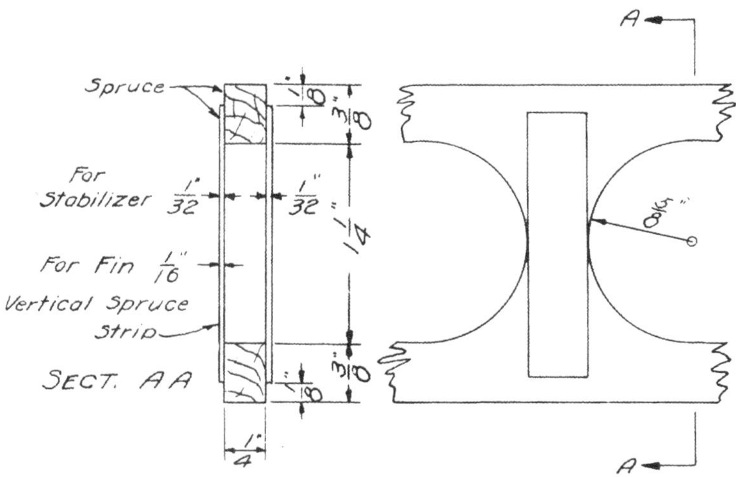


Fig. 140. Stabilizer Rib on Curtiss Kirkham

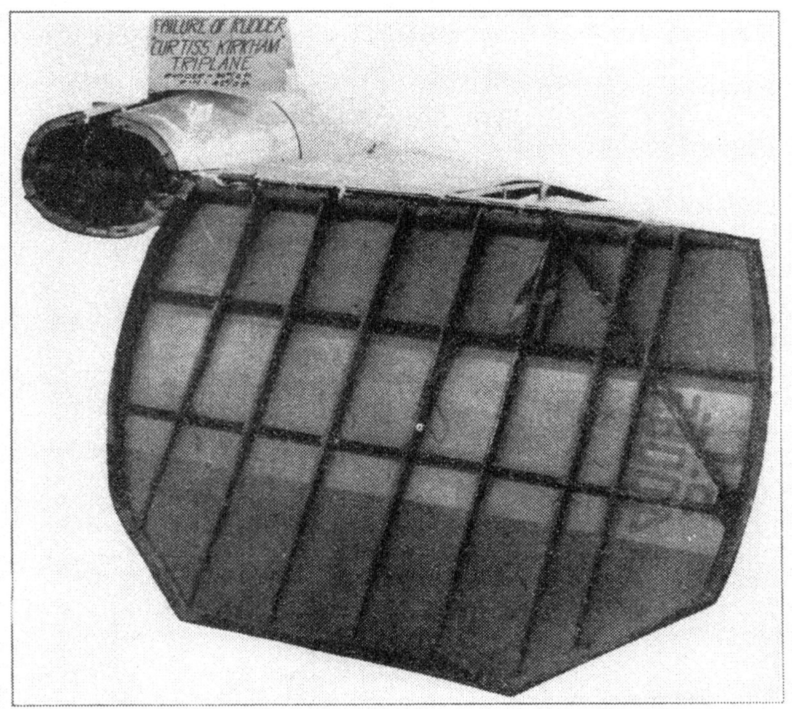


Fig. 141. Internal Construction of Curtiss Kirkham 18-T Rudder

a small part of the weight of a surface, the addition of two or three of them will often considerably increase its strength and rigidity without adding appreciably to the weight. The diagonal tube running up from the torque tube to the upper edge of the rudder serves to prevent excessive deflection of this portion of the rudder.

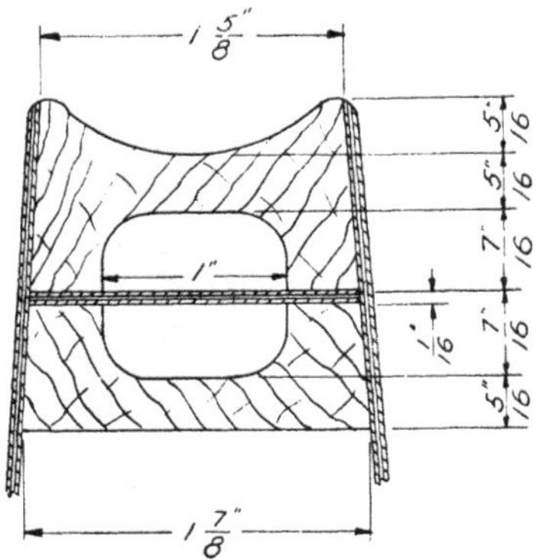

Fig. 142. Section of Stabilizer Spar on Curtiss Kirkham Half Way Out

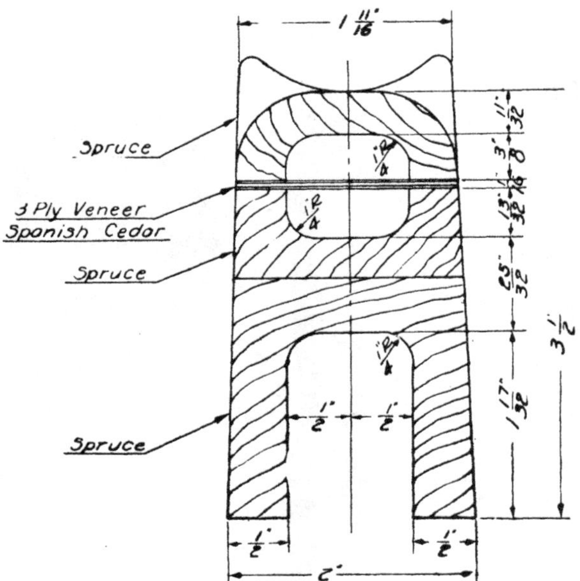


Fig. 143. Maximum Section of Stabilizer Spar on Curtiss Kirkham 18-T

The entire load on the horizontal and vertical surfaces is carried in bending by the spars in the fixed surfaces acting as cantilevers in combination with the torque tubes in the elevator and rudder. The stabilizer and fin spars increase in section toward the fuselage. Fig. 142 is a section through the stabilizer spar half way out, and Fig. 143 the maximum section at the fuselage. The fin spar is of the same type, but slightly smaller. For the torque tubes on rudder and elevator, 1½ in. O.D. tubing was employed, of .035 in. gage for the elevator and .032 in. gage for the rudder.

The hinge used on these surfaces has some excellent features. Fig. 139 shows it to be of the streamline type that allows very little leakage. Fig. 144 illustrates the design in detail. The guide collar and the hinge must be slipped on the torque tube before the ribs and frame are welded to it. Aside from the great difficulty of replacing a broken strap, this hinge is probably the simplest and best kind in use. Fig. 145 illustrates a hinge that overcomes this objection.

Type 3. The type of surface illustrated in Fig. 146 is well suited to good sized surfaces on two-seater airplanes. The construction is very simple and strong. Although the ultimate strength of this design is given in Table XXIII as only 20 lbs. per sq. ft., neither the elevator nor stabilizer was injured at this load, but the test was discontinued on account of the complete failure of the control system. The weight of the

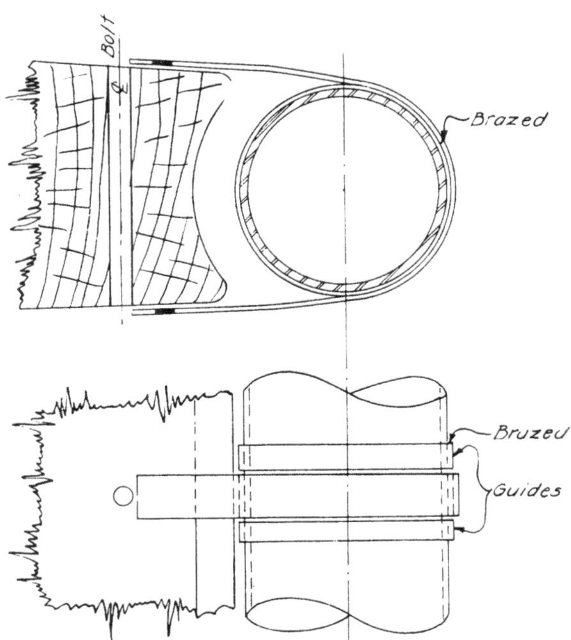


Fig. 144. Elevator Hinge on Curtiss Kirkham 18-T

ribs may be decreased by using 3/32 in. 3-ply spruce lightened by circular holes. The secondary spar is a 1/2 in. spruce rod. A strip of .065 in. gage aluminum forms the trailing edge, with bent ash for the curved portion at the tips. The main spar is constructed of spruce and 3-ply mahogany. Its width is 6 in. As is evident in Fig. 146, the spar offers a broad base for the control masts, and is also capable of carrying a large torque. Under the masts, the spar should be reinforced by blocks of hardwood, about 2 in. long. This will prevent a crushing of the spruce web at the point of attachment of the mast. These webs may be routed except at mast and strut points, and tapered slightly in plan view to advantage. The horizontal surfaces are braced by two 11/16 in. O.D., .049 in. gage struts running from the bottom of the fuselage, one to the end of the rear stabilizer spar and one to the forward stabilizer spar at a point 3/5 way out from the fuselage.

Type 4. The thin surface of tubular construction illustrated in Fig. 147 has been used to a considerable extent. From the aerodynamic point of view these surfaces are not as efficient as the thicker, double-cambered surfaces, i.e., their $L/D$ is less and also their maximum $K_y$ or unit lift. One point in their favor is the rapidity with which the lift builds up. They are more sensitive than thicker sections; in other words, a less angle of incidence is required to give the needed force. Thin surfaces are perhaps better adapted to slow or moderate speed airplanes than to pursuit airplanes, because the high speed of the latter insures sensitive controls with either type of surface. Furthermore, as it is desirable to eliminate external bracing in fast airplanes, thin surfaces, which must be braced, are not suitable for such airplanes. The simplicity of the tubular construction makes it good production work.

One of the principal problems in tubular work is the joint between the torque tube and the ribs. Formerly, this was made by slotting each member, inserting a gusset plate, and brazing the parts together. Slot-

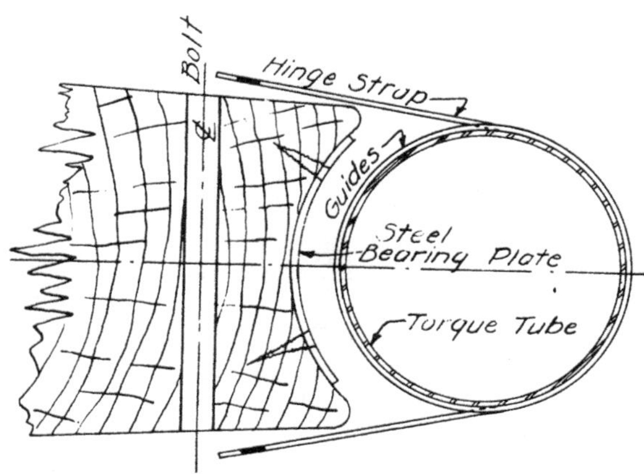


Fig. 145. Elevator Hinge with Removable Strap

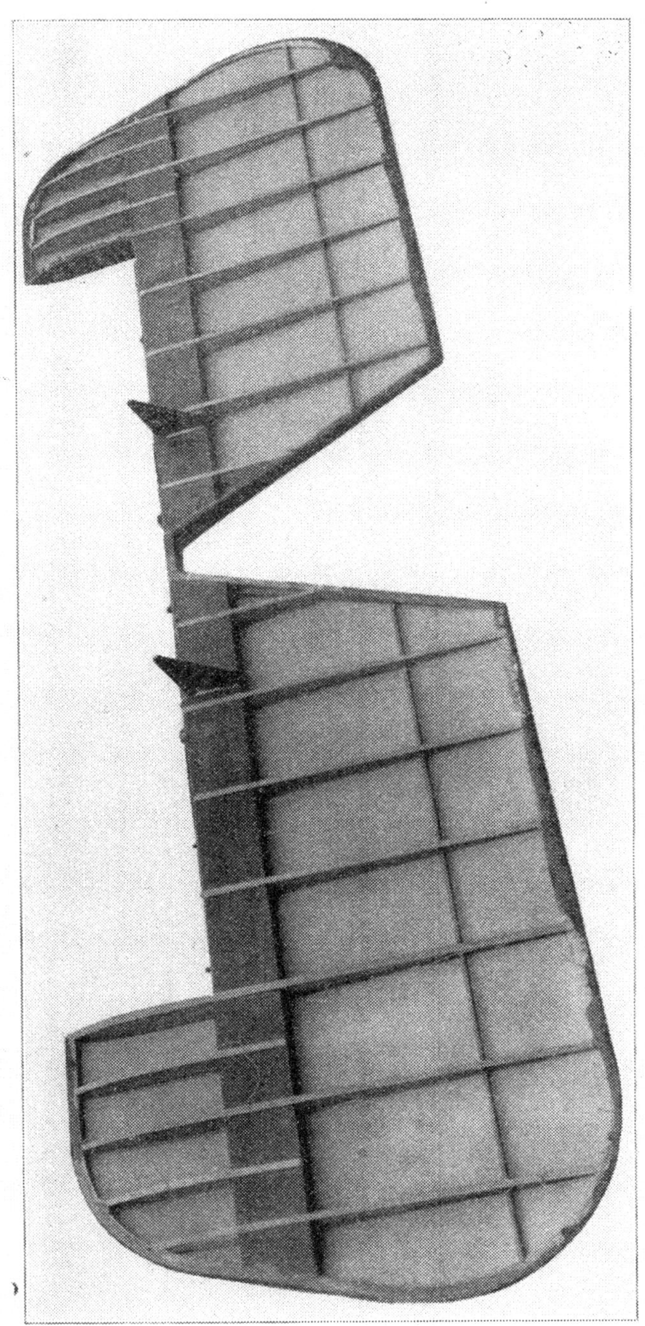

Fig 146. Elevator of LePere Pursuit Airplane

ting the main tube and rib weakened them materially. A much more satisfactory joint is shown in Fig. 148. It will be observed from Fig. 147 that where a rib of smaller diameter than the torque tube is joined to the latter the rib is flattened so that its long axis is vertical, while at the junction of the rib and trailing edge the rib is flattened so that its long axis is horizonal.

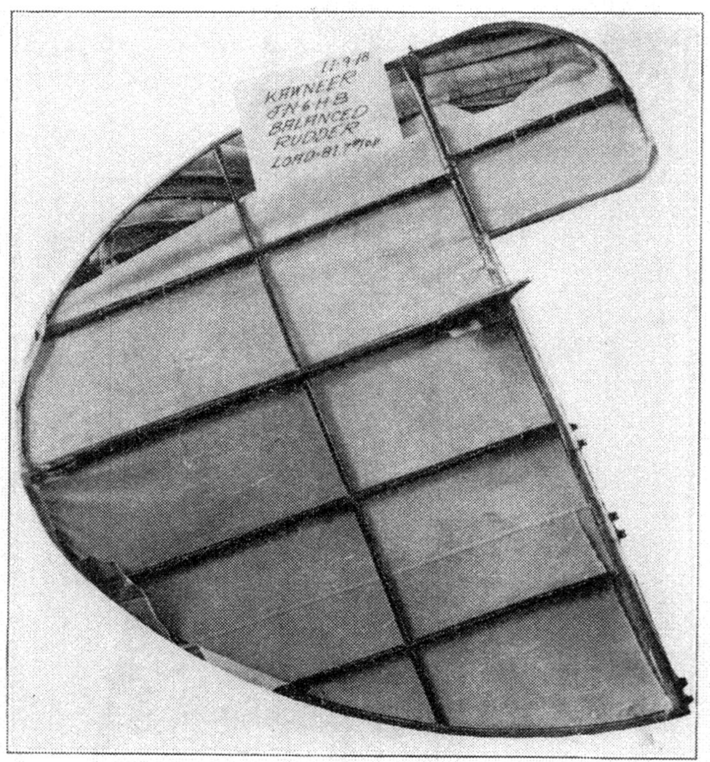


**Fig. 147. Rudder of Tubular Construction**

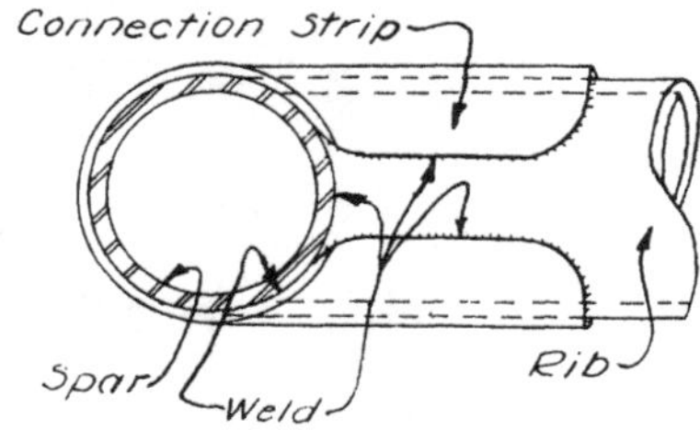


**Fig. 148. Method of Joining Ribs and Main Spars**

Owing to their low section modulus, neither the spar nor rib tubes can carry large bending moments. It is, therefore, necessary to brace the main spar from the fuselage, and the trailing edge or auxiliary spar from the mast. For this reason, the mast rib, and also the trailing edge or secondary spar, should be fairly stout. The importance of putting the mast and mast rib in line with the control cable should again be emphasized. It is well to locate hinges as near the ends of the elevator or rudder torque tubes as possible to avoid excessive deflections of the tips. If a tip cannot be properly braced by ribs at right angles to the torque tube, a diagonal bracing tube is run from the latter out to the end of the tip, as in Fig. 141.

Because of the large excess strength of the Curtiss JN-6 HB rudder, its weight can be cut materially by reducing the gage of the torque tube and mast rib from .049 in. and the gage of the other ribs from .035 in. The gage of the trailing edge and of the auxiliary spar should not be decreased. In fact, since with this type of surface the auxiliary spar largely increases the strength and rigidity of the structure, it might be well to increase slightly the diameter of this member which is now 7/16 in. In changing to a smaller gage it should be remembered that the ultimate unit strength of the metal is reduced.

Type 5. A design which retains most of the desirable features of the steel, tubular construction, but which has the aerodynamic efficiency of the double-cambered surface, is illustrated in Fig. 131. The wood ribs are secured to the torque tube and trailing edge by means of light straps welded to the tubes and nailed to the ribs. The heavy mast rib and secondary spar should be noted.

Type 6. For the thick double-cambered surfaces of large two-seater airplanes, the general construction shown in Fig. 149 has proved very efficient. The ribs and spars are of light gage, pressed steel. All joints are welded. A detail sketch of the mast box is shown in Fig. 132. The manner of bracing the stabilizer with a horizontal truss may be noted. For the compression struts 1/2 in. tubing is used. Instead of a tube, a composite spar made up of a steel member, similar to the front stabilizer spar, and a spruce member is employed for the stabilizer trailing edge. The construction of the hinge is evident from the general photograph. It is noted that the width of the hinge is greater than necessary. A wide hinge strap produces excessive friction. An excellent example of this type of hinge is illustrated in Fig. 150. In passing, attention should be called to the wire or light strap, welded to the elevator tube, which forms a loop about the hinge strap. The fabric on the elevator is sewed to this loop. The surfaces are externally braced by wires from the rudder post and fuselage to the end of the stabilizer spar.

In this construction the elevator ribs are of deep enough section at the torque tube to act efficiently as cantilevers with no support from the trailing edge or a secondary spar. It is, therefore, more economical to omit an auxiliary spar such as is shown in Fig. 129 and brace the ribs laterally with piano wire, spot welded to their chords. The weight thus

Fig. 149. Truss Type of All-Steel Control Surface Construction

Fig. 150. Aileron Hinge

saved should go into extra ribs. In the test, the first failure occurred by buckling of the chords of the elevator ribs. As a rule, the spacing of ribs should not exceed 12 in. Final failure took place through shearing of the bolts securing the lug of the brace wire from the fin. As can be seen from the photograph the diagonals on the stabilizer ribs are too slender. A slight increase in their section adds very slightly to the total weight. A good value for the gage of all ribs is .020 in. This is sufficiently heavy, and if lighter material is used the construction is not stiff enough. As the ribs form only a small portion of the total weight, it is not worth while to cut down their gage too much.

When a tube is used for the trailing edge of a stabilizer the type of hinge shown in Fig. 151 has given satisfaction. With this kind of hinge the strap around the elevator tube can be easily replaced. In conjunction with this hinge a light support, or former, Fig. 152, to which the fabric is attached, is secured to the stabilizer tube every 6 in. by little legs. This arrangement reduces the leakage and streamlines the hinge.

For main spars which are subjected to torsional stresses in excess of 10,000 lbs. per sq. in., and preferably for all spars of steel tubing, seamless tubes must be used instead of "locked seam" tubes which are decidedly weak in torsion.

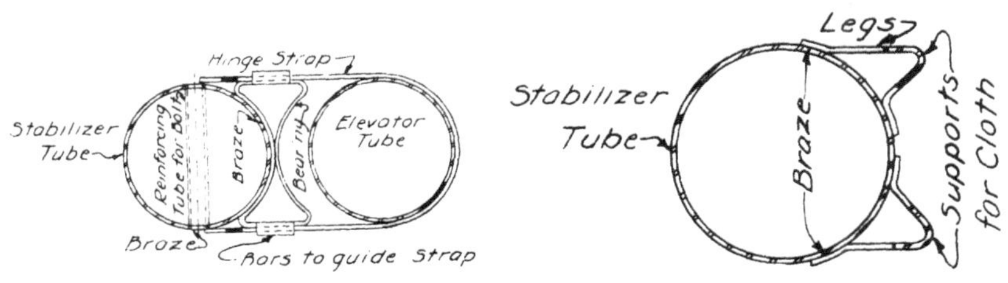


Fig. 151. Aileron Hinge

Fig. 152

## *STRESS ANALYSIS*

148. *Loading*—The loading which is made the basis of an analysis should be that given in the Government Specification for tail surfaces. The distribution of load between the fixed and movable surfaces is shown in Fig. 107, its intensity for the different types of airplanes being stated in the Specifications.

149. *Unbraced Surfaces*—The calculation of the external moments on an unbraced empennage is a simple, determinate problem. In most cases there is no front stabilizer spar, and the entire bending moment is

carried by the rear stabilizer spar acting with the elevator spar. In case auxiliary elevator or stabilizer spars are present, their effect in relieving the main spars may be neglected, or some arbitrary allowance made for them. The moment of the loads on both surfaces should first be calculated about the outer point of support of these spars. It may be noted that in this computation the same moment will be obtained if the load per sq. ft. on the elevator is taken as uniform instead of triangular. This moment, divided by the ultimate strength of the material, gives the total necessary section modulus for both spars. In case the stabilizer spar is wood, and the elevator torque tube steel, the proportion of the total load carried by each member must be determined by the method explained in Art. 25. The principle upon which this method is based is that both spars deflect together. When the moment resisted by each spar is thus computed the member is designed as before. In addition to bending stresses, the elevator torque tube is subjected to torsional stresses, which may be calculated by the formulas of Art. 26. The torsional stresses so obtained should then be combined with the bending stress according to the formulas given in the above article. For wood members the strength of the material in horizontal rather than in direct shear will probably be the limiting factor. In order to secure lightness, spars as a rule taper toward the outer end. It is well to check the stresses at one or more points to guard against excessive taper. If steel tubes are employed for the members, especially where the tubes are long and where the torsional stresses are high, a reduction in weight may be secured by splicing the tubes as shown in Fig. 154, using for the outer section a tube of the same diameter as the main section, but of lighter gage. Splicing should be resorted to for special cases only. The ultimate stresses for torsion given in Fig. 153 should not be exceeded. This curve is an arbitrary one based on a limited number of tests, but it will give reasonable and safe results. Ratios of $D/t$ greater than 125 should never be used, and a ratio of 100 is a better limiting value. The same methods of analysis apply to the fin and rudder unit also.

150. *Ribs*—Except for steel ribs which are in the form of a truss or tube, no accurate analysis can be made of the stresses in a rib. For indeterminate cases the design must be purely empirical and based on good practice as outlined in Art. 83. With a truss rib the usual assumptions are made that the members are pin jointed and carry only direct stresses. The distributed load on the rib must be divided up into concentrated loads applied at the panel points. The graphical method of solution explained in Art. 12 will be found convenient. The design of compression members in the rib can be based on the parabolic column formula of Art. 24. In selecting the proper value the members should be assumed to be hinged at the ends. Where a rib is continuous over a spar, as in the case of a stabilizer rib supported by a front stabilizer spar, the best method of solution is to assume a diagonal in place of the spar. The reaction of the rib on the spar may be considered as two concentrated loads acting at the panel points adjacent to the spar. With tubular ribs the analysis and design is simple. The maximum

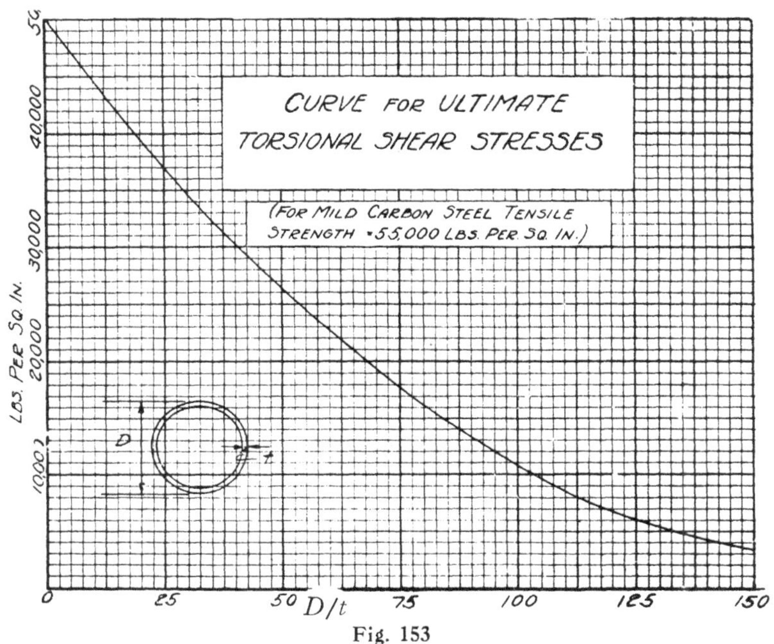


Fig. 153

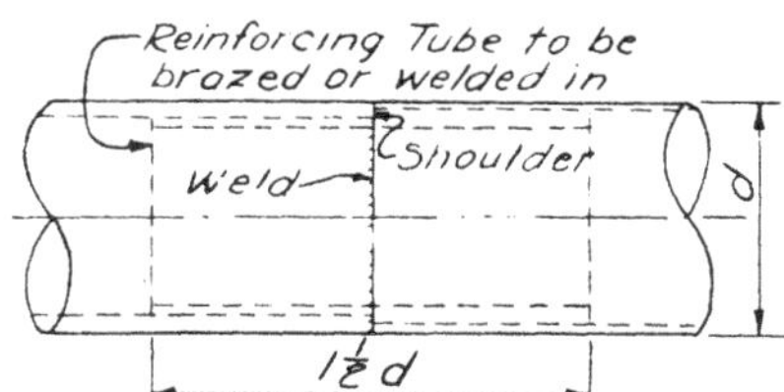


Fig. 154. Splice in Tubular Spar

stress on a cantilever rib occurs at the torque tube, and equals the moment of the distributed load carried by the rib, divided by the section modulus of the tube at that point. A mast rib which transmits torque from the mast to the trailing edge or to the auxiliary spar of a surface is an indeterminate structure, whether the mast is braced or not. If a guy wire is used, the rib is subjected to a heavy compression equal to the horizontal component of the stress in the wire, and to a bending moment equal to the algebraic sum of the moments of the vertical component in the wire, the distributed load on the rib, and the reaction of the auxiliary spar and trailing edge. If a secondary spar is present the worst stress in the rib probably occurs at the point at which the reaction from this spar goes into the rib. When the mast is unbraced the rib carries only bending moment. An estimate of the reactions of a secondary spar at the mast rib can be only approximate as they are dependent upon the relative deflections of the secondary spar or trailing edge and the various ribs. These deflections in turn vary with the width and length of the surface, as well as with the moment of inertia of the ribs, auxiliary spar, trailing edge, and main torque tube. The angular deflection of the last member affects the deflection of the ribs. Another important factor in the problem is the location of the mast rib. If, for instance, this rib should be placed near the inner end of the surface, a much larger proportion of the torque from the mast would be transmitted directly to the ribs by the main torque tube instead of going into the mast rib.

151. *Braced Surfaces*—When a surface is braced by stay wires from the fuselage and fin to the stabilizer spar, it becomes a statically indeterminate structure if the spars are continuous over the fuselage. Should the spars be hinged at the fuselage the vertical component of the stress in the stay wire equals the moment of the distributed load on the stabilizer and elevator about the point at which the spars are hinged, divided by the distance between this hinge point and the point of attachment of the wire. With the spars continuous, this vertical component of the wire stress equals the reaction at the point of attachment of the wire which would be obtained from a solution of the "three moment equation" written for the entire length of the spars. Owing to the fact that the horizontal surfaces are symmetrical about the centerline of the airplane a very simple expression can be written for this reaction. Referring to Fig. 155, $R_1 = -w\,(.375\,L_1 + L_0 + .75\,L_0^2/L_1)$, in which $w$ is the total

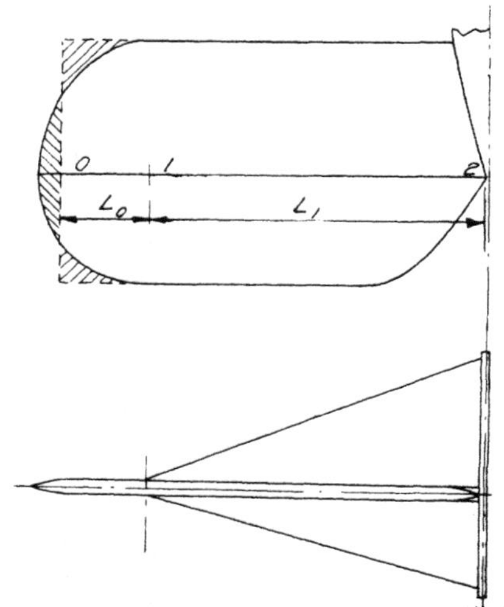


**Fig. 155. Example for Solution of Braced Surfaces**

load per in. run on the surfaces. The rounding of the ends of the surfaces is allowed for by decreasing the span so that the shaded areas in Fig. 155 are equal, an approximation that is permissible in most cases. The assumption that the spars are supported at a single point at the fuselage, instead of at two points as is usual, introduces a slight error. The fact that the stress in the wire produces in it a certain amount of elongation, thus allowing point 1 to fall slightly below the level of point 2, is another source of error. To neglect this increases the stress in the wire about 3 or 4 per cent. The effect of this deflection can be calculated, if desired, by using the "three moment equation," $2m_1L_1 + 4\,m_2L_1 = wL_1^3/2 + 12\,EIv_1/L_1$, in which $v_1$ is the deflection of point 1 with reference to point 2, to recompute the bending moment at 2, and from that to correct the reaction.

The horizontal component of the wire stress equals the compression that is put into the stabilizer spar which may be assumed to carry the entire compression. The center of the span is probably the critical section for the stabilizer spar and should be investigated for combined bending and compression.

In determining the stresses in a braced rudder post the method of deflection is perhaps best. It consists of computing the deflection due to a uniform load of the cantilever rudder post, at the point of attachment of the brace wire, as if no wire were present. Then the concentrated load, applied at this same point and perpendicular to the axis of the rudder post, which would be required to produce an equal and opposite deflection, is calculated. This load equals the component in the stay wire perpendicular to the rudder post. For a uniformly loaded cantilever beam the deflection at any point equals

$$d = \frac{w}{2\,EI}\left(\frac{x^4}{12} - \frac{L^3x}{13} + \frac{L^4}{4}\right)$$

$x$ being measured from the outer limit of loading, and $L$ being the loaded length. For a cantilever beam with a concentrated load applied at a distance $L_1$ from the support, the deflection under the load equals $d = WL_1^3/3\,EI$. These two deflections are then equated and a solution made for $W$. Since the value of $EI$ is common to both formulas it may be cancelled for the deflection.

## CHAPTER VIII

# FUSELAGE

152. *Truss Type*—Fuselages may be classified under three main types: Truss, Monocoque and Semi-monocoque. The truss type consists of a box frame made up of four longerons, running the length of the fuselage, connected by vertical and horizontal members so as to form four trusses. Truss fuselages may be divided into two sub-types; one in which the sides of the fuselage are of cloth and serve only to decrease the air resistance, another in which the sides are of plywood and reinforce the trusses. The first sub-type is frequently known as the "stick and wire" type, as most fuselages built in this manner consist of Pratt trusses with wooden longerons and struts, and steel wire diagonals. The forward portion of the fuselage, however, is often constructed as a Warren truss. The name stick and wire is not precise since some fuselages of this type, for example the Fokker D-7, are of metal construction.

The chief advantage of the stick and wire fuselage is the ease with which it can be repaired. The chief disadvantages are: the continual adjustment or truing up necessary to maintain the fuselage in condition for flying; the complexity of the structure owing to the large number of tie-rods and fittings; and its "vulnerability to attack" due to the fact that the failure of any one member results in serious crippling of the fuselage. The first and second disadvantages may be largely overcome, as in the Fokker D-7, by using a welded metal framework. The third is of importance only in certain types of military airplanes.

In the plywood covered type the diagonals and struts are wood members glued and nailed to the plywood sides, which act very much as gusset plates. This construction eliminates the numerous fittings found in the type previously discussed. It has the further advantages of being cheaper and easier to build, requiring no "truing up," and being less vulnerable to attack. It is, however, somewhat more difficult to repair than the stick and wire type.

153. *Monocoque Type*—The distinguishing feature of this type is the single-piece shell of plywood which forms the fuselage. This shell is built up in a mold which is removed after the shell has been constructed. After the removal of the mold the shell is usually reinforced by bulkheads. The building of the shell is carried on in different ways by various manufacturers. One firm which makes a specialty of this type of construction uses the following method. A collapsible form is built up and covered with unbleached muslin. The first layer of veneer strips is then lightly tacked on and held firmly in place by straps and special rolls. These rolls are so designed that a single strip of veneer can be taken off at a time. Each strip is taken off the mold, coated with glue and replaced. The next layer is of unbleached muslin tape, which is followed, in turn, by two layers of veneer and the outside layer of unbleached muslin. The veneer is in long strips 2 to 3 in. wide and usually 1/16 in. thick. The first layer is wound in a spiral making an angle of about 30 degs. with the axis of the fuselage. The

second layer is also in a spiral but in the opposite direction. The strips of the third layer are laid parallel to the axis of the fuselage.

The main advantages of the monocoque type are: its combination of lightness and strength; its strength under attack, in that small injuries do not render the fuselage unsafe; and the ease with which it can be made with a good streamline shape. The most serious disadvantage of this type is the difficulty experienced in making satisfactory repairs. Other disadvantages urged against it are that unusual skill and excessive time are required in its construction and that it is much more costly. The mold is expensive, but in quantity production its cost is divided up, so that very little need be charged for it against any one fuselage, making it a serious disadvantage only in experimental work where but a few fuselages of a given design are to be made. The disadvantages of the skill and time required have already been reduced by proper methods of construction so that this type of fuselage can be constructed as cheaply and quickly as the truss type when it is produced in quantity.

154. *Semi-Monocoque*—In this type the shell is reinforced by bulkheads and longerons, the former being spaced at intervals of about 1½ to 2 ft. In large fuselages of this type the longerons are assumed to carry all of the direct stresses and the plywood shell to bind them together so that they will act as a unit and carry the shear. In smaller examples the shell is assumed to carry some of the direct stresses also. The longerons are usually of spruce or ash; the plywood of spruce, poplar, redwood, or mahogany; and the bulkheads of metal or plywood.

155. *Airplane Considered for Analysis and Loading Conditions*—The fuselage of a Vought VE-7 two-seater training biplane will be taken for analysis. This fuselage is of the truss or "stick and wire" type and is standard.

Four conditions of loading are considered:

1. *Flying Condition with Elevator Up*—In this condition the following forces are acting; forces due to the weight of the airplane and its equipment; propeller forces, torque and traction; and the air load on the horizontal tail surfaces.
2. *Flying Condition with Rudder Turned*—The only forces considered are the air loads on the vertical tail surfaces and the reactions at the wings.
3. *Landing with Tail Up*—The propeller axis is considered horizontal, and the forces acting are the weight of the airplane and its equipment, the reactions from the chassis struts, and a balancing air load on the tail surfaces.
4. *Three-Point Landing*—In this condition the air load on the tail surfaces is replaced by the reaction at the tail skid.

In all these cases the weight of the wings is omitted as it is not carried by the fuselage in flying and not after the airplane has touched the ground, until it has slowed down considerably.

When the critical loads occur, the effect of the weight of the different parts of the airplane is increased by the fact that the airplane has a large angular acceleration or is subject to shock. To take care of this dynamic effect and also for defects of workmanship and material the dead loads must be multiplied by a factor called sometimes a load factor and sometimes a dynamic factor, though the former term is preferable. For air loads 5 lbs. per sq. ft. is considered the equivalent of a load factor of one. If a fuselage is designed for a load factor of 7, the ultimate air load will be considered equal to 35 lbs. per sq. ft.

156. *Computation of Loads for Flying Conditions*—The loads due to the weight of the airplane and its equipment, their location, and their center of gravity are shown in the balance drawing, Fig. 156. Table XXIV gives the same information in tabular form.

*TABLE XXIV*

*BALANCE TABLE*

| Item of Group | Weight (Lbs.) | Moment Arm About Rear Face Propeller Hub (In.) | Moment (In. Lbs.) |
|---|---|---|---|
| Propeller | 30 | —3.0 | —90 |
| Radiator, Water and Piping | 100 | 5.0 | 500 |
| Engine | 520 | 21.0 | 10,920 |
| Oil Tank and Oil Radiator, 30 lbs. Oil | 45 | 27.0 | 1,215 |
| Gas Tank and Gas (Front) | 130 | 50.0 | 6,650 |
| Chassis | 95 | 52.0 | 4,940 |
| Passenger | 170 | 74.0 | 12,580 |
| Gas Tank and Gas (Rear) | 75 | 101.0 | 7,575 |
| Pilot | 170 | 110.0 | 18,700 |
| Tail Skid | 10 | 236.0 | 2,360 |
| Empennage | 50 | 256.0 | 12,800 |
| Fuselage (inc. Inst., Controls, Cowl., Seats, Piping, Wiring, Etc.) | 335 | 69.0 | 23,115 |
| Wing Cell | 305 | 65.0 | 19,825 |
| Total | 2035 | 59.5 | 121,090 |
| Without Wing | 1730 | 58.5 | 101,265 |

Fig. 157 indicates the loads and their position relative to the vertical fuselage trusses as taken from the balance drawing. In computing the stresses in the fuselage it is considered as composed of two trusses loaded at the panel points. Fig. 158 shows one of these trusses with the panel loads as computed from the balance drawing. The computations to obtain these panel loads are as follows:

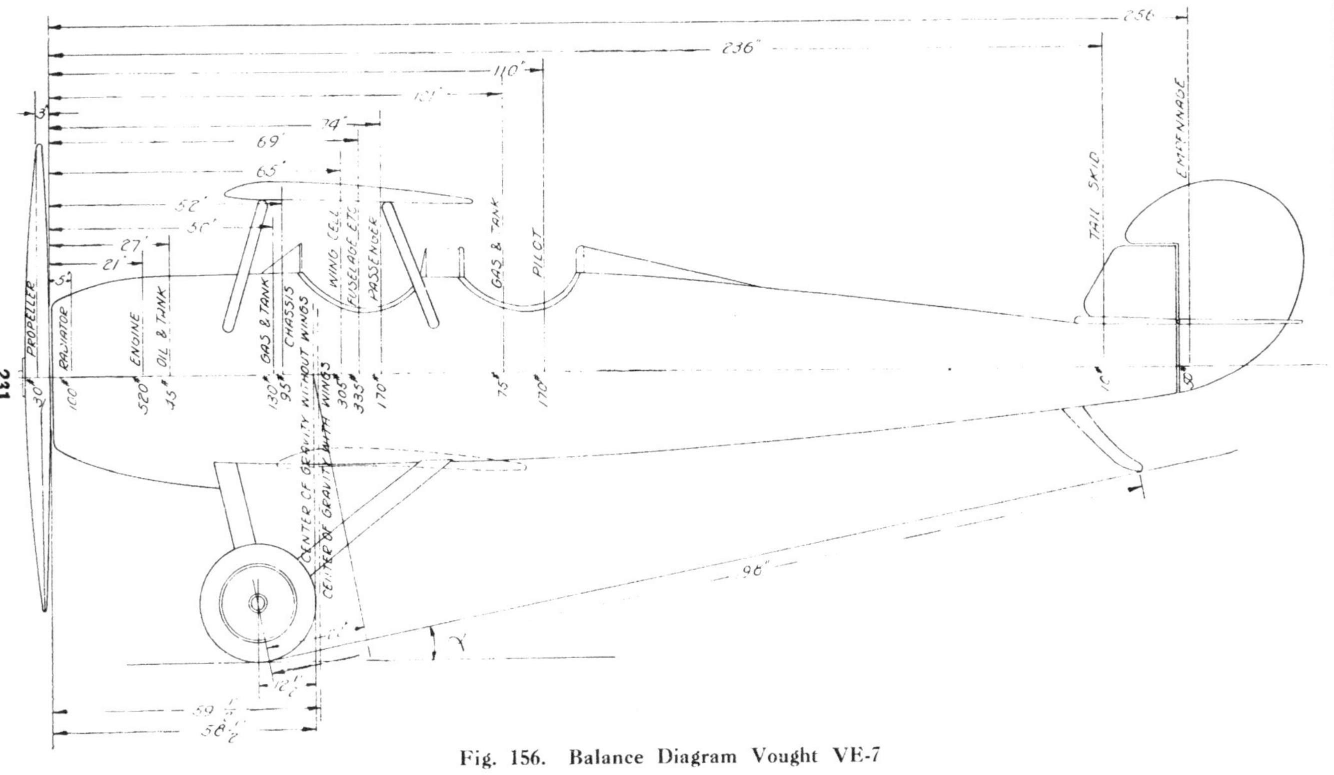


Fig. 156. Balance Diagram Vought VE-7

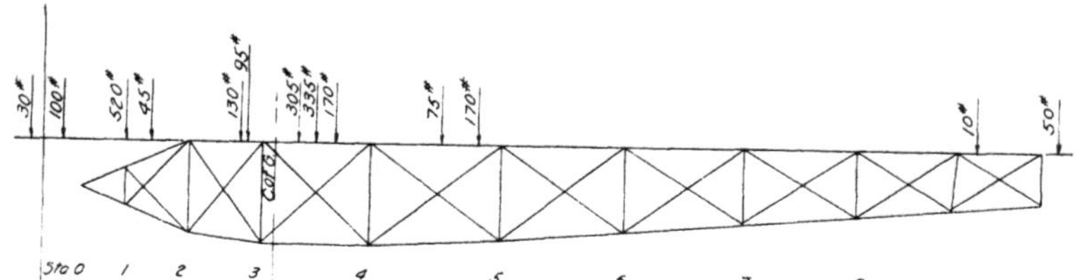


Fig. 157. Load Positions Relative to Fuselage Trusses

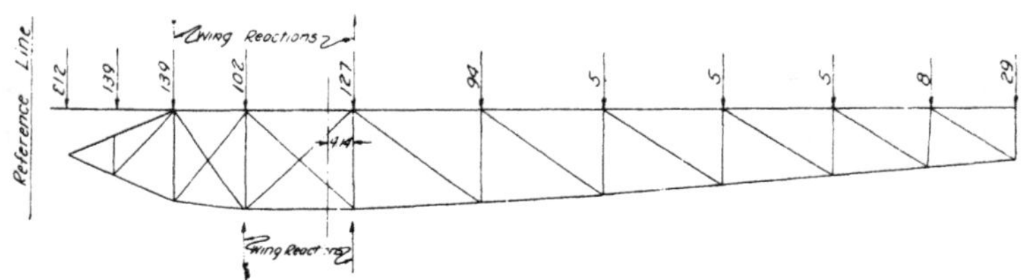


Fig. 158. Panel Loads Due to Dead Weight

**Station 0**

| | |
|---|---|
| Engine, Radiator, Propeller .... | 379 |
| Fuselage ...... | 45 |
| For two trusses | 424 |
| For one truss.. | 212 |

**Station 1**

| | |
|---|---|
| Engine, Radiator, Propeller .... | 197 |
| Oil and tank $45 \times \frac{11}{16}$ ..... | 31 |
| Fuselage ...... | 50 |
| For two trusses | 278 |
| For one truss.. | 139 |

**Station 2**

| | |
|---|---|
| Engine, Radiator, Propeller .... | 74 |
| Oil and tank $45 \times \frac{5}{16}$ ..... | 14 |
| Gas and tank $130 \times \frac{7}{19}$ ..... | 48 |
| Chassis $95 \times \frac{34}{48}$ ..... | 67 |
| Fuselage ...... | 75 |
| For two trusses | 278 |
| For one truss.. | 139 |

Station 3

| | |
|---|---|
| Gas and tank | |
| $130 \times \frac{12}{19}$ ..... | 82 |
| Passenger | |
| $170 \times \frac{12}{29}$ ..... | 71 |
| Fuselage ...... | 51 |
| For two trusses | 204 |
| For one truss.. | 102 |

Station 4

| | |
|---|---|
| Chassis | |
| $93 \times \frac{14}{48}$ ..... | 28 |
| Passenger | |
| $170 \times \frac{17}{29}$ ..... | 99 |
| Gas and tank | |
| $75 \times \frac{18}{33}$ ..... | 41 |
| Pilot | |
| $170 \times \frac{9}{33}$ ..... | 46 |
| Fuselage ...... | 40 |
| For two trusses | 254 |
| For one truss.. | 127 |

Station 5

| | |
|---|---|
| Gas and tank | |
| $75 \times \frac{15}{33}$ ..... | 34 |
| Pilot | |
| $170 \times \frac{24}{33}$ ..... | 124 |
| Fuselage ...... | 30 |
| For two trusses | 188 |
| For one truss.. | 94 |

Station 9

| | |
|---|---|
| Tail skid | |
| $10 \times \frac{18}{20}$ ..... | 9 |
| Fuselage ...... | 8 |
| For two trusses | 17 |
| For one truss.. | 8 |

Station 10

| | |
|---|---|
| Tail skid | |
| $10 \times \frac{2}{20}$ ..... | 1 |
| Empennage ... | 50 |
| Fuselage ...... | 6 |
| For two trusses | 57 |
| For one truss.. | 29 |

At stations 6, 7, and 8 a load of 5 lbs. on each truss is taken for the weight of the fuselage.

The loads should be divided between the panel points in such a manner that the center of gravity of the panel loads will be the same as the center of gravity of the loads considered acting. In order to obtain this result most of the concentrated loads should be divided between the two adjacent panel points in inverse proportion to the distances from the loads to the panel points. The weight of the fuselage and the engine and any weights in front of the first panel point cannot be treated in this manner. The weight of the fuselage should be distributed among all the panel points in such a manner as to have the center of gravity of the panel weights coincident with the center of gravity of the fuselage. This can be done only by trial. When the center of gravity of the fuselage is unknown it can be assumed to be between 40 and 45 per cent to the rear of the forward end. Approximately 60 per cent of the weight of the fuselage will be forward of its center of gravity and the remaining 40 per cent to the rear. The quickest way to distribute the fuselage load is to assume all the panel loads but two, and compute these two, knowing the center of gravity and the sum of the two unknown loads. If the values obtained are reasonable they should be accepted. If they are not, another trial should be made.

Where a radial or rotary engine is used and is fixed to an engine plate forming the front of the fuselage, the weight of the engine and propeller unit can be cared for by replacing it with a down load and a couple at the first panel point of the fuselage. In most cases the power unit is carried on a pair of engine bearers supported at two or more points. When the bearers are supported at only two points they can be computed as simply supported beams and the reactions easily found. If they are supported at three or more points they should be treated as continuous beams. Between adjacent supports the load due to the engine may be considered uniformly distributed, and the load per inch run in each span should be such that the location of the center of gravity of the engine remains unchanged. The propeller may be considered as a concentrated load on the cantilever end of the engine bearers. The reaction at each of the supports of the engine bearers can then be computed.

157. *Air Loads on Horizontal Tail Surfaces*—At Station 10 an air load of 5 lbs. per sq. ft. should be assumed. This value represents the increase of tail load occurring when the angular acceleration of the airplane is increased so that the dynamic effect of each component mass in the airplane is increased by the weight of that mass. The maximum tail load occurs while pulling out of a dive and may reach very high values. Instead of trying to design the fuselage and tail to carry the maximum possible values of the tail load, they are designed to be only as strong as the wings. A discussion of the strength required of the wings in the diving condition can be found in Art. 40.

There are several methods of calculating the tail load for which the fuselage should be designed in order to have the same strength as the wings. Three of these will be mentioned. U. S. Air Service Specification No. 1003 gives the formula, $F = 0.0026\,AV^2$, where $F$ is the force on the tail, $A$ is the area of the horizontal tail surfaces, and $V$ is the normal high speed of the airplane. For the case under consideration the formula becomes: $F = 0.0026 \times 32.7 \times 108^2 = 993$ lbs. The intensity of this load will be $993/32.7 = 30.3$ lbs. per sq. ft.

A British Admiralty report gives the formula, $F = 1.5\,WC/L$, where $W$ is the weight of the airplane, $C$ is the chord of the wings, and $L$ is the distance between the center of gravity of the airplane and the center of pressure of the horizontal tail surfaces.

$$F = \frac{1.5 \times 2035 \times 55}{197.5} = 850 \text{ lbs.}$$

or $\dfrac{850}{32.7} = 26.0$ lbs. per sq. ft. for the maximum intensity of air load on the tail.

Air Service Specifications for airplanes of this type call for a breaking load of 25 lbs. per sq. ft. on the horizontal tail surfaces. This seems rather low when compared with the loads as derived by the formulas above, but observation has shown that the structures designed to meet these loads are at least safe, and so in the analysis a basis of 25 lbs. per sq. ft. for the breaking load on the horizontal tail surfaces is used. Experimental investigation covering this subject is now under way from which it is hoped that considerable information will be derived. This load, however, is an ultimate load, while the panel loads just computed are working loads.

Air Service Specifications call for an allowable load at failure corresponding to a load factor of 5. Then the air load on the tail corresponding to the panel loading given by Fig. 158 will be equal to 25 lbs. per sq. ft. divided by 5, or 5 lbs. per sq. ft. This will increase the panel load on one truss at Station 10 by $(5 \times 32.7)/2 = 81$ lbs., making a total panel load of $81 + 29 = 110$ lbs.

158. *Traction and Torque*—The propeller traction is a function of the horsepower and the speed of the airplane. In pounds it equals 550 times the horsepower divided by the velocity in feet per second. In the case under consideration the horsepower is 150 and the speed 114 M.P.H.

$$\text{The traction} = \frac{150 \times 550 \times 60}{114 \times 88} = 494 \text{ lbs.}$$

Half of this total traction, or 250 lbs., will be taken by each truss.

The loading from the propeller torque is found from the formula

$$\text{Torque reaction} = \frac{63000\ P}{n\ d}$$

where $P$ is the horsepower of the engine, $n$ the revolutions per minute, and $d$ the distance between engine bearers in inches.

Substituting in the formula the various known quantities we have,

$$\text{Torque reaction} = \frac{63000 \times 150}{1600 \times 13} = 455 \text{ lbs.}$$

This torque tends to lighten the load on one truss and to increase it on the other. The panel loads at Stations 0, 1, and 2 are, therefore, increased to take care of the torque loads. The revision of these panel loads is shown below, and the revised loads for the flying condition are shown in Fig. 159. As the values obtained for the torque and traction are maximum values, the load factor of 5, which must be applied to the static loads due to the weight of the airplane, need not be applied to them. In order, therefore, to obtain torque and traction loads comparable to the loads shown in Fig. 158 these values are divided by the dynamic factor of 5. The torque is distributed in the same proportions as the weight of the power unit.

Station 0

From Art. 156 ............... 212

$$\text{Torque } \frac{455}{5} \times \frac{379}{650} \text{ ............. } 53$$

265 lbs.

Station 1

From Art. 156 ............... 139

$$\text{Torque } \frac{455}{5} \times \frac{197}{650} \text{ ............. } 27$$

166 lbs.

Station 2

From Art. 156 ............... 139

$$\text{Torque } \frac{455}{5} \times \frac{74}{650} \text{ ............. } 11$$

150 lbs.

159. *Supporting Forces*—The loads shown in Fig. 159 are supported at four points by the wing spars. As the distribution of the total load between these four points is indeterminate, assumptions must be made. In this case it is assumed that 25 per cent of the load is carried by the upper wing and 75 per cent by the lower. The lower spar connections are considered to carry the majority of the load on account of the lift wires being connected to the fuselage at the lower spar hinges. Where the wings are internally braced, as in the Fokker D-7, the upper wing is assumed to take half or slightly more than half of the load. The load carried by each wing is assumed to be divided between the front and rear spars in inverse proportion to the distance from the center of gravity of the loads to the respective panel points as illustrated by Fig. 160.

The propeller traction and torque are not at their maximum simultaneously with the maximum tail load, but as the loading on the tail has very little effect on those members forward of the cockpit, and, as the loading on the forward part of the fuselage has very little effect on the members rear of the cockpit, both of these maximum loading conditions may be considered simultaneously.

Sometimes the center of gravity of the loads falls to the rear of the lower rear spar, and in this case the forward and rear portion of the fuselage must be figured separately. In computing the rear portion, enough load should be added to the forward panel points to bring the center of gravity forward of the rear spar. For the forward portion enough load should be assumed to be subtracted from the tail load to obtain the same result.

Another method of compensating for the displacement of the center of gravity of the loads by the air load on the tail is preferred by some

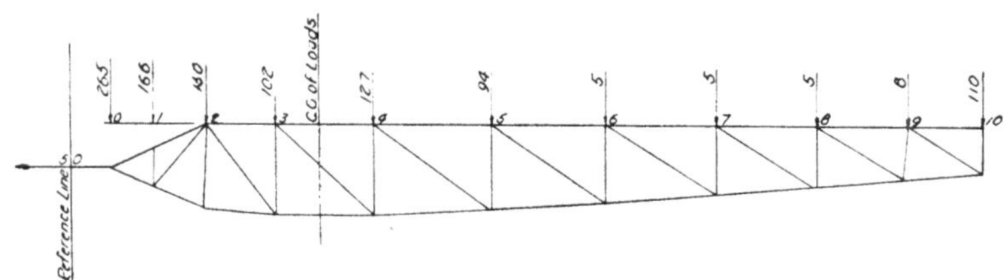

Fig. 159. Panel Loads in Maximum Air Load Condition

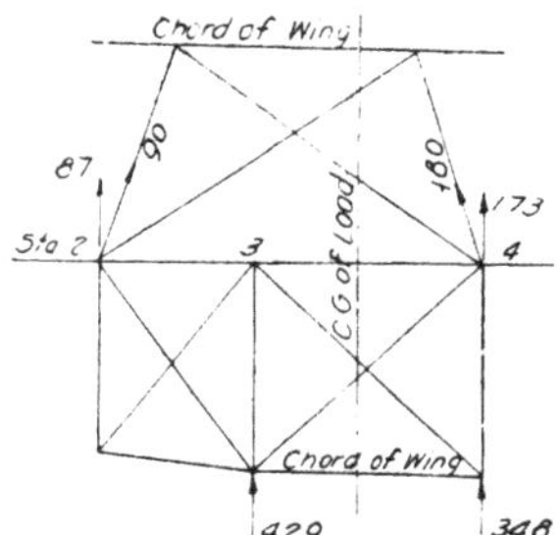


Fig. 160. Wing Spar Reaction in Maximum Air Load Condition

designers. This method is to decrease the panel loads to the rear of the center of gravity and increase those forward of it so that the center of gravity of the air load and the modified panel loads will be at the center of gravity of the airplane. In following this method the changes in the panel loads should be directly proportional to the distance from the center of gravity. This method is theoretically more correct than the one followed in this chapter, though the latter is more conservative.

160. *Stresses for Flying Conditions*—Fig. 161 shows a line diagram of the rear portion of the fuselage with the panel loads indicated, and also the stress diagram.

Fig. 162 shows the stresses in the forward part of the fuselage due to the loads shown. It is assumed that half of the traction is taken out at each of the two points where the front wing spars are connected to the fuselage. Force *EF* is increased to balance the horizontal component of force *GH*. The cabane wires will exert this force.

161. *Flying Condition with Rudder Turned*—The only forces that need be considered in this condition are the air load on the fin and rudder and the reaction at the wing spars. The air load may be considered as a concentrated load acting on the tail post at the level of the center of gravity of the vertical tail surfaces. This is not exact, as the actual distribution of load on these surfaces is unknown, but it is a reasonable

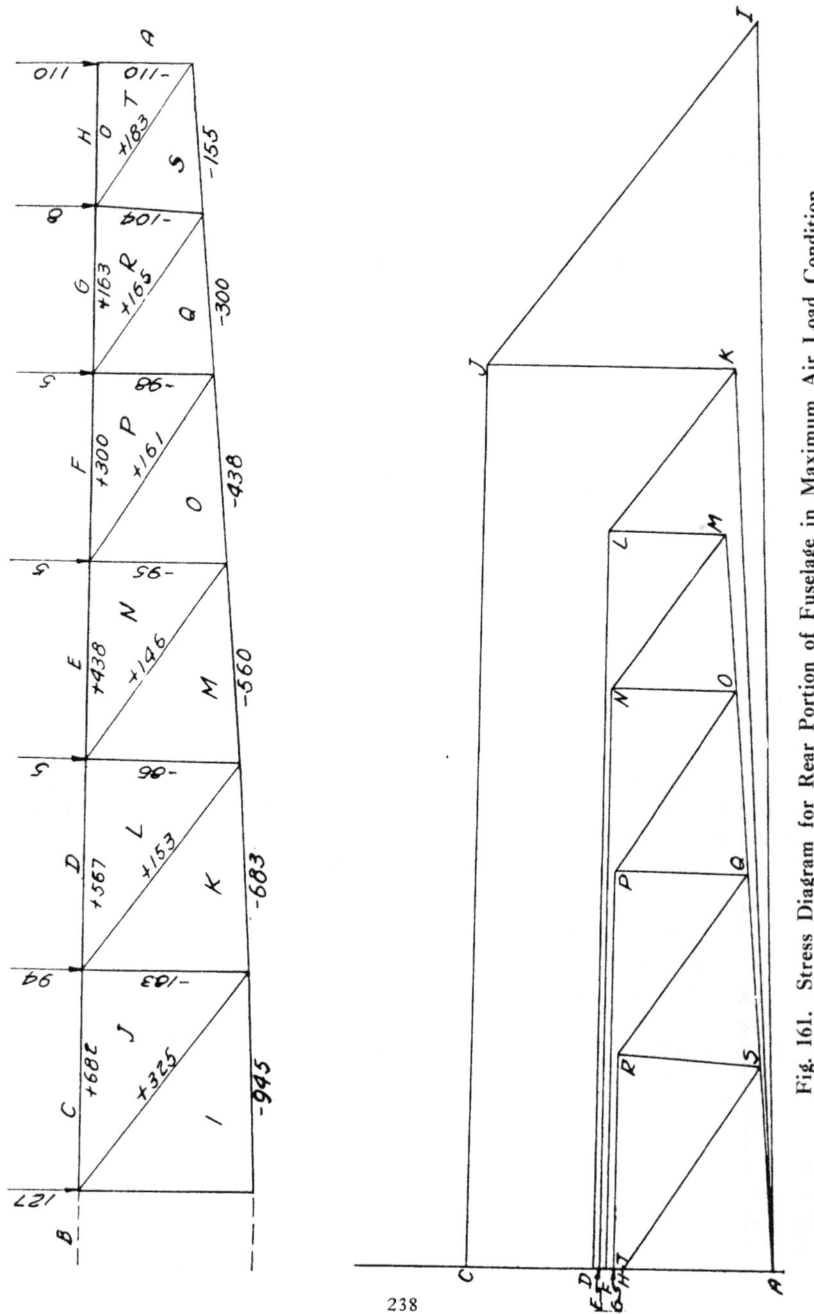

Fig. 161. Stress Diagram for Rear Portion of Fuselage in Maximum Air Load Condition

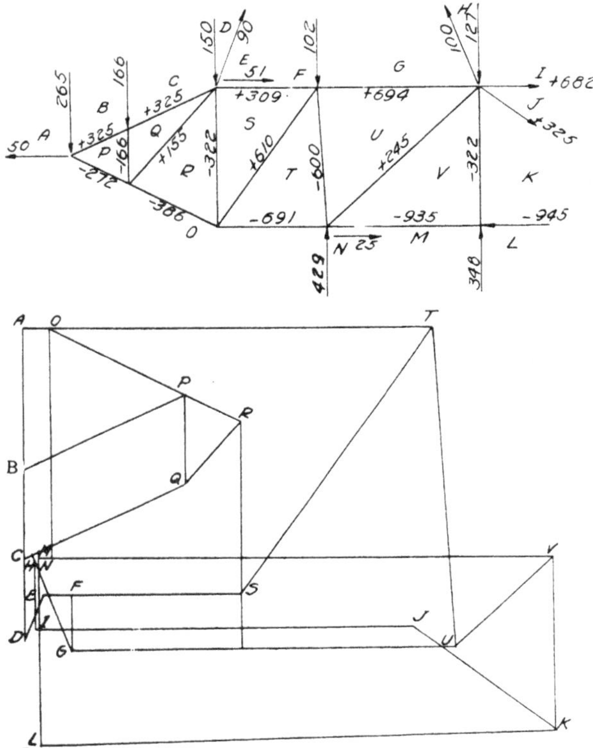


Fig. 162. Stress Diagram for Front Portion of Fuselage in Maximum Air Load Condition

assumption and a simple one. The tail post will then act as a beam loaded at the level of the center of gravity of the vertical surfaces, and supported at two points by the longerons. It is then a simple matter to compute the horizontal load on each pair of longerons. Very often the center of gravity of the vertical tail surfaces will be above the upper longerons, in which case these longerons will carry a load greater than the horizontal air load, and the load on the lower longerons will be in the opposite direction. A stress diagram needs be drawn only for the horizontal truss containing the pair of longerons carrying the heavier load, as the web members of both trusses will usually be identical. These web members are the only ones in the fuselage determined by this condition of loading since the longerons are determined by the stresses due to the heavier vertical loads. The stresses in the horizontal trusses will be less than those computed owing to the action of the diagonal cross wires at the panel points which transfer part of the stress from the upper truss to the lower.

In the case under consideration the longerons are 13 in. apart at the tail post and the center of gravity of the fin and rudder 15 in. above the upper longerons. The upper horizontal truss will then carry 28/13 = 2.16 times the horizontal air load. Air Service Specifications require an ultimate horizontal tail loading of 25 lbs. per sq. ft. Therefore, the upper horizontal truss will have to carry a load at the tail post of 2.16 × 25 × 10.1 = 545, say 550 lbs. Fig. 163 shows the line and stress diagram of the rear portion of the airplane for this load. It is not necessary to carry these stresses further forward as the stresses are small and other considerations will govern the design of the members in front of those computed.

162. *Landing with Tail Up*—The propeller traction and torque are negligible quantities, and the load on the tail surfaces is assumed to be just enough to keep the line of propeller thrust horizontal. The panel loadings for all panels except No. 10 are as shown by Fig. 158. At that panel there is an upward tail load balancing the airplane. This can be found by taking moments about the point of contact of the landing wheel and the ground.

$$\text{T.L.} \times 208 = 865 \times 12.5$$

$$\text{Tail load} = \frac{865 \times 12.5}{208} = 52 \text{ lbs.}$$

$$\text{Chassis reaction} = 865 - 52 = 813 \text{ lbs.}$$

The reaction at panels 2 and 4 is found by taking moments about the center of gravity.

$$0 - 52 \times 208 - R_4 \times 27 + R_2 \times 21 = 0$$

$$R_2 + R_4 = 813$$

$$R_2 \times 21 - 27 \times 813 + 27\,R_2 - 52 \times 208 = 0$$

$$48.0\,R_2 = 21{,}951 + 10{,}816 = 32{,}767$$

$$R_2 = 683 \text{ lbs.}$$

$$R_4 = 130 \text{ lbs.}$$

It would be more exact to subtract the weight of the chassis from both the panel loads and the chassis reaction, but this would require relocating the center of gravity and would entail unnecessary computation with little or no real gain in accuracy. Fig. 164 is the stress diagram for this loading. The diagram for the rear portion of the fuselage is drawn to a larger scale than the front portion on account of the small forces acting.

In the foregoing the attempt is made to represent dynamic conditions by static loads in equilibrium. The vertical reactions figured at the chassis strut points do not represent functions of the stresses in the chassis struts under the landing condition with the tail up, but are only the loads located at the points shown, which will maintain the structure in equilibrium.

The stresses shown in the bottom longerons between the reaction points are not, then, as large as they would be under actual conditions

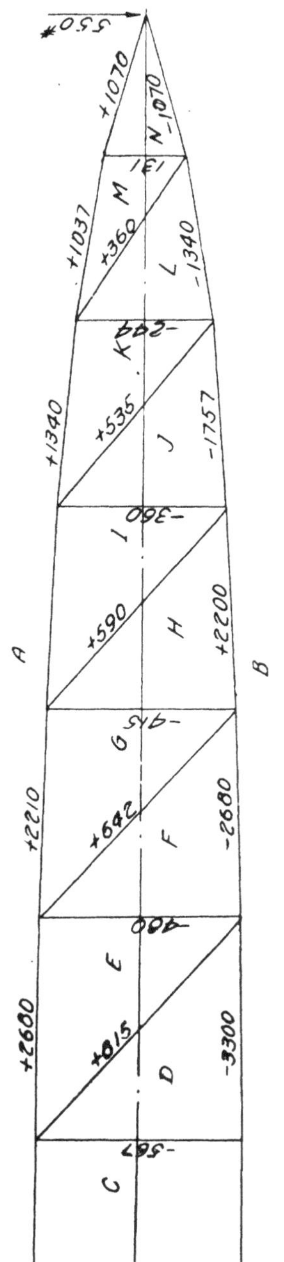


Fig. 163. Stress Diagram for Horizontal Fuselage Truss

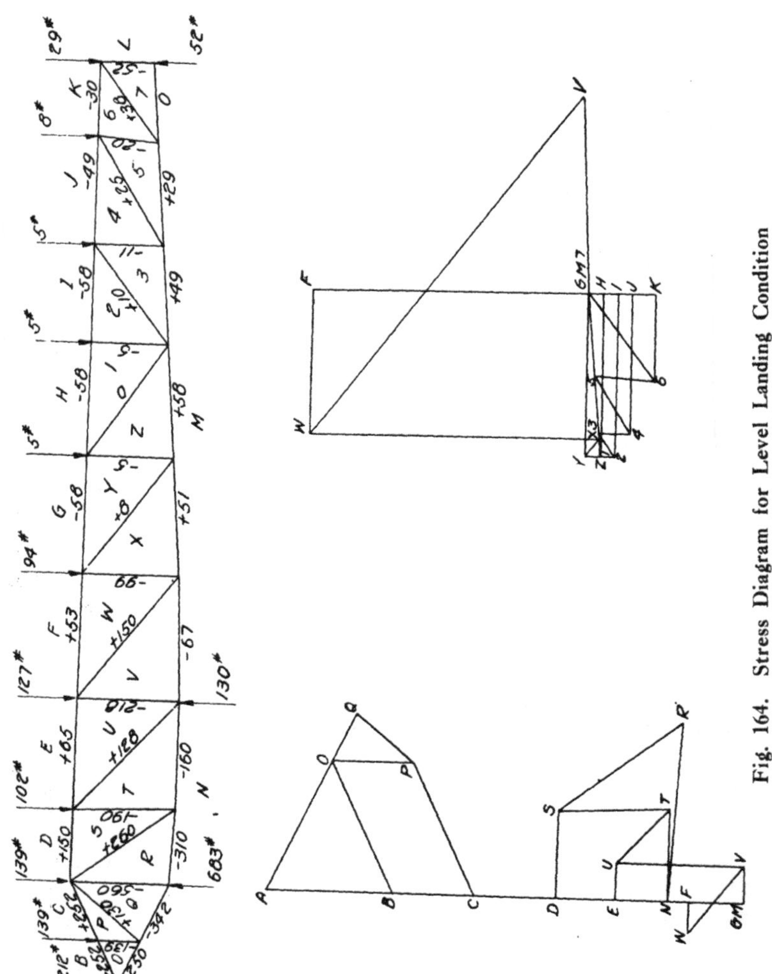

Fig. 164. Stress Diagram for Level Landing Condition

of landing with the tail up. If, however, the connections between the fuselage and chassis are made strong enough it is felt that the stresses in the bottom longerons will not become troublesome. There are a number of strengthening members in the fuselage which are not shown by the diagram. These tend to make the problem of distribution of stresses in this part of the fuselage indeterminate, but they also strengthen it. Hence the conclusion is that to make the connections rigid enough to transmit the stresses from the fuselage to the chassis is all that is required.

163. *Computation of Loads and Stresses for Three Point Landing Condition*—When landing with the tail down, the upward reaction on the tail skid causes compression in the upper longerons and is the determining condition for the design of these members and also for some of the web members.

The most reasonable assumption as to the distribution of the weight of the airplane between the chassis and the tail skid is to assume the airplane to be a simple beam supported at two points. The weight of the airplane should be assumed to act at the projection of the center of gravity on the line joining the points of support. No air load on the tail needs be considered, and the weight of the wings may be subtracted from the total weight of the airplane, as they probably sustain at least their own weight for some time after the airplane has reached the ground. The tail skid is in contact with the ground at a point behind where it is attached to the fuselage, and its load should be represented at the point of attachment by a force and a couple. The chassis reactions should be divided between the points of attachment of the chassis struts in such a manner that the center of gravity of the upward reactions will coincide with the center of gravity of the airplane. Allowance should be made for the fact that the chassis reaction is decreased before reaching the fuselage by the weight of the landing gear.

When the tail skid touches the ground in a three-point landing, the reaction of the earth can be divided into two components. The vertical component resists that part of the weight of the airplane being carried by the tail skid. The horizontal component is the friction between the skid and the ground. This frictional force tends to make the line of action of the resultant tail skid load perpendicular to the line of the propeller thrust. The force acting on the wheels may also be divided into a vertical component and a horizontal component due to friction or inequalities of the ground. But the horizontal component will not be as large proportionately as for the tail skid owing to the smaller coefficient of friction. These frictional forces, together with the drag, etc., are used in slowing down the airplane. Each individual weight in the airplane absorbs a portion of the total decelerating force. The portion absorbed by any particular unit may be considered as a small force opposing the decelerating force of friction and may be called the "inertia load" on the unit. These small inertia loads may be represented by one load at the center of gravity.

Fig. 165 is a diagram showing the loads in this type of landing. $N_c$

and $N_{ts}$ are the vertical components of the chassis and tail skid reactions respectively. $F_c$ and $F_{ts}$ are the frictional components of these reactions. $W$ is the weight of the airplane considered as acting at the center of gravity and $I$ is the sum of the inertia loads, also acting at the center of gravity. The ratios $F_{ts}/N_{ts}$, $F_c/N_c$, and $I/W$ are undoubtedly different and it would be very difficult if not impossible to determine them. They all tend to make the three resultant forces more nearly perpendicular to the line of propeller thrust than the vertical forces $N_c$, $N_{ts}$ and $W$ are. The assumption sometimes made is that these ratios are the same and equal to the tangent of the angle $a$ between the ground and the line of propeller thrust. This makes all the loads perpendicular to the thrust line. When this is done the tail skid load is equal to $\frac{W \times L_3}{L_3 + L_4}$. If the horizontal loads are neglected, we get the tail skid load, $N_{ts} = \frac{W \times L_1}{L_1 + L_2}$, which will always be a larger value.

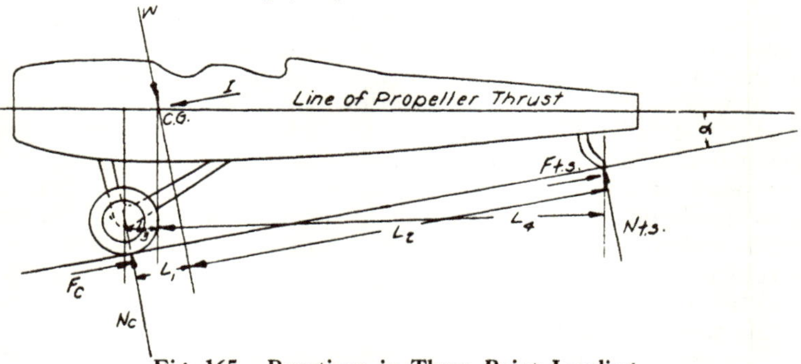


**Fig. 165. Reactions in Three Point Landing**

The true value is somewhere between the two. As the directions of the resultant forces are unknown, and as the assumption that they are perpendicular to the line of propeller thrust gives loads that are on the unsafe side, it is recommended that the second assumption, neglecting the horizontal components, be used. Since it gives a higher tail skid load, it is the safer assumption. This will be done in the present example, and all loads will be considered to act at the angle, 90°—$a$, to the line of the propeller thrust. Inspection of Fig. 165 shows that this assumption also gives a higher load on the front chassis strut, for a given value of $N_c$, than the assumption that the loads are perpendicular to the thrust line. In most cases it will give a higher actual value in spite of the smaller value of $N_c$. The value of the stress in the rear chassis strut will be much too low, but this is unimportant as the maximum value of stress in that strut comes in the level landing condition.

The forces $N_{ts}$ and $N_c$ are first computed, and $N_c$ is divided between the front and rear chassis struts in such a manner as to locate the center of gravity of the supporting forces at the same point as the center of gravity of the down loads. The panel loads are assumed to act at the line of the propeller thrust. This is allowed for by dividing the panel loads between the two longerons. The computation of the supporting forces follow.

The center of gravity of the airplane without wings is 58.5 inches to the rear of the propeller hub (see Table XXIV). It may be assumed to be on the line of the propeller thrust. This is not exact, but is a reasonable assumption for an airplane of this type. It is almost impossible to compute the exact location of the center of gravity of an airplane from the drawings, the only practical procedure being to compute its location as closely as possible and check the computations by experiment on the finished airplane. The line joining the points of contact with the ground of the landing wheels and the tail skid is 198 inches long (See Fig. 156) and the projection of the center of gravity on this line is 22 inches from the landing wheels. From this we have,

Tail skid load $= 1730 \times \frac{22}{198} =$ 192 lbs. for both trusses
96 lbs. for one truss

Chassis reaction $= 1730 \times \frac{176}{198} =$ 1538 lbs. for both trusses
769 lbs. for one truss

The chassis reaction must be divided into reactions at Stations 2 and 4.

$R_2 + R_4 = 769$

Taking moment about $R_4$

$-865 \times 23 - 96 \times 153 + R_2 \times 46 = 0$

$$R_2 = \frac{19{,}895 + 14{,}688}{46} = \frac{34583}{46} = 752 \text{ lbs.}$$

Taking moments about $R_2$

$865 \times 23 - 96 \times 199 - R_4 \times 46 = 0$

$46\,R_4 = 19{,}895 - 19{,}104 = 791$

$R_4 = 17$ lbs.

Check: $752 + 17 = 769$

Fig. 166 is the stress diagram for the rear portion and Fig. 167 for the front portion of the fuselage due to this loading. The panel loads are divided between the upper and lower ends of the struts, as shown in the diagrams and the weight of the chassis is subtracted from both the panel loads and the chassis strut reactions.

164. *Summary of Stresses in Fuselage*—Table XXV shows the stresses in the fuselage due to the loads assumed. For the nomenclature of the members see Fig. 168. The column in Table XXV headed "Section," refers to the types of sections shown in Fig. 169. Where no dimension is shown in Fig. 169, it is indicated in the table.

It should be noted that the stresses given in the above table for the horizontal members 4—4, etc., are computed from a load to which a load factor of 5 has been applied, while the stresses in the other mem-

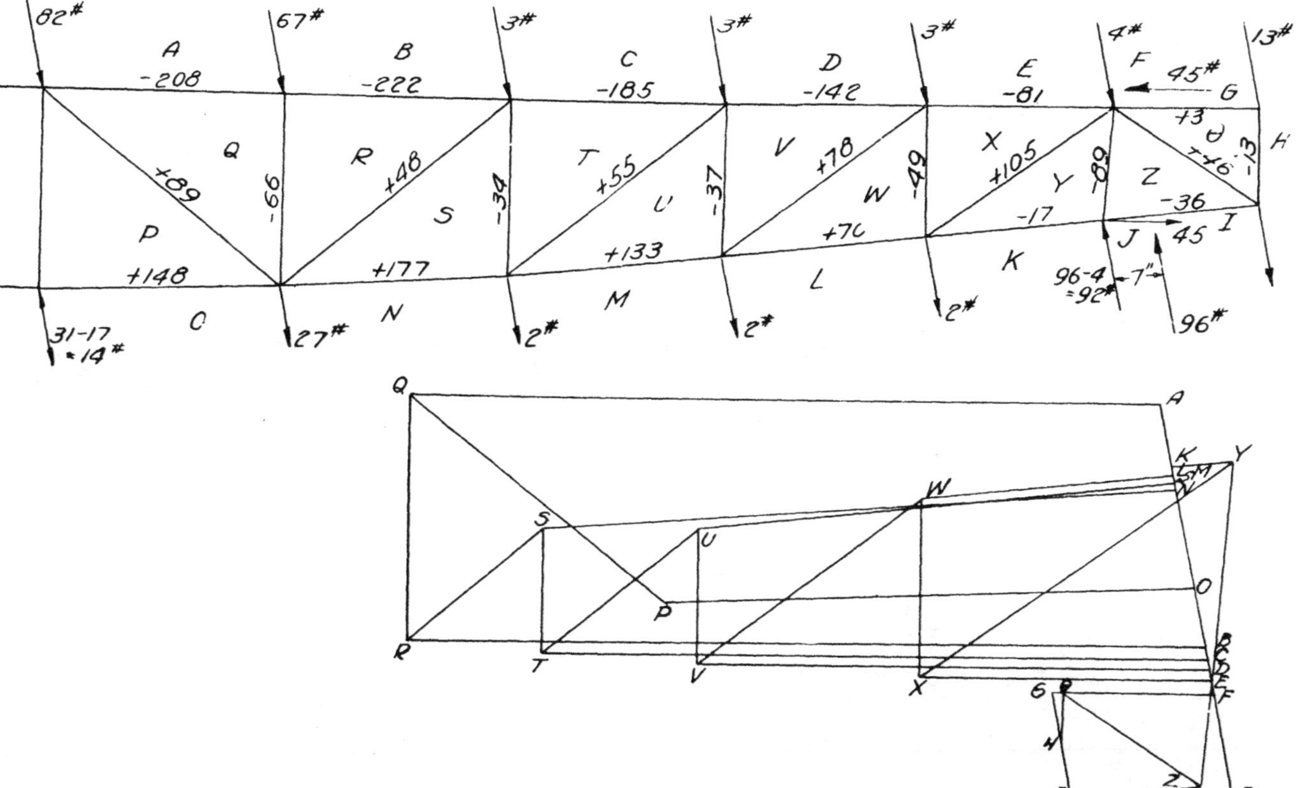


Fig. 166. Stress Diagram for Rear Portion of Fuselage in Three Point Landing Condition

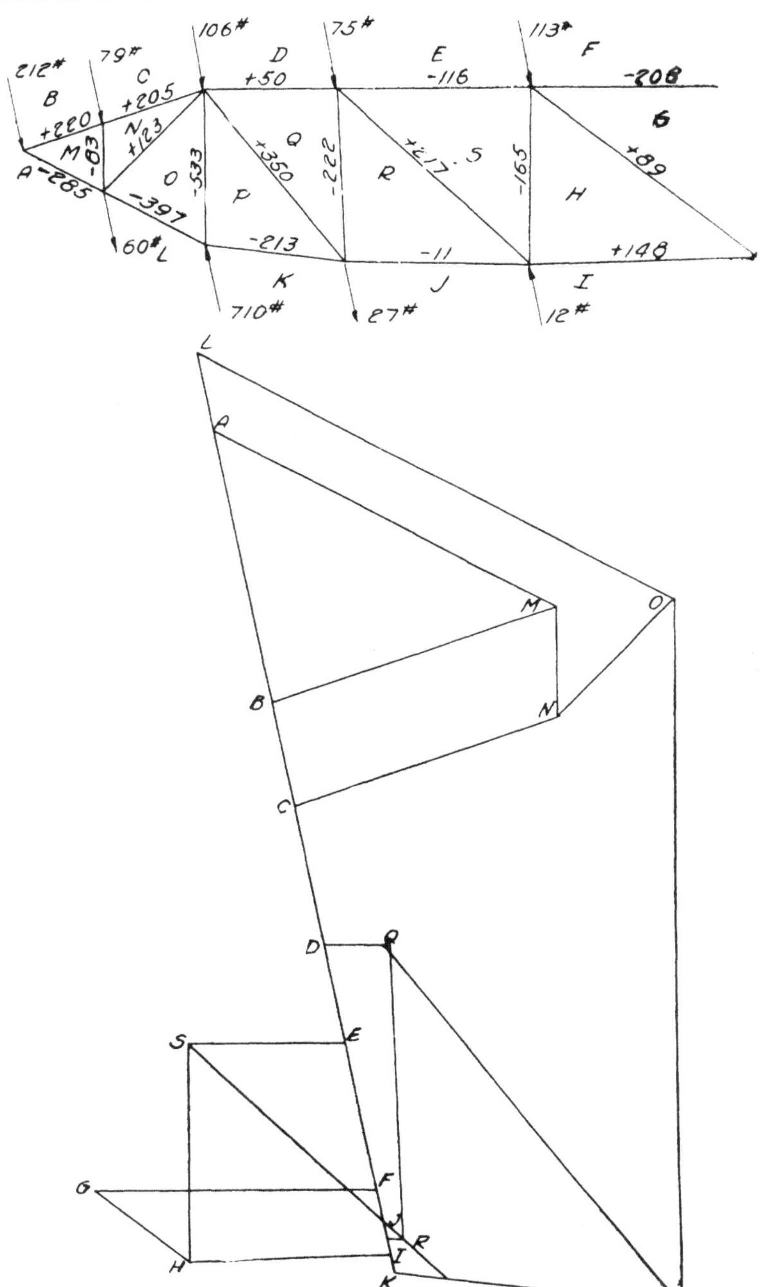


Fig. 167. Stress Diagram for Front Portion of Fuselage in Three Point Landing Condition

## TABLE XXV

### STRESSES AND PROPERTIES OF SECTIONS

| Member | Flying Conditions | Landing Tail Up | Three Point Landing | Material | Section | Dimensions | Area sq. in. | Length | I | $\rho$ | $L/\rho$ |
|---|---|---|---|---|---|---|---|---|---|---|---|
| 0—1 | +325 | +252 | +220 | Steel | E | 1-1/4"—0.035g | 0.233 | 7" | 0.0412 | 0.420 | 16.7 |
| 1—2 | +325 | +252 | +205 | Steel | E | 1-1/4"—0.035g | 0.233 | 18" | 0.0412 | 0.420 | 42.9 |
| 2—3 | +309 | +150 | + 50 | Ash | B | 1-1/2 x 1-1/8 | 1.436 | 19" | 0.1254 | 0.315 | 64.4 |
| 3—4 | +694 | + 65 | —116 | Ash | B | 1-1/2 x 1-1/8 | 1.436 | 27" | 0.1254 | 0.315 | 91.5 |
| 4—5 | +682 | — 53 | —208 | Ash | B | 1-1/2 x 1-1/8 | 1.436 | 33" | 0.1254 | 0.315 | 112.0 |
| 5—6 | +567 | — 58 | —222 | Spruce | A | 1-1/16 x 1-1/16 | 1.130 | 31" | 0.1068 | 0.308 | 100.5 |
| 6—7 | +438 | — 58 | —185 | Spruce | A | 1 x 1 | 1.000 | 31" | 0.0833 | 0.289 | 107.3 |
| 7—8 | +300 | — 58 | —142 | Spruce | A | 15/16 x 15/16 | 0.880 | 29" | 0.0645 | 0.270 | 107.3 |
| 8—9 | +163 | — 49 | — 81 | Spruce | A | 7/8 x 7/8 | 0.766 | 26" | 0.0488 | 0.252 | 103.1 |
| 9—10 | 0 | — 30 | + 3 | Spruce | A | 7/8 x 7/8 | 0.766 | 21" | 0.0488 | 0.252 | 83.4 |
| 0—11 | —272 | —250 | —285 | Steel | E | 1-1/4"—0.035g | 0.233 | 7" | 0.0412 | 0.420 | 16.7 |
| 11—12 | —386 | —342 | —397 | Steel | E | 1-1/4"—0.035g | 0.233 | 18" | 0.0412 | 0.420 | 42.7 |
| 12—13 | —691 | —310 | —213 | Ash | B | 1-1/2 x 1-1/8 | 1.436 | 19" | 0.1254 | 0.315 | 64.4 |
| 13—14 | —935 | —160 | — 11 | Ash | B | 1-1/2 x 1-1/8 | 1.436 | 27" | 0.1254 | 0.315 | 91.5 |
| 14—15 | —945 | — 67 | +148 | Ash | B | 1-1/2 x 1-1/8 | 1.436 | 33" | 0.1254 | 0.315 | 112.0 |
| 15—16 | —683 | + 51 | +177 | Spruce | A | 1-1/16 x 1-1/16 | 1.130 | 31" | 0.1068 | 0.308 | 100.5 |
| 16—17 | —560 | + 58 | +133 | Spruce | A | 1 x 1 | 1.000 | 31" | 0.0833 | 0.289 | 107.3 |
| 17—18 | —438 | + 49 | + 70 | Spruce | A | 15/16 x 15/16 | 0.880 | 29" | 0.0645 | 0.271 | 107.1 |
| 18—19 | —300 | + 29 | — 17 | Spruce | A | 7/8 x 7/8 | 0.766 | 24" | 0.0488 | 0.252 | 95.4 |
| 19—20 | —155 | 0 | — 36 | Spruce | A | 7/8 x 7/8 | 0.766 | 23" | 0.0488 | 0.252 | 91.3 |
| 1—11 | —166 | —139 | — 83 | | Bracket | Bracket | | | | | |
| 2—12 | —322 | —560 | —533 | Ash | D | 1-3/4 x 1-3/8 | 1.848 | 25" | 0.2287 | 0.352 | 71.0 |
| 3—13 | —600 | —190 | —222 | Ash | D | 1-3/4 x 1-1/8 | 1.766 | 26" | 0.1275 | 0.268 | 97.0 |
| 4—14 | —322 | —218 | —165 | Ash | A | 1-3/4 x 1-3/8 | 2.406 | 26" | 0.3790 | 0.397 | 65.5 |

*TABLE XXV (Continued)*

| Member | Flying Conditions | Landing Tail Up | Three Point Landing | Material | Section | Dimensions | Area sq. in. | Length | I | $\rho$ | L/$\rho$ |
|---|---|---|---|---|---|---|---|---|---|---|---|
| 5—15 | —183 | — 99 | — 66 | Spruce | C | 1-1/8 x 1-1/8 | 0.9947 | 25″ | 0.0983 | 0.314 | 79.4 |
| 6—16 | — 86 | — 5 | — 34 | Spruce | C | 1-1/16 x 1-1/16 | 0.8580 | 23″ | 0.0751 | 0.298 | 77.2 |
| 7—17 | — 95 | — 5 | — 37 | Spruce | C | 1 x 1 | 0.7291 | 20″ | 0.0561 | 0.278 | 72.0 |
| 8—18 | — 98 | — 11 | — 49 | Spruce | C | 15/16 x 15/16 | 0.6080 | 18″ | 0.0408 | 0.260 | 69.3 |
| 9—19 | —104 | — 20 | — 89 | Spruce | C | 15/16 x 15/16 | 0.6080 | 16″ | 0.0408 | 0.260 | 61.5 |
| 10—20 | —110 | — 52 | — 13 | Steel | F | 7/8″—0.035g | 0.0924 | 13″ | 0.00816 | 0.297 | 44.4 |
| 2—11 | +155 | +130 | +123 | 3400 lb. Wire | | | | | | | |
| 3—12 | +610 | +260 | +350 | 3800 lb. Wire | | | | | | | |
| 4—13 | +245 | +128 | +217 | 3800 lb. Wire | | | | | | | |
| 4—15 | +325 | +150 | + 89 | 3800 lb. Wire | | | | | | | |
| 5—16 | +153 | + 8 | + 48 | 1900 lb. Wire | | | | | | | |
| 6—17 | +146 | 0 | + 55 | 1900 lb. Wire | | | | | | | |
| 7—18 | +161 | + 10 | + 78 | 1900 lb. Wire | | | | | | | |
| 8—19 | +165 | + 25 | +105 | 1900 lb. Wire | | | | | | | |
| 9—20 | +183 | + 38 | + 46 | 1900 lb. Wire | | | | | | | |
| | | | | Horizontal Truss Members | | | | | | | |
| 4—4 | —567 | | | Spruce | C | 1-1/8 x 1-1/8 | 0.9947 | 30″ | 0.0983 | 0.314 | 95.5 |
| 5—5 | —480 | | | Spruce | C | 1-1/8 x 1-1/8 | 0.9947 | 29″ | 0.0983 | 0.314 | 92.4 |
| 6—6 | —415 | | | Spruce | C | 1-1/16 x 1-1/16 | 0.8580 | 27″ | 0.0751 | 0.298 | 90.6 |
| 7—7 | —360 | | | Spruce | C | 1 x1 | 0.7291 | 24″ | 0.0561 | 0.278 | 86.4 |
| 8—8 | —244 | | | Spruce | C | 15/16 x 15/16 | 0.6080 | 19″ | 0.0408 | 0.260 | 73.0 |
| 9—9 | —131 | | | Steel Tube | | 3/4″—0.047g | 0.1035 | 11″ | 0.0064 | 0.249 | 44.2 |
| | | | | Veneer Frame | | | | | | | |
| 4—5 | +815 | | | 1900 lb. Wire | | | | | | | |
| 5—6 | +642 | | | 1900 lb. Wire | | | | | | | |
| 6—7 | +590 | | | 1900 lb. Wire | | | | | | | |
| 7—8 | +535 | | | 1900 lb. Wire | | | | | | | |
| 8—9 | +360 | | | | | | | | | | |

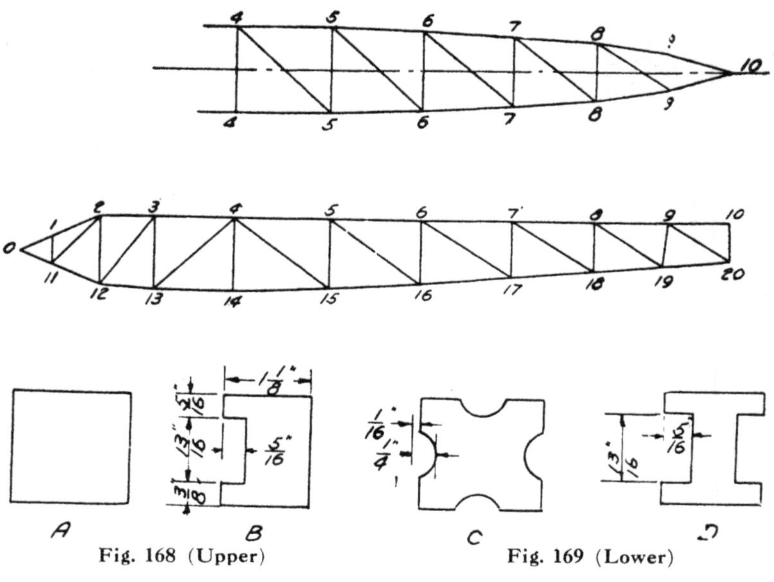


Fig. 168 (Upper) Fig. 169 (Lower)

bers must be multiplied by the load factor to give the ultimate stress. (See Art. 161.)

The table shows the material and dimensions of each member as shown in the figures. Using the stresses obtained above it is possible to compute the factors of safety of the different members. On design work the ultimate stresses are figured and the members designed to carry those stresses.

165. *Computation of Factors of Safety*—The following values have been assumed for the properties of the materials used.

| | E Lbs. per sq in. | Ultimate Tensile Strength | Ultimate Compressive Strength |
|---|---|---|---|
| Steel | 28,000,000 | 55,000 | 36,000 |
| Spruce | 1,600,000 | 12,000 | 5,500 |
| Ash | 1,600,000 | 16,000 | 7,800 |

The ultimate compressive strength in columns is found from the formulas $P/A = \frac{f^2}{8\pi^2 E}\left[\frac{L}{\rho}\right]^2$ where $P/A$ is greater than $f/2$, and $P/A = \frac{2\pi^2 E}{(L/\rho)^2}$ where it is less than $f/2$. The constants in these formulas are those for a column with restrained ends, the degree of restraint being represented by the coefficient 2, where the coefficient for a pin-ended column would be 1 and a fixed ended column 4.

Table XXVI shows the ultimate strength of the various members in the vertical trusses and also the factors of safety under the loads figured above.

*TABLE XXVI*

| Member | Ultimate Strength | | Flying Condition | | Three Point Landing | | Landing Tail Up | |
|---|---|---|---|---|---|---|---|---|
| | T | C | Stress | F.S. | Stress | F.S. | Stress | F.S. |
| 0—1 | 12,800 | | +325 | 39.4 | | | | |
| 1—2 | 12,800 | | +325 | 39.4 | | | | |
| 2—3 | 17,200 | 8,340 | +309 | 55.7 | | | | |
| 3—4 | 17,200 | 5,400 | +694 | 24.8 | —116 | 46.5 | | |
| 4—5 | 17,200 | 3,600 | +682 | 25.2 | —208 | 17.3 | | |
| 5—6 | 13,500 | 3,560 | +567 | 23.8 | —222 | 16.0 | | |
| 6—7 | 12,000 | 2,750 | +438 | 27.4 | —185 | 14.9 | | |
| 7—8 | 10,500 | 2,420 | +300 | 35.0 | —142 | 17.0 | | |
| 8—9 | 9,200 | 2,270 | +163 | 56.5 | — 81 | 28.0 | | |
| 9—10 | | 2,960 | | | | | — 30 | 98.6 |
| 0—11 | | 8,300 | | | —285 | 29.2 | | |
| 11—12 | | 8,100 | | | —397 | 20.4 | | |
| 12—13 | | 8,340 | —691 | 12.1 | | | | |
| 13—14 | | 5,400 | —935 | 5.8 | | | | |
| 14—15 | | 3,600 | —945 | 3.81 | | | | |
| 15—16 | | 3,560 | —683 | 5.2 | | | | |
| 16—17 | | 2,750 | —560 | 4.91 | | | | |
| 17—18 | | 2,420 | —438 | 5.5 | | | | |
| 18—19 | | 2,550 | —300 | 8.5 | | | | |
| 19—20 | | 2,700 | —155 | 17.4 | | | | |
| 2—12 | | 8,030 | | | | | —560 | 14.3 |
| 3—13 | | 3,850 | —600 | 6.4 | | | | |
| 4—14 | | 10,600 | —322 | 33.0 | | | | |
| 5—15 | | 3,980 | —183 | 21.8 | | | | |
| 6—16 | | 3,560 | — 86 | 41.4 | | | | |
| 7—17 | | 3,140 | — 95 | 33.0 | | | | |
| 8—18 | | 2,680 | — 98 | 27.3 | | | | |
| 9—19 | | 2,830 | —104 | 27.2 | | | | |
| 10—20 | | 3,200 | —110 | 29.0 | | | | |
| 2—11 | 3,400 | | +155 | 22.0 | | | | |
| 3—12 | 3,800 | | +610 | 6.2 | | | | |
| 4—13 | 3,800 | | +245 | 15.5 | | | | |
| 4—15 | 3,800 | | +325 | 11.7 | | | | |
| 5—16 | 1,900 | | +153 | 12.4 | | | | |
| 6—17 | 1,900 | | +146 | 13.0 | | | | |
| 7—18 | 1,900 | | +161 | 11.8 | | | | |
| 8—19 | 1,900 | | +165 | 11.5 | | | | |
| 9—20 | 1,900 | | +183 | 10.4 | | | | |

Table XXVII shows the ultimate strengths of the various web members in the upper horizontal truss and their factors of safety. The stresses shown are ultimate stresses based on a load factor of 5 so the value given as the factor of safety is equal to five times the ultimate strength divided by the stress due to the loads used.

*TABLE XXVII*

| Member | Stress | Ultimate Allowable | F. S. |
|---|---|---|---|
| 4—4 | —567 | 3310 lbs. | 29.0 |
| 5—5 | —480 | 3470 lbs. | 36.0 |
| 6—6 | —415 | 3050 lbs. | 36.5 |
| 7—7 | —360 | 2730 lbs. | 38.0 |
| 8—8 | —244 | 2620 lbs. | 53.5 |
| 9—9 | —131 | 2580 lbs. | 98.5 |
| 4—5 | +815 | Veneer Frame | |
| 5—6 | +642 | 1900 lbs. | 14.8 |
| 6—7 | +590 | 1900 lbs. | 16.1 |
| 7—8 | +535 | 1900 lbs. | 17.7 |
| 8—9 | +360 | 1900 lbs. | 26.4 |

166. *Discussion of the factors of Safety Computed*—The ultimate strengths given in the tables in the last article are all computed on the gross sections of the members. For tension members, however, the ultimate strength should be computed on the net sections. Inspection of the factors of safety, however, shows that none of the tension members, except some wires, are critically stressed. In the case of wires, however, the ultimate strength was figured on the net area. It is not worth while in this case, therefore, to compute the net area of the tension members and correct the values of the ultimate strength and the factor of safety.

Air Service Specifications call for an allowable load factor of 5 for this type of airplane. Although the formulas proposed to compute the probable values of the dynamic loads on an airplane vary considerably, experience has shown that an airplane of this type designed for a factor of 5 is safe. Inspection of the tables of factors of safety reveals two members in which the factor is less than 5. In member 14—15 the factor is 3.81, and in member 16—17 it is 4.91. These factors raise the question of revision of the size of these members. They do not necessarily indicate that the airplane is unsafe, as there are several features of the design which tend to make the strength of the fuselage greater than indicated by the figures above. There are a number of redundant members which increase the rigidity of the framework. The degree of fixity may be greater than that assumed. The degree assumed was that corresponding to a value of $C = 2$ in the column formulas (See Art. 24). The air load on the tail will probably never reach the assumed value of 25 lbs. per sq. ft., and if it should the entire airplane would be moving in a curved path, which would induce inertia loads in the fuselage which would decrease the stress in the lower longeron. If the degree of fixity of the longerons is such that $C=2.25$ in the member 16—17, that member will show a factor of safety of 5.45. To raise the safety factor of member 14—15 to 5.0 the degree of fixity of the member will have to be raised so that $C=2.63$. A value of 2.25 for $C$ in member 16—17 is reasonable and no fear needs be felt for the safety of the member. A value of $C=2.63$ for member 14—15, however, is quite high, 2.5 being considered the maximum value for the longerons in this type of construction. It would seem advisable to revise the size of this member. If the routing were omitted the member would have an ultimate strength of 5160 lbs. and a factor of safety of 5160/970 = 5.46. Although member 14—15 appears unsafe from this analysis, the Vought VE-7 has been built and flown very successfully, and no trouble has been experienced with this member. This may be partially accounted for by the considerations discussed above, which may have resulted in the member having an actual fixity coefficient of 2.63.

167. *Veneer Covered Truss Type Fuselages*—A modern development of the truss type of fuselage is the use of plywood instead of cloth for covering the fuselage and the use of wood diagonals instead of wires. When this is done only one diagonal is placed in each bay in small fuselages, though in large airplanes a second diagonal is often needed to stiffen the plywood. The principal advantage of this type of construction from a structural point of view is the greater degree of fixity of the compression members. It can be assumed that the joint between a member and the plywood prevents the member from slipping along the plywood. The longerons, therefore, if they have plywood on two sides at right angles to each other, may be considered fixed in both directions along their entire length and may be figured to carry the ultimate compressive stress for the material, or 5500 lbs. per sq. in. for spruce. The members which have plywood on one side only may be considered as fixed along their entire length in one direction, and as columns partially fixed at the ends in the other. The column constant to be used in this case will vary between 2 and 3, depending on the details of the construction. Owing to the character of the restraint on the members, it is advisable to make the longerons with plywood on both sides square, and the other members rectangular with the narrow face in con-

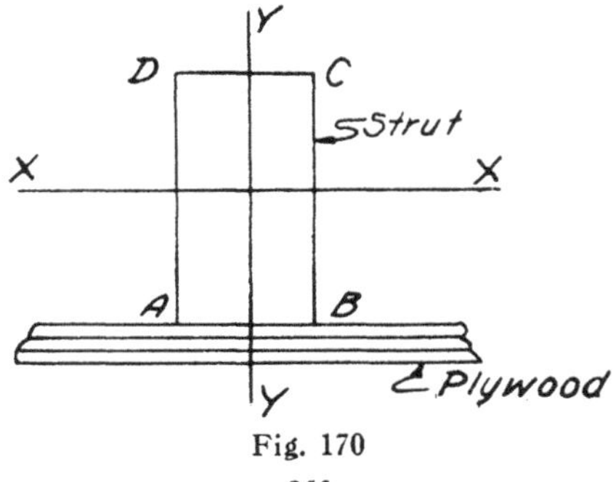


Fig. 170

tact with the plywood. Consideration of Fig. 170 will show why this is the case.

In Fig. 170, *ABCD* is a strut glued and screwed to the plywood along the face *AB*. The strength of the joint will prevent slipping, so the strut will not deflect in the plane *XX*. The plywood, however, offers little resistance to deflection in the *YY* plane owing to its flexibility, and that is the plane in which deflection will occur. It is not important, therefore, to have a large radius of gyration about the *YY* axis, though it is necessary to have it about the *XX* axis. For a given area the greatest value of $\rho$ about *XX* is obtained with a rectangle in which *AD* is greater than *AB*. If this were the only consideration a very thin strip with the edge against the plywood would be the ideal shape for the strut, but the danger of local failure and the necessity of having glue joints large enough to fix the strut in the plane parallel to the plywood, limit the ratio of length to breadth of the section. A ratio of about 2 to 1 for small struts and about 2.5 to 1 for large ones is satisfactory.

168. *Semi-Monocoque Type*—The semi-monocoque fuselages differ from fuselages of the truss type in that the loads are not applied to the fuselage at panel points, and the loads and their points of application may be taken directly from the balance diagram after suitable provision has been made for the distribution of the weight of the fuselage, engine, and any other loads acting over a considerable length of the fuselage. It is best to assume these loads as distributed loads, either uniform or uniformly varying, rather than as concentrated loads.

In semi-monocoque fuselages the bulkheads are assumed to stiffen the fuselage so that failure will result more from bending as a whole than from local buckling. The solution for a fuselage of this type is indeterminate, but a good approximate analysis can be made. The chief assumption is that the fuselage will act as a beam. To find the strength at any section, therefore, the bending moment should be calculated as usual, and also the section modulus of the effective section. In very large fuselages where the ratio of plywood thickness to fuselage diameter is very small, the effective section consists of the longerons and part of the shell near them, the shell being relied on mainly to fix the longerons and insure their acting together. In construction with multiple longerons, they are placed more closely near the vertical axis of the cross section of the fuselage than near the horizontal axis, as the vertical loads are more severe than the horizontal. In the smaller fuselages the entire shell may be included in the effective area. The plywood layers in the shell make different angles with a line parallel to the axis of the fuselage and, on this account, their efficiency in carrying stress varies. It seems a reasonable assumption that the effective area of each ply be assumed equal to the cross sectional area of the ply multiplied by the cosine of the angle it makes, with a line parallel to the axis of the fuselage. If the plies were parallel and perpendicular to such a line the effective area would be the area of the parallel plies. If each ply made an angle of 30 degs. with such a line the effective area would be 0.866 times the area of the plies. The remarks in Art. 167 regarding the shape and degree of fixity of members apply also to the longerons in the semi-monocoque type of construction. However, owing to the curved section of the shell, where multiple longerons are used they may be assumed to carry the full compressive stress of 5500 lbs. per sq. in.

169. *Monocoque Type*—Where the shell is not braced by longerons and where the bulkheads are too widely spaced, failure will not occur as a result of simple bending. The tube will buckle on the compression side due to the excessive secondary stresses set up by the primary bending stresses. The action is similar to that which takes place in a long column under compression. There is this difference, however, that in the case of the long column the primary stresses are usually caused by the end compression and the secondary stresses by the moment resulting from the end loads acting with the side deflection of the column as an arm; whereas in the case of the monocoque fuselage there usually is little or no end compression due to external loads. The initial loadings cause deflection of the fuselage and put the fuselage under bending stresses. The secondary stresses caused by the moment of these initial stresses acting through the deflection cause ultimate failure.

The calculation of the stresses in a fuselage of the monocoque type is an indeterminate problem. The strength of a new fuselage may be predicted by comparison with standard fuselages of similar type. The best way to analyze this type of structure is to consider it a beam acting under bending forces, and to use a low value for the ultimate allowable stress, owing to the local buckling. Just what value should be used has not yet been determined, but static tests on full size fuselages of this type should show what values are correct. In the light of such tests as have already been made, a value of 2000 to 2500 lbs. per sq. in. for spruce or poplar plywood, and 3000 to 3500 lbs. per sq. in. for birch are recommended for the ultimate allowable compressive stresses in the plywood shells of both monocoque and semi-monocoque fuselages. However, where only narrow strips of the shell next the longerons are considered part of the effective section, they may be assumed to carry as high a stress as the longerons themselves.

170. *Prediction of High Speed at Ground*—For the purpose of making a stress analysis of an airplane it is necessary to determine its maximum speed at the ground. The chart shown in Fig. 171 affords an accurate and ready means for doing this. It is based on the principle that performance is dependent on the pounds per horsepower and pounds per square foot of an airplane, modified by the fineness or general cleanness of the design. What is known as the "fineness factor" of an airplane is a number proportional to the cube root of its total $L/D$ at any value of $K_y$. From actual official performance tests on a number of airplanes their fineness factors were determined. On the basis of the factors so obtained, the following ranges are recommended for the various general types of airplane.

| Type | Fineness Factor |
|---|---|
| Single-seater pursuit | 105 to 115 |
| Two-seater fighters and observation airplanes | 100 to 110 |
| Training, small mail or passenger airplanes | 100 to 110 |
| Day bombers, intermediate size mail or passenger airplanes with side engine nacelles | 93 to 96 |
| Large bombers or large commercial airplanes with engine nacelles | 88 to 93 |

To fall within the classes indicated the designs can in no case be poor. To illustrate this point the actual fineness factors for several well-known airplanes are given.

| | |
|---|---|
| Curtiss JN-4D-2 | 97 |
| Vought VE-7 | 106 |
| U.S. D-4 | 100 |
| U.S. D-9A | 100 |
| American SE-5 | 108 |
| Fokker D-7 | 103 |
| Le Pere Triplane | 92.5 |
| Martin Bomber (with bombs) | 93 |
| Orenco "D" | 108 |
| Thomas-Morse MB-3 | 113 |

The most important elements in determining the fineness of an airplane are the disposition of the wing surface (whether monoplane, biplane, or triplane), the number of bays in the wing cellule, the number of exposed wires (control cables, empennage bracing wires, etc.), whether streamline wires, faired or unfaired cables, number of engine nacelles, and, especially, the shape of the fuselage.

For a more detailed explanation of this method of predicting airplane performance attention is invited to McCook Field Report A.D.M. No. 489 (Serial 1221).

171. *Strength Factors for Wings, Fuselage and Tail Surfaces.*

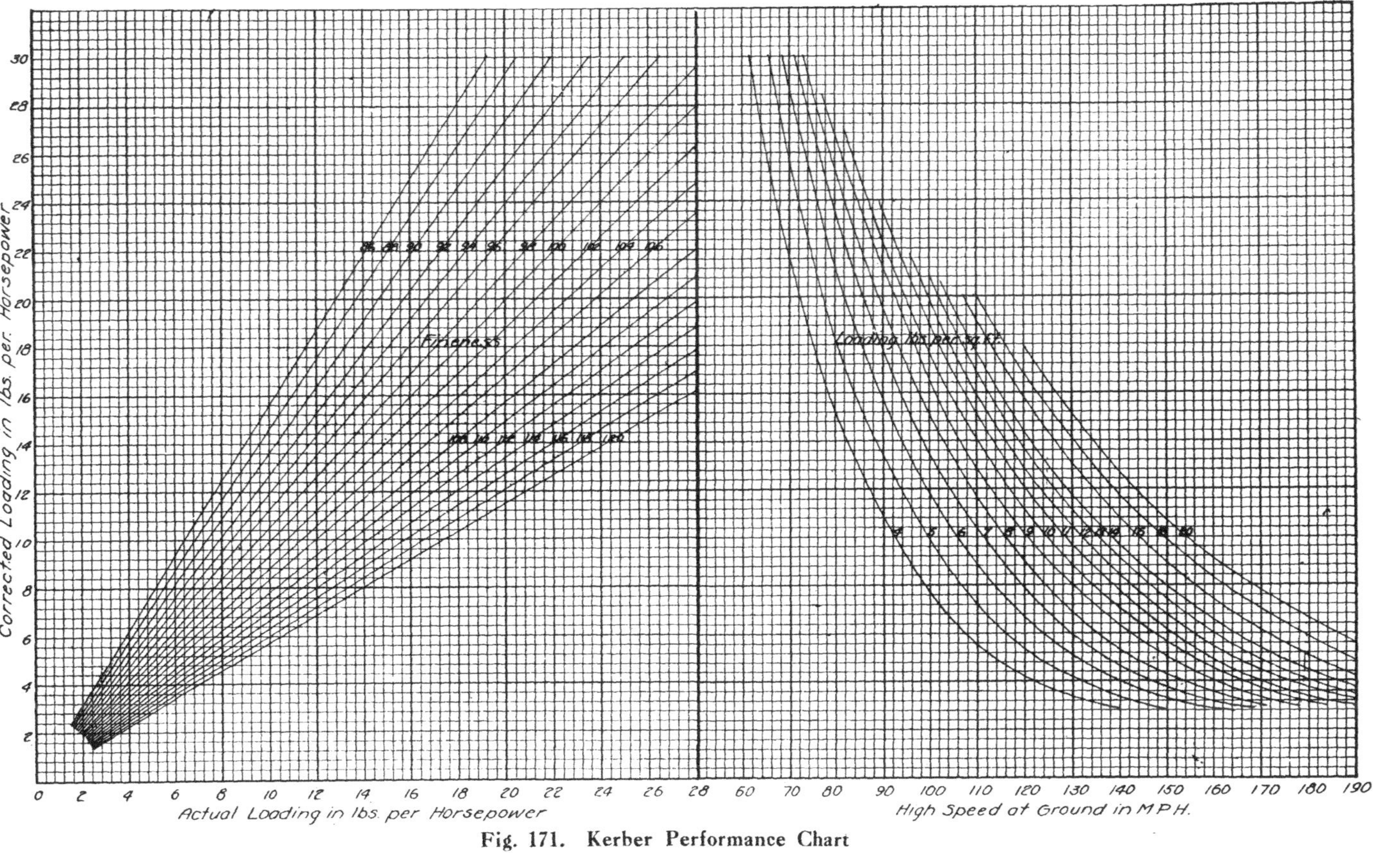


**Fig. 171. Kerber Performance Chart**

*TABLE XXVIII*

| Type | | Load Factors | | | | | |
|---|---|---|---|---|---|---|---|
| | | (1) | (2) | (3) | (4) | (5) | (6) |
| I. | Single Seater Pursuit for Day Work—Water-Cooled Engine | 8.5 | 5.5 | 3.5 | 35 | 30 | 7 |
| II. | Single Seater Pursuit for Night Work—Air or Water-Cooled Engine ............ | 7.5 | 5.0 | 3.5 | 30 | 25 | 6 |
| III. | Single Seater Pursuit for Day Work—Air-Cooled Engine .. | 8.5 | 5.5 | 3.5 | 35 | 30 | 7 |
| IV. | Single Seater Pursuit for Ground Work—Air or Water-Cooled Engine (armored).. | *7–3 | *4.5–2.5 | 3.0 | 30 | 25 | 6 |
| V. | Two-Seater Pursuit—Air or Water-Cooled Engine ...... | 7.5 | 5.0 | 3.5 | 30 | 25 | 6 |
| VI. | Three Seater for Ground Attack—Air or Water-Cooled Engine (armored) ......... | 7–3 | 4.5–2.5 | 3.0 | 25 | 20 | 5 |
| VII. | Two Seater Infantry Liaison —Air or Water-Cooled Engine (armored) ........... | 6–3 | 4.0–2.5 | 3.0 | 25 | 20 | 5 |
| VIII. | Two Seater Night Observation — Air or Water-Cooled Engine .................. | 6.5 | 4.5 | 3.0 | 25 | 20 | 5 |
| IX. | Three Seater Army and Coast Artillery Observation and Surveillance ............... | 6.5 | 4.5 | 3.0 | 25 | 20 | 5 |
| X. | Two Seater Corps Observation ...................... | 6.5 | 4.5 | 3.0 | 25 | 20 | 5 |
| XI. | Day Bombardment ........ | 5.5 | 3.5 | 2.5 | 25 | 20 | 5 |
| XII. | Night Bombardment, Short Distance .................. | 4.5 | 3.0 | 2.5 | 20 | 15 | 4 |
| XIII. | Night Bombardment, Long Distance .................. | 4.0 | 2.5 | 2.0 | 15 | 10 | 3 |
| XIV. | Training—Air-Cooled Engine (radial or rotary) ......... | 8.0 | 5.5 | 3.5 | 35 | 30 | 7 |
| XV. | Training—Water-Cooled Engine ...................... | 8.0 | 5.5 | 3.5 | 35 | 30 | 7 |

In all cases, both computations and static test are based on full military load.

Column 1 Gives the load factor for high incidence condition with center of pressure at its most forward position.

Column 2 Gives load factor for low incidence condition with center of pressure at location corresponding to maximum ground speed.

Column 3 Gives the required negative load factor to be carried by the main plane structure in static test, with center of pressure at .25 of chord.

Column 4 Gives the required average load per sq. ft. to be carried by the horizontal tail surfaces as determined by the static test.

Column 5 Gives the required average load per sq. ft. to be carried by the vertical tail surfaces as determined by static test.

Column 6 Gives the required load factor to be carried by the fuselage as determined by static test.

For Types 1, 2, 3, 4, 5, 14 and 15 for the condition of straight dive with down load on the front truss and up load on the rear truss a load factor of 1¾ is required for biplanes and triplanes in which there is adequate incidence bracing or its equivalent, and a factor of 3.0 where such bracing is not present as in monoplanes. For the other types of airplanes no computations need be made for this diving condition except to determine the drag bracing. In computations for this condition, incidence bracing is to be neglected.

*Note—Where split factors are given, as, for example, Type IV, Columns 1 and 2, the first factor is to be given by the structure as a whole; the second factor is to be given by the structure with any one structural member removed.

172. *Equations for Continuous Beams*—In this section are given three-moment equations and deflection formulas for continuous beams for all those conditions of loading which are ordinarily used in the stress analysis of airplanes. For the derivation of these equations and formulas reference can be made to the report issued by the Engineering Division of the Air Service entitled "Equations for Continuous Beams," Serial Number 759.

Great care must be used in the application of these formulas to give the proper algebraic signs to the moments and shears. Art. 2 explains the conventions which are followed in this book.

The nomenclature given below will be used throughout in all the formulas.

$m_1$ = the bending moment at support 1.

$m_2$ = the bending momen at support 2.

$m_3$ = the bending moment at support 3.

$s_2$ = the shearing force at a section adjacent to and at the right side of the support 2.

$s_{-2}$ = the shearing force at a section adjacent to and at the left side of the support 2.

$i_2$ = the slope at 2, the angle being measured between the tangent and the axis to the right of 2.

$i_{-2}$ = the slope at 2, the angle being measured between the tangent and the axis to the left of 2.

s = the shearing force, $m$ = the bending moment, $i$ = the slope and $v$ = the deflection at any cross section at a distance $x$ from 2.

$v_1$ = the difference in level between the supports 1 and 2.
$v_3$ = the difference in level between the supports 3 and 2.

For ordinary conditions all the supports are assumed to be on the same level so that all terms in the formulas involving $v_1$ and $v_3$ equal zero. When the supports are not on the same level the values of $v_1$ and $v_3$ are positive when the supports 1 and 3 are higher than the support 2, and negative when 1 and 3 are lower than 2.

All these formulas are derived for upward vertical loads which are considered as positive. In the case of downward loads it is only necessary to change the algebraic sign of the loads $W_1$, $W_2$, $w_1$ or $w_2$, as the case may be.

In the ordinary stress analysis of an airplane structure the only deflections required are those within each span above or below the line passing through the supports at the ends of the span. Cases 1A—5A give formulas for such deflections. Their derivation is similar to the corresponding cases with two spans. Deflections are positive when upward and negative when downward. The deflection which is required in an analysis is that occurring at the point of maximum moment in the span, where the stresses are a maximum. The point of maximum moment is the point of zero shear (considering all the loads and moments) and can be easily found from the shears. For combinations of loads the deflections for the different cases should be added together to obtain the final deflection.

The reactions at the supports are found by adding the shears on either side of the support, taking account of the algebraic signs given for continuous beams. A positive result indicates an upward reaction and a negative one the reverse.

Case 1—Concentrated Loads.

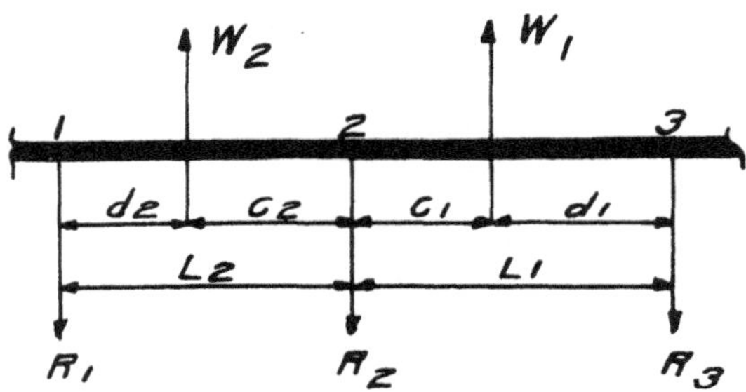


Origin at Support 2.

$$m_1L_2+2m_2(L_1+L_2)+m_3L_1=\frac{W_2\,d_2}{L_2}(L_2^2-d_2^2)$$

$$+\frac{W_1\,d_1}{L_1}(L_1^2-d_1^2)+\frac{6EIv_3}{L_1}+\frac{6EIv_1}{L_2}$$

Case 1A—Deflection in One Span for Concentrated Loads.

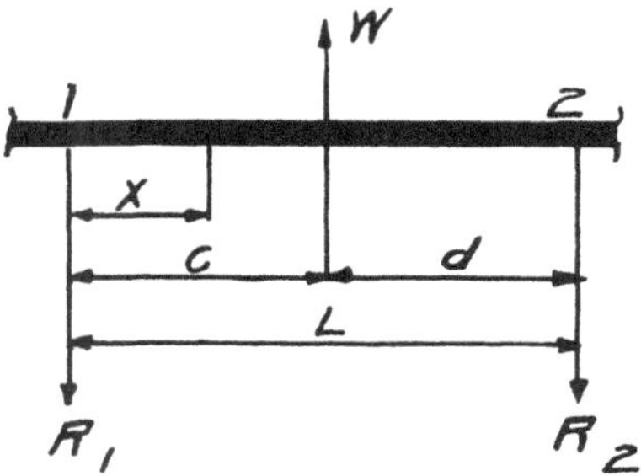


Origin at Support 1.

Deflection for Part of Beam between x = 0 and C.

$$EIv = \frac{m_1 x^2}{2} + \frac{s_1 x^3}{6} - \left( \frac{m_1 L}{2} + \frac{s_1 L^2}{6} + \frac{W d^3}{6L} \right) x$$

Deflection for part of beam between x = C and L.

$$EIv = \frac{m_1 x^2}{2} + \frac{s_1 x^3}{6} + \frac{W(x—C)^3}{6} - \left( \frac{m_1 L}{2} + \frac{s_1 L^2}{6} + \frac{W d^3}{6L} \right)$$

Case 2—Uniform Loads.

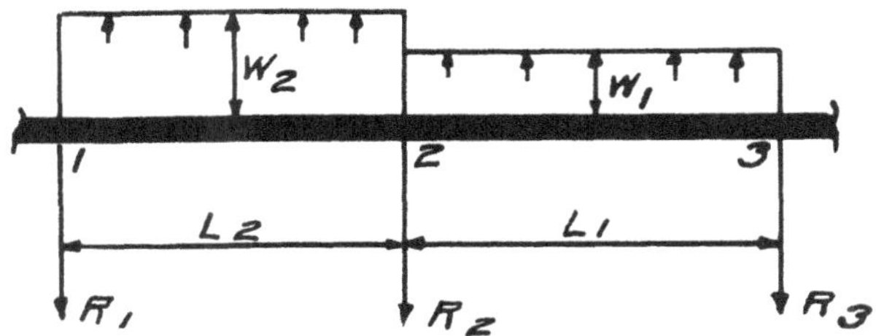


Origin at Support 2.

$$m_1 L_2 + 2m_2(L_1 + L_2) + m_3 L_2 = \frac{w_2 L_2^3}{4} + \frac{w_1 L_1^3}{4} + \frac{6EIv_3}{L_1} + \frac{6EIv_1}{L_2}$$

Case 2A—Deflection in One Span for Uniform Loads.

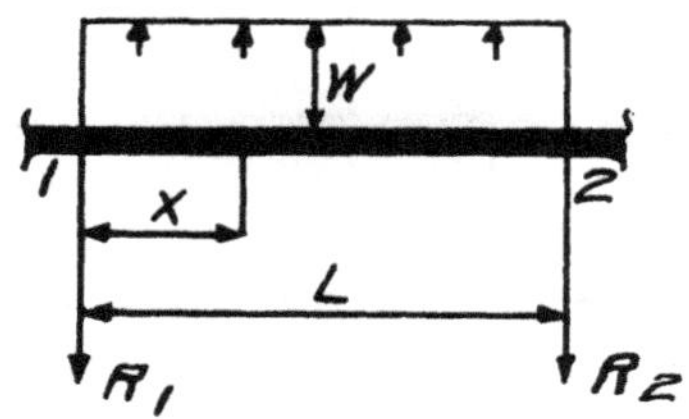


Origin at Support I.

$$EIv = x(x - L)\left[\frac{m_1}{2} + \frac{s_1}{6}(x+L) + \frac{w}{24}(x^2+xL+L^2)\right]$$

Case 3—Uniformly Varying Loads.

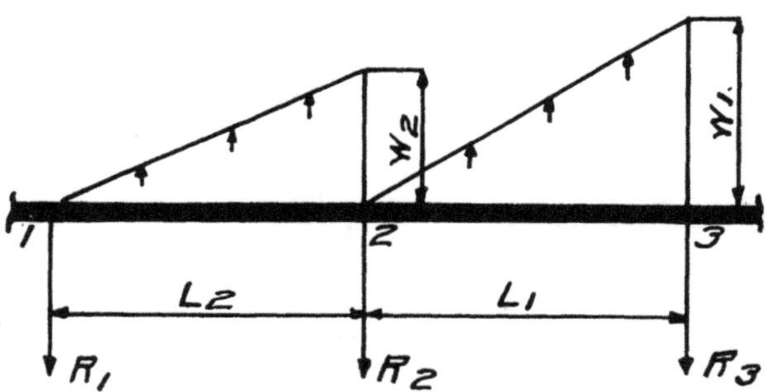


Origin at Support 2.

$$m_1L_2 + 2m_2(L_1+L_2) + m_3L_1 = \frac{7w_1L_1^3}{60} + \frac{2w_2L_2^3}{15} + \frac{6EIv_3}{L_1} + \frac{6EIv_1}{L_2}$$

Case 3A—Deflection in One Span for Uniformly Varying Loads.

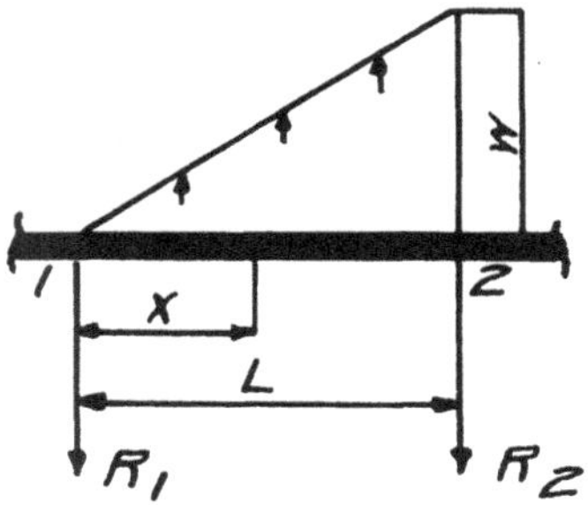


Origin at Support 1.

$$EIv = x(x-L)\left(\frac{m_1}{2}+\frac{s_1(x+L)}{6}+\frac{w(x+L)\,(x^2+L^2)}{120L}\right)$$

$$s_1 = \frac{m_2-m_1}{L}-\frac{wL}{6}$$

Case 3B—Deflection in One Span for Uniformly Varying Loads.

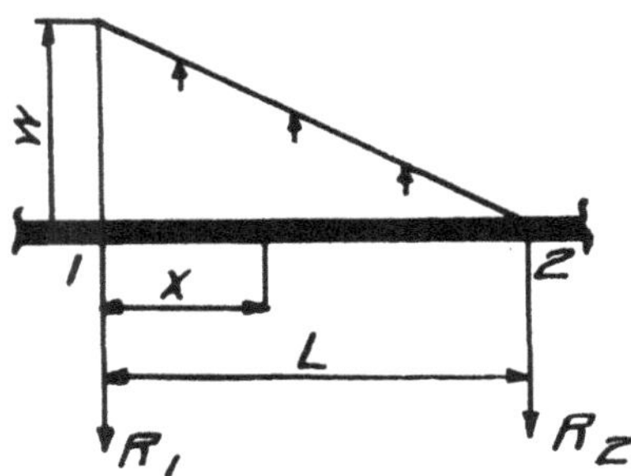


Origin at Support. 1.

$$EIv=x(x-L)\left(\frac{m_1}{2}+\frac{s_1}{6}(x+L)\right)+\frac{wx}{120L}\times(5x^3L-x^4-4L^4)$$

$$s_1 = \frac{m_2-m_1}{L}-\frac{wL}{3}$$

Case 4—External Moments Within Spans.

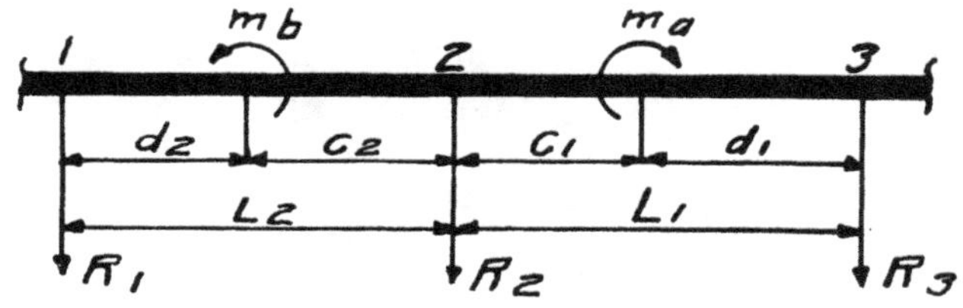


Origin at Support 2.

$$m_1L_2+2m_2(L_1+L_2)+m_3L_1 = m_a\left(6C_1-\frac{3C_1^2}{L_1}-2L_1\right)$$

$$+m_b\left(6C_2-\frac{3C_2^2}{L_2}-2L_2\right)+\frac{6EIv_3}{L_1}+\frac{6EIv_1}{L_2}$$

The algebraic signs of the moments $m_a$ and $m_b$ may be determined by the following rules:

(1) Moments in the span to the right of the origin are positive when clockwise and negative when counter-clockwise.

(2) Moments in the span to the left of the origin are positive when counter-clockwise and negative when clockwise.

Case 4A—Deflection in One Span for External Moments Within Span.

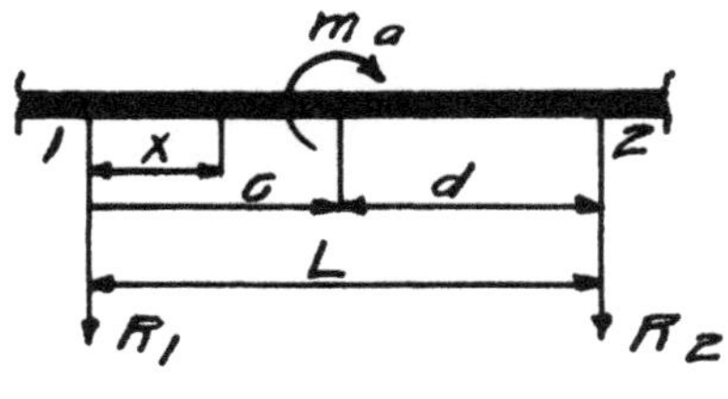

Origin at Support 1.

Deflection for Part of Beam between x = 0 and C.

$$EIv = x(x-L)\left[\frac{m_1}{2}+\frac{s_1(x+L)}{6}\right]-\frac{m_a d^2 x}{2L}$$

Deflection for Part of Beam between x = C and L.

$$EIv = x(x-L)\left[\frac{m_1}{2}+\frac{s_1(x+L)}{6}\right]+\frac{m_a}{2}\left[(x-C)^2-\frac{d^2x}{L}\right]$$

Case 5—External Moments at Supports.

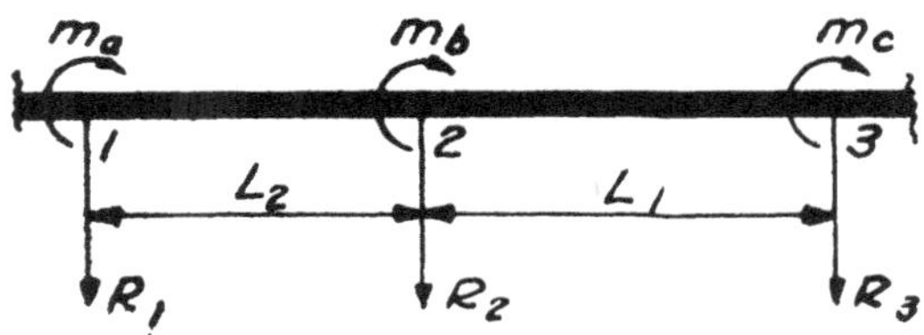


Origin at Support 2.

$$m_1L_2+2m_2(L_1+L_2)+m_3L_1 = -m_aL_2-2m_bL_1+\frac{6EIv_3}{L_1}+\frac{6EIv_1}{L_2}$$

The moments $m_1$ and $m_2$ are positive when clockwise and negative when counter-clockwise.

The derivation of this formula is given below.
$m_1$, $m_2$ and $m_3$ are the moments to the left of supports 1, 2 and 3 respectively.

Case 5A—Deflection in One Span for External Moments at Supports.

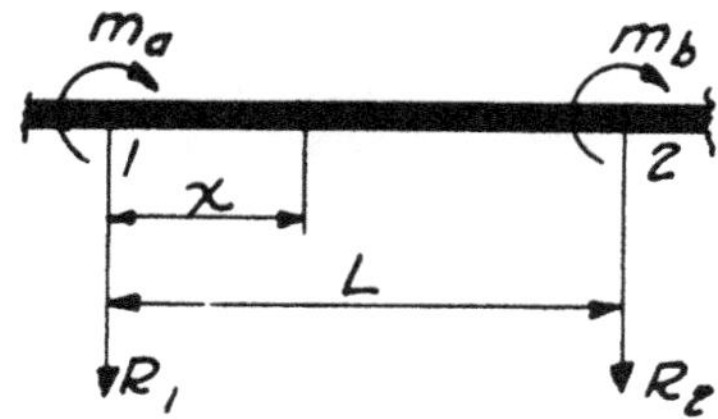


Origin at Support 1.

$$EIv = \frac{x\,(x-L)}{6\,L}\left[(m_1+m_a)\ (2L-x)+m_2(L+x)\right]$$

$m_1$ and $m_2$ are the moments calculated from Case 5.

Derivation of Three-Moment Equation for Case 3.

Consider Span 2—3.

(1) $$s = s_2 + \frac{w_1x^2}{2\,L_1}$$

(2) $$m = m_2 + s_2x + \frac{w_1x^3}{6\,L_1}$$

(3) $$EIi = m_2x + \frac{s_2x^2}{2} + \frac{w_1x^4}{24L_1} + k.$$ When $x=0$, $k=EIi_2$.

(4) $$EIv = \frac{m_2x^2}{2} + \frac{s_2x^3}{6} + \frac{w_1x^5}{120L_1} + EIi_2\,x + k_1.$$ When $x=0$, $v=0$ and $k_1 = 0$.

Let $x = L_1$ in equation (2) and solve for $s_2$, noting that $m$ becomes $m_3$.

(5) $$s_2 = \frac{m_3-m_2}{L_1} - \frac{w_1L_1}{6}$$

Let $x=L_1$ in equation (4) using the value of $s_2$ from equation (5) and solve for $EIi_2$.

(6) $EIi_2 = \frac{EIv_3}{L_1} - \frac{m_2L_1}{3} - \frac{m_3L_1}{6} + \frac{7w_1L_1^3}{360}$

Consider Span 1—2.

(7) $m = m_2 + s_{-2}x + \frac{w_2x^2}{2} - \frac{w_2x^3}{6L_2}$

(8) $EIi = m_2x + \frac{S_{-2}x^2}{2} + \frac{w_2x^3}{6} - \frac{w_2x^4}{24L_2} + k_2$. When $x = 0$, $k_2 = EIi_{-2}$.

(9) $EIv = \frac{m_2x^2}{2} + \frac{s_{-2}x^3}{6} + \frac{w_2x^4}{24} - \frac{w_2x^5}{120L_2} + EIi_{-2}x + k_3$. When $x = 0$, $v = 0$ and $k_3 = 0$.

Let $x = L_2$ in equation (7) and solve for $s_{-2}$.

(10) $s_{-2} = \frac{m_1 - m_2}{L_2} - \frac{w_2L_2}{3}$

Let $x = L_2$ in equatiòn (9) using the value for $s_{-2}$ from equation (10) and solve for $EIi_{-2}$.

(11) $EIi_{-2} = \frac{EIv_1}{L_2} - \frac{m_2L_2}{3} - \frac{m_1L_2}{6} + \frac{8w_2L_2^3}{360}$

Since the slopes given by equations (6) and (11) are of opposite signs, $EIi_2 + EIi_{-2} = 0$.

Adding these equations and transposing

(12) $m_1L_2 + 2m_2 \times (L_1 + L_2) + m_3L_1 = \frac{7w_1L_1^3}{60} + \frac{2w_2L_2^3}{15} + \frac{6EIv_3}{L_1} + \frac{6EIv_1}{L_2}$

Derivation of Three-Moment Equation for Case 4.

Consider Span 2—3.

For Values of x from 0 to $C_1$.

(1) $m = m_2 + s_2x$

(2) $EIi = m_2x + \frac{s_2x^2}{2} + k$ When $x = 0$, $k = EIi_2$.

(3) $EIv = \frac{m_2x^2}{2} + \frac{s_2x^3}{6} + EIi_2x + k_1$ When $x = 0$, $v = 0$ and $k_1 = 0$.

For Values of x from $C_1$ to $L_1$.

(4) $m = m_2 + s_2x + m_a$

(5) $EIi = m_2x + \frac{s_2x^2}{2} + m_ax + k_2$

When $x = C_1$, equations (2) and (5) are equal because of the continuity.

Equating and solving for $k_2$, $k_2 = EIi_2 - m_aC_1$

(6) $EIv = \frac{m_2x^2}{2} + \frac{s_2x^3}{6} + \frac{m_ax^2}{2} + EIi_2x - m_aC_1x + k_3$

When $x = C_1$ equations (3) and (6) are equal.

Equating and solving for $k_3$, $k_3 = \frac{m_aC_1^2}{2}$

When $x = L_1$, $EIv = EIv_3$.

Let $x = L_1$ in equation (6) and solve for the value of $EIi_2$

(7) $EIi_2 = \frac{EIv_3}{L_1} - \frac{m_2L_1}{2} - \frac{s_2L_1^2}{6} - \frac{m_aL_1}{2} + m_aC_1 - \frac{m_aC_1^2}{2L_1}$

Let $x = L_1$ in equation (4) and solve for $s_2$.

(8) $s_2 = \frac{m_3 - m_2 - m_a}{L_1}$

Substituting this value of $s_2$ in equation (7)

(9) $EIi_2 = \frac{EIv_3}{L_1} - \left(\frac{2m_2 + 2m_a + m_3}{6}\right)L_1 - \frac{m_aC_1^2}{2L_1} + m_aC_1$

Treating span 1—2 in the same way

(10) $EIi_{-2} = \frac{EIv_1}{L_2} - \left(\frac{2m_2 + 2m_b + m_1}{6}\right)L_2 - \frac{m_bC_2^2}{2L_2} + m_bC_2$

Since the slopes given by equations (9) and (10) are of opposite signs $EIi_2 + EIi_{-2} = 0$

(11) $$0 = \frac{EIv_3}{L_1} + \frac{EIv_1}{L_2} - \frac{m_3L_1}{6} - \frac{2m_2(L_1 + L_2)}{6} - \frac{m_1L_2}{6}$$
$$+ m_a\left(C_1 - \frac{C_1^2}{2L_1} - \frac{L_1}{3}\right) + m_b\left(C_2 - \frac{C_2^2}{2L_2} - \frac{L^2}{3}\right)$$

$$(12)\quad m_1L_2+2m_2(L_1+L_2)+m_3L_1 = m_a\left(6C_1-\frac{3C_1^2}{L_1}-2L_1\right)$$

$$+m_b\left(6C_2-\frac{3C_2^2}{L_2}-2L_2\right)+\frac{6EIv_3}{L_1}+\frac{6EIv_1}{L_2}$$

Derivation of Three-Moment Equation for Case 5.

Consider Span 2—3.

$(1)\quad m = m_2 + m_b + s_2x$

$(2)\quad EIi = (m_2 + m_b)x + \dfrac{s_2\,x^2}{2} + k.$ When $x = 0$, $k = EIi_2$.

$(3)\quad EIv = (m_2 + m_b)\dfrac{x^2}{2} + \dfrac{s_2x^3}{6} + EIi_2x + k_1.$ When $x=0$, $v=0$ and $k_1 = 0$.

Let $x = L_1$ in equation (1) and solve for $s_2$.

$$(4)\quad s_2 = \frac{m_3-m_2-m_b}{L_1}$$

Let $x = L_1$ in equation (3) and substitute for $s_2$ its value from equation (4).

Solving for $EIi_2$.

$$(5)\quad EIi_2 = \frac{EIv_3}{L_1}-\frac{m_2L_1}{3}-\frac{m_bL_1}{3}-\frac{m_3L_1}{6}$$

Consider Span 1—2.

$(6)\quad m = m_2 + s_{-2}x$

$(7)\quad EIi = m_2x+\dfrac{s_{-2}x^2}{2}+k_2.$ When $x=0$, $k_2=EIi_{-2}$

$(8)\quad EIv = \dfrac{m_2x^2}{2}+\dfrac{s_{-2}x^3}{6}+EIi_{-2}x+k_3.$ When $x=0$, $v=0$ and $k_3=0$.

$$(9)\quad s_{-2} = \frac{m_1-m_2+m_a}{L_2}$$

Let $x = L_2$ in equation (8) and substitute for $s_{-2}$ its value from equation (9).

Solving for $EIi_{-2}$:

$$(10) \quad EIi_{-2} = \frac{EIv_1}{L_2} - \frac{m_2L_2}{3} - \frac{m_1L_2}{6} - \frac{m_aL_2}{6}$$

Since the slopes given by equations (5) and (10) are of opposite signs

$EIi_2 + EIi_{-2} = 0$

Adding these equations and transposing

$$(11) \quad m_1L_2 + 2m_2(L_1 + L_2) + m_3L_1 = -m_aL_2 - 2m_bL_1 + \frac{6\,EIv_3}{L_1} + \frac{6\,EIv_1}{L_2}$$

173. *Methods of Determining Deflection of Beams with Varying Load and Section*—All methods, whether analytical or graphical, depend on the proposition that the deflection curve is developed from the loading curve by four successive integrations, or, if the moment curve is known, from the moment curve by two successive integrations. The principles upon which the special methods treated here rest are: the deflection of any point $x_1$, measured from a line tangent to the beam at another point $x_0$, is equal to the moment about $x_1$ of the portion of the $M/I$ curve of the beam between $x_0$ and $x_1$, (p. 154, Boyd's Strength of Materials); the moment about any point of any number of co-planar forces is equal to the product of the intercept on a line through the point and parallel to the resultant of the forces, between the strings holding the resultant in equilibrium, the pole distance of the force polygon, and a factor depending on the scales of the force and funicular polygons, (p. 343, Spoffords, Theory of Structures).

Fig. 172 shows the variation in the moment of inertia of a simply supported beam whose deflection curve is desired. It adds but slightly to the work if this curve is irregular instead of being a straight line. Fig. 173 shows the loading curve for the beam. The force polygon, Fig. 174, is constructed from the loading curve by dividing the latter into a convenient number of parts and considering the area of each part as a concentrated vertical force. These different areas are plotted to scale on the line 1—10 and the various rays drawn to the pole 0, at a perpendicular distance, *h*, from the resultant of the forces 1—10. In the usual manner, the equilibrium or funicular polygon, Fig. 175, is constructed from the force polygon of Fig. 174, each string being parallel to its respective rays in the force polygon. Each ordinate in the funicular polygon is proportional to the bending moment at that section. To obtain the curve of Fig. 176 each ordinate of the moment curve is divided by the ordinate of the moment of inertia curve at the corresponding point in the beam, and the quotient plotted to a suitable scale. The force polygon of Fig. 177 is obtained from this modified moment curve in the same way that Fig. 174 was obtained from Fig. 173. The area of each section is treated as a concentrated load or force applied at the center of the section. From the force polygon for the $M/I$ curve is drawn the final funicular polygon, Fig. 178. A curve inscribed in this polygon represents the elastic curve of the beam and will be referred to as the elastic curve.

The proof of this construction is as follows: Assume the number of sections of the $M/I$ curve used in constructing the force and funicular polygons to increase indefinitely. As the area of the individual sections approaches zero as a limit, the funicular polygon becomes a curve inscribed in the polygon actually constructed. Each tangent to this curve is the string of the funicular polygon between the lines of action of the differential areas of the $M/I$ curve each side of the point of tangency, and is parallel to the rays of the force polygon holding in equilibrium the differential area at the point of tangency. Consider any tangent to the curve. The intercept on any vertical line between this tangent and the elastic curve represents the deflection of the point where the vertical cuts the beam, measured from the tangent. This is true because the intercept considered is that between the strings of the funicular polygon holding in equilibrium the portion of the $M/I$ curve between the two points in question. In other words, the ordinates from the curve $AB$ to any tangent to it when multiplied by the proper scale are ordinates equal to the deflections of the beam measured from the tangent in question. Similarly the ordinates from the curve to any line are equal to the deflections from that line when multiplied by the vertical scale. In short, the smooth curve inscribed in the funicular polygon, Fig. 178, is a graph of the elastic curve plotted to inclined co-ordinates, and with the vertical scale of ordinates differing from the scale of lengths. If the direction of the axis of lengths is found, it is easy to plot the position of any line and measure the deflections of the beam from it directly from the drawing.

In the case of a simply supported beam the deflections are desired from a line passing through the two supports. To obtain them, draw a line through the two points on the elastic curve representing the supports, and the ordinates from this line to the curve will give the deflections desired. This case is illustrated in Fig. 178. In the case of a cantilever the deflections are desired from a tangent to the elastic curve at the support, and can be found by drawing such a tangent and measuring the deflections. This case is shown in Fig. 179. In general, there will be either two points at which we know the deflections or one at which we know the direction of the tangent, making it possible to draw a line from which the desired deflections can be measured. In the case of a wing spar like that of the upper wing of the Fokker D-7, the tangent to the elastic curve at the center line of the airplane is horizontal and the deflection at the cabane struts is zero. The deflections desired are those measured to the curve of the funicular polygon from a line intersecting that curve at the cabane strut point and parallel to the tangent at the center line.

In all of this work great care must be used to multiply the ordinates measured from the drawing by the correct scale. The calculation of the scale is as follows:

E is in lbs. per sq. in.
I is in in.$^4$
1 in. = q in. for the linear scale.
1 in. = p lbs. per inch run.

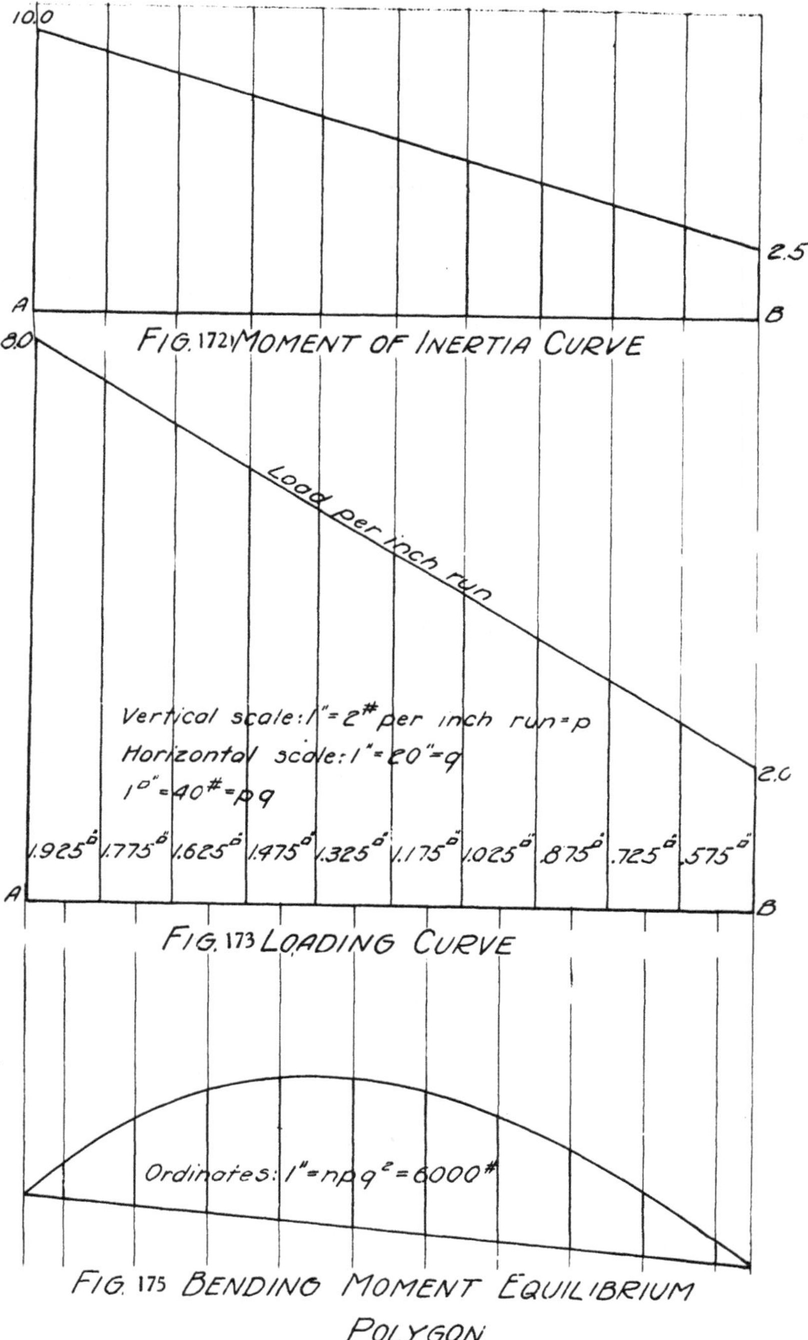

10.0
2.5
A
B
FIG. 172 MOMENT OF INERTIA CURVE
8.0
Load per inch run
Vertical scale: 1" = 2# per inch run = p
Horizontal scale: 1" = 20" = q
1□" = 40# = pq
1.925□ 1.775□ 1.625□ 1.475□ 1.325□ 1.175□ 1.025□ .875□ .725□ .575□
2.0
A
B
FIG. 173 LOADING CURVE
Ordinates: 1" = npq² = 6000#
FIG. 175 BENDING MOMENT EQUILIBRIUM POLYGON

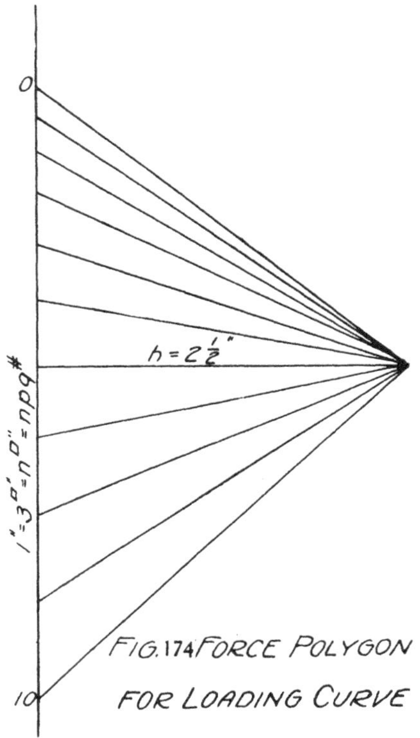


FIG. 174 FORCE POLYGON FOR LOADING CURVE

For Fig. 173, 1 sq. in. = p $\times$ q lbs.

For Fig. 174, let 1 in. = n sq. in. from Fig. 173, whence 1 in. in Fig. 174 = n $\times$ p $\times$ q lbs.

If the pole distance for Fig. 174 is *h* in., the scale for the bending moment polygon is 1 in. = n $\times$ p $\times$ $q^2$ $\times$ h in. lbs.

If the ordinates of the moment polygon are now divided by the ordinates to the moment of inertia curve, and the quotients plotted to a scale of 1 in. to m in. of modified bending moment ordinates, the scale for Fig. 176 is 1 in.=$mnpq^2h$ lbs. per $in.^3$. Therefore, 1 sq. in. for Fig. 176 = $mnpq^3h$ lbs. per $in.^2$.

For Fig. 177, let 1 in. = r sq. in. from Fig. 176. Then 1 in. in Fig. 177 = $rmnpq^3h$ lbs. per $in.^2$. If the pole distance for Fig. 177 is $h_1$ in. the scale of ordinates for the deflection polygon is 1 in. = $rmnpq^4hh_1$ lbs. per in. The value of the ordinates multiplied by this scale must be divided by the modulus of elasticity of the material in the beam to get the true deflection. A dimensional equation shows that the above scale is correct.

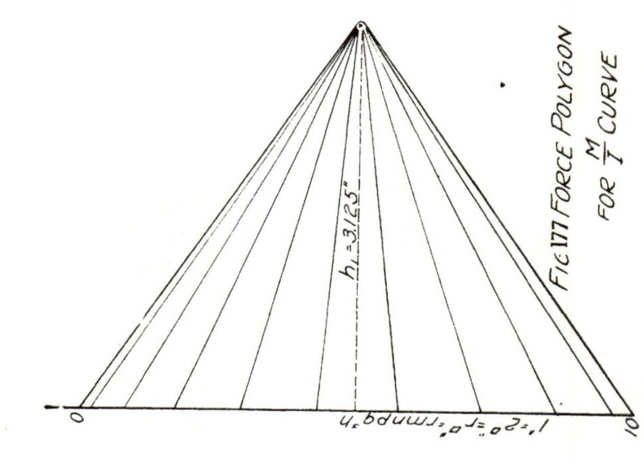


FIG 177 FORCE POLYGON FOR $\frac{M}{I}$ CURVE

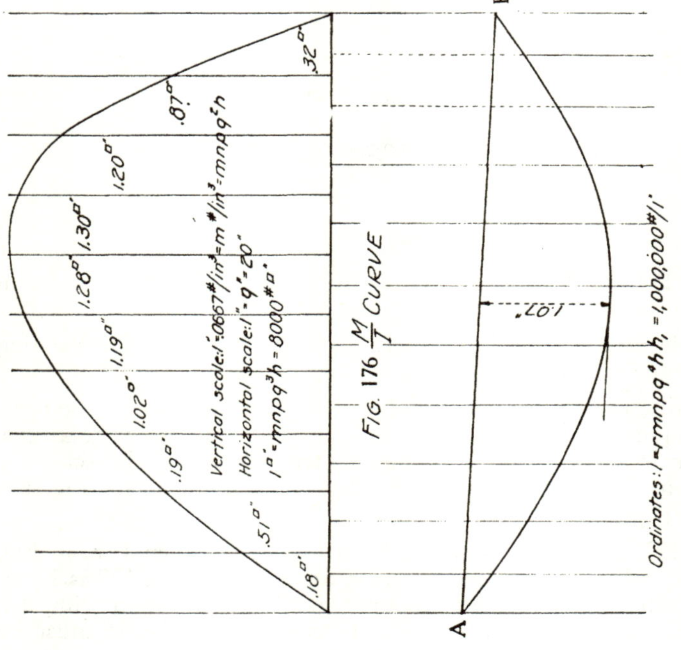


FIG. 176 $\frac{M}{I}$ CURVE

FIG 178 DEFLECTION EQUILIBRIUM POLYGON

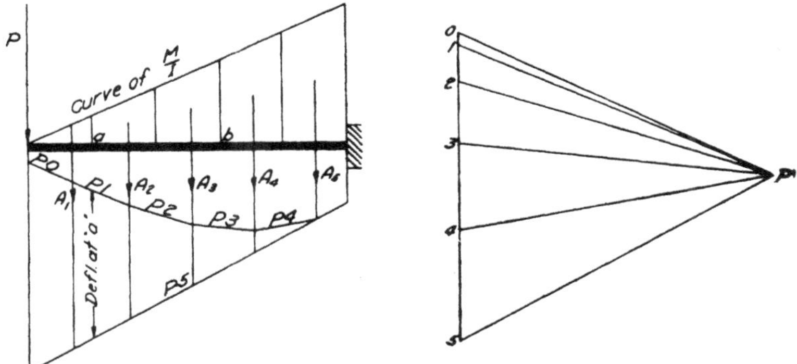


Fig. 179. Deflection of a Cantilever Beam

Since it is usually more convenient to compute the values of $M/I$ than to obtain them graphically, the $M/I$ curves may be plotted from calculated values. The scale, $s$, used in plotting this curve equals the vertical scale for Fig. 176 1 in. $= mnpq^2h = s$. The remainder of the work is the same as that shown in Fig. 178, but it should be remembered that the value of $mnpq^2h$ is to be replaced in subsequent formulas by the value of the scale, $s$, for $M/I$. The scale of ordinates for the deflection polygon becomes 1 in. $= rsq^2h_1$.

The graphical method may be checked roughly by using the average moment of inertia of the beam and computing the deflection by the ordinary deflection formulas.

Any difficulty there may seem to be in determining the correct scale factor is removed if it is understood that in all the scaling and plotting the ordinates and areas dealt with are true lengths and actual areas in inches and square inches, respectively. As each diagram is drawn the scale that was used for it is noted, and not till the deflection is obtained is it necessary to compute the scale, which is, therefore, done in a single operation.

If the example given for a singly supported beam had been solved analytically, using an average load and moment of inertia, the resulting deflection is in error by somewhat less than 3% on the unsafe side.

$$d = \frac{5\,wl^4}{384\,EI} = \frac{5 \times 5 \times 100^4}{384 \times 6.25\,E} = \frac{1{,}040{,}000}{E}$$

In the case of a beam with the same loading and length but with the moment of inertia varying from 10 to 1 the deflection by the analytical method, with averaged loads and moments of inertia was again too small by about 8%.

In airplanes such as the Fokker D-7, the upper and lower wings are internally braced cantilevers. The deflections of the spars are equalized by the use of an "N" outboard strut. In the Fokker D-7 the spars are not the same size, and hence, without the outboard strut would not deflect equally. Therefore, as a strut is used, some of the spars take part of the load from the others. Before the final design can be made of the spars the distribution of load between the spars must be determined. This can be done as follows:

1. Determine by the graphical method the deflection of points *c* and *d*, Fig. 180, due to the distributed load and the reaction at *d*, and the deflections at *a* and *b* due to the distributed load and the reaction at *a*. These deflections should be measured from tangents to the elastic curves at *e* and *f* as these are both horizontal by symmetry. The deflections are measured from these tangents because they are the only ones to the two curves which are known to be horizontal. The points *e* and *f*, however, are not fixed vertically, so the deflections at the fixed points *d* and *a* must be subtracted from those at *c* and *b* to find the absolute deflections of the latter. This may be done graphically as described above by measuring the deflections from lines through *d* and *a* and parallel to the tangents at *c* and *f*. When the center section of the lower spar is enclosed in a sleeve allowance must be made in the $M/I$ curve for the increased moment of inertia of the spar unit.

2. Subtract the absolute deflection of *b* from that of *c* to find the relative movement of the two points.

3. Apply a concentrated load of unity to each beam at the strut point, and determine the deflection of each beam at this point due to the unit load.

4. The relative deflection of the beams of either truss at the strut point produced by the distributed load, divided by the numerical sum of their deflections due to the load of unity equals the value of the stress in the strut, or the amount of support that one beam affords the other.

5. Find the relative deflection of the front and rear trusses and the forces necessary to equalize them by the same method as above.

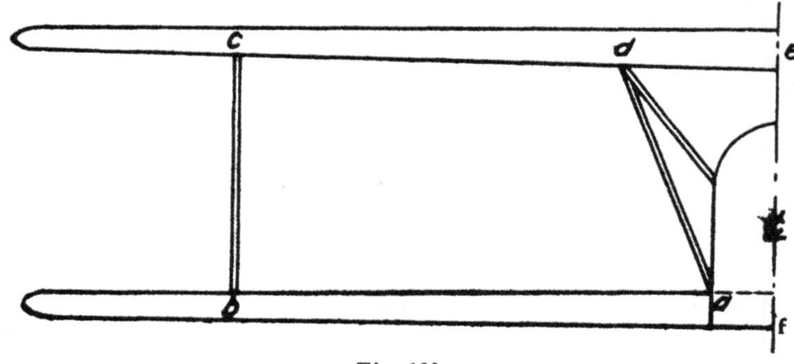


Fig. 180

6. Figure the resultant forces acting on each spar at the $N$ strut and correct the total bending moment in the spars by the moments due to the forces.

There are several designs that have been built or proposed in which the wings are unbraced cantilevers. As far as strength is concerned, such wings can be readily designed, but in the opinion of many engineers this type of construction is limited by the large deflections of the wing tips. Whether or not this deflection is detrimental to the flying qualities of the airplane, the determination of the deflection curve of the wing spars is an essential part of the stress calculation.

In addition to the graphical solution, the deflections may be obtained analytically. Essentially the graphical method consisted in the determination of the moment curve of a beam under a loading represented by the $M/I$ or $M/EI$ curve. This can be done by any of the analytical methods of finding bending moment from the load curve. Care must be taken, however, to make sure that the deflections obtained are those desired. In the case of a cantilever the moment desired is that of the area of the curve between the given point and the point where the axis of the beam is horizontal, about point $x$. In the case of a simply supported beam the moment of the area of the $M/EI$ curve between a point and a support will give the deflection from the tangent at that support. To find the deflection from a line passing through the supports it is necessary to know the slope of this tangent which can be found by dividing the deflection of the other support from it by the span. Knowing these values it is a simple matter to find the deflection desired.

If the work is done analytically it is best to keep the work in tabular form as far as possible and to find the moments by the formula in Art. 19. By using this formula it is a simple, though somewhat tedious, matter to find the moment curve of any $M/I$ curve no matter how irregular. The $M/I$ curve should be divided into a convenient number of parts, each one of which may be considered as a trapezoid or even a rectangle with very little loss of precision, and the moments computed at each division point. In order to illustrate the method of using the formula of Art. 19, the moments of the example in Art. 17 are worked out below.

| 1 | 2 | 3 | 4 | 5 | 6 | 7 | 8 |
|---|---|---|---|---|---|---|---|
| Sta. | S | x | Sx | F | a | Fa | M |
| A | —100 | | | | | | 0 |
| | | 25 | —2500 | —150 | 12.5 | —1875 | |
| B | 387.6 | | | | | | —4375 |
| | | 20 | 7752 | —120 | 10.0 | —1200 | |
| C | 217.6 | | | | | | 2177 |
| | | 25 | 5440 | —150 | 12.5 | —1875 | |
| D | — 22.4 | | | | | | 5742 |
| | | 40 | — 896 | — 40 | 20.0 | — 800 | |
| E | —262.4 | | | | | | 4086 |
| | | 15 | —3936 | — 15 | 7.5 | — 112.5 | |
| F | 0 | | | | | | — 2.5 |

Col. 2 gives the shear to the right of the station in col. 1.

Col. 3 gives the distance between consecutive stations.

Col. 4 gives the product of the distance in col. 3 on the same line and the shear in col. 2 in the next line above.

Col. 5 gives the amount of load in the distance represented by the figure in col. 3.

Col. 6 gives the distance from the center of gravity of this load to the station at which the moment is being computed.

Col. 7 is the product of cols. 5 and 6.

Col. 8 gives the moment at all the stations on the same line in col. 1. It is the algebraic sum of the value just above in cols. 8, 4 and 7, and is obtained from the formula of Art. 19: $M_x = M_b + S_b x + F \cdot a$.

These values do not agree exactly with those in Art. 17 because the values of the reactions are found to one more significant figure. The value of the moment at *F* should equal zero. That it does not is due to the fact that the reactions were figured only to the nearest tenth of a pound. If they had been figured only to the nearest pound the error would have been 37.5 in. lbs. instead of —2.5 in. lbs.

In the case of figuring deflections, there are no concentrated loads or reactions and the value of the shear at any point is the sum of the elementary areas up to that point, i.e., $S = \Sigma F$. The work can easily be arranged in similar tabular form. For a cantilever this method is probably as easy as the graphical method, though for simply supported beams the advantage may be with the latter. The point to be stressed is that both methods are basically the same and the work may be carried on, if desired, by almost any combination of the two methods, some steps being taken graphically and some analytically. Up to and including the computation of the values of $M/I$ or $M/EI$, the analytical method is almost always preferable.

The terms $M/I$ curve and $M/EI$ curve have been used interchangeably in the above discussion. This is allowable if it be assumed, as it nearly always is, that $E$ is a constant. In that case exactly the same deflections will be obtained by using the $M/I$ curve and dividing its moment by E as by using the $M/EI$ curve. Two advantages of the former method are that the numbers used are generally greater than unity, instead of small decimals, making them more convenient to figure with, and that fewer divisions by $E$ are usually necessary, as in nearly all cases the ordinates of more points on the $M/I$ curve must be determined than deflections.

174. *Properties of Woods at 10 Per Cent Moisture*—The properties of wood, which have been in common use, have been based on a moisture content of 15 per cent. As most specimens of woods in actual use on land airplanes average slightly less than 10 per cent moisture content, the strength properties given in Table XXIX were reduced to this value. For seaplanes and flying boats, however, all strength properties should be based on 15 per cent moisture. The woods included in this table were taken from Signal Corps Specification 15020-B. Only those species listed in Column 2 of Table XXIX are permitted by the above specifi-

cation. The strength properties are based on the values given in Bulletin No. 556 of the Forest Service of the U. S. Department of Agriculture which, as far as data were available, are adjusted to 10 per cent moisture content by the formula $s_2 = s_1 \cdot 10^{a\,(m_1 - m_2)}$ in which $s_1$ is the strength value as found from tests at $m_1$ per cent moisture, and $s_2$ equals the strength value adjusted to 10 per cent moisture; $a$ is a constant derived from experiment. This formula is the result of work done at the Forest Products Laboratory. In the cases where values for $a$ could not be obtained the method of adjustment given in Bulletin No. 556 was followed.

With the exception of elm all the values in Table XXIX are averages of the properties of each of the species, listed in Column 2, of the wood in question. In the case of elm the properties of rock elm only were considered. As the average specific gravities of slippery and white elm are .54 and .51, respectively, while the specific gravity requirement in Column 3 for elm is .60, any specimens of slippery or white elm meeting this requirement will have strength values equal to those of rock elm. It should be further noted that the values for the moduli of rupture in Column 7 are for rectangular sections only. With routed sections a large decrease in the modulus of rupture occurs, which varies with the degree of routing. Tests have been conducted to determine quantitatively the value of this decrease. Three series were run, the first with very slight routing, the second with normal routing, and the third with excessive routing. There were ten routed specimens in each series except series two, in which there were fifty, and for each routed specimen there was a matched solid specimen. All specimens were of the same outside dimensions, 2¼ x 1½ in. As a result of a study of the data from these tests, the following empirical formula was obtained, which is believed to give reasonably accurate values for the modulus of rupture of spruce or fir spars of a normally routed I section:

$$F_r = \frac{1.12\, F_s\, I_s}{I_r}$$

$F_r$ = modulus of rupture of routed section.
$F_s$ = modulus of rupture of solid section.
$I_r$ = moment of inertia of routed section.
$I_s$ = moment of inertia of a solid rectangular section of same area and center height as the routed section.

Further, experimental work is being done to verify this formula, and also to determine the effect of routing on strength in horizontal shear, which is reduced probably even more than the modulus of rupture. The value of 750 lbs. per sq. in. that is given below for spruce is for routed sections. In calculating the shear in solid sections, such as is produced by bolts which are subjected to pull parallel to the grain of the wood, as in a longeron splice, the values of the shear strength listed in Col. 12, Table XXIX may be used.

For the reasons that spruce is so largely employed for highly stressed members in airplane structures, and that each of the three species of spruce are in common use, the working values for the various properties recommended for design work are not the average of these properties for the three species, but, in the case of the ultimate compressive strength parallel to the grain, are the lowest of the three values, and in the case of the modulus of elasticity and ultimate horizontal shear are values which experience has shown to be suitable.

*STRENGTH VALUES FOR SPRUCE*

| Modulus of Rupture | Modulus of Elasticity | Ultimate Compressive Strength Parallel to Grain | Horizontal Shear |
|---|---|---|---|
| | Lbs. per sq. in. | | |
| 10,300 | 1,600,000 | 5,500 | 750 |

In Fig. 181 is given a curve showing the reduction in stress that is necessary to allow for column action in spruce columns. Similar curves for other woods may be constructed by the aid of the formulas in Fig. 181. Interplane struts as a rule fall in the class of slender columns, and their strength may, therefore, be computed by Euler's formula for pin-ended columns. This formula is not applicable to spruce or fir struts with an $L/\rho$ less than 80 to 90. For center section struts use Johnson's formula or the curve in Fig. 181 for pin-ended columns.

Where members, such as wing beams, are subjected to both bending and compression, the following method of calculating ultimate allowable stresses is recommended.

$$f_a = \left\{ \frac{f_b}{f_b + f_c} (F - C) \right\} + C$$

$f_a$ = ultimate allowable stress
$f_b$ = calculated stress due to bending.
$f_c$ = calculated stress due to compression.
$F$ = modulus of rupture.
$C$ = maximum allowable $P/A$ as determined from the curve for pin-ended columns in Fig. 181.

When considering a point near mid-span the length of the column may be conservatively taken as equal to the span. At inner strut points where column action is present the column length may be taken as equal to .25 of the span on one side of the strut plus .25 of the span on the other. At cantilever strut points $C$ equals the full compressive strength of the material.

The weights in Col. 4 of Table XXIX are calculated by using the specific gravities, given in Bulletin No. 556, which are averages for the species and are based on oven dry weight and air dry volume, to compute the weight of a cubic foot of dry wood. This weight is then increased by 10 per cent to allow for the moisture.

# TABLE XXIX

## PROPERTIES OF WOODS AT 10 PER CENT MOISTURE CONTENT

| Wood | Species permitted under specification. | Minimum allowable specific gravity.[1] | Average weight per c.f. at 10 per cent moisture. | Static bending. Work to maximum load. | Static bending. Fiber stress at elastic limit. | Static bending. Modulus of rupture. | Static bending. Modulus of elasticity. | Impact bending. Fiber stress at elastic limit. | Maximum compression. Parallel to grain. | Maximum compression. Perpendicular to grain at elastic limit. | Shear parallel to grain. |
|---|---|---|---|---|---|---|---|---|---|---|---|
| 1 | 2 | 3 | 4 | 5 | 6 | 7 | 8 | 9 | 10 | 11 | 12 |
| *Hardwoods* | | | *Lbs.* | *In Lbs.* | | | | *Lbs. per sq. in.* | | | |
| Ash | White, green, blue, and Biltmore | 0.56 | 39.5 | 13.8 | 9,050 | 14,600 | 1,620,000 | 17,500 | 7,830 | 1,630 | 1,950 |
| Do | Black | .48 | 34.5 | 15.4 | 8,500 | 14,000 | 1,690,000 | 12,400 | 7,000 | 1,100 | 1,760 |
| Birch | Sweet, yellow | .61 | 44.5 | 18.7 | 12,250 | 18,600 | 2,140,000 | 23,200 | 9,880 | 1,500 | 2,200 |
| Basswood | | .36 | 26.5 | 8.0 | 6,600 | 9,500 | 1,500,000 | 10,500 | 5,430 | 520 | 1,160 |
| Cherry | Black | .48 | 35.0 | 11.1 | 10,450 | 13,400 | 1,530,000 | 14,300 | 8,000 | 970 | 1,870 |
| Elm | Rock, white and slippery | .60 | 45.0 | 18.7 | 8,550 | 15,800 | 1,580,000 | 17,400 | 7,940 | 1,690 | 2,010 |
| True hickory | Pignut, mockernut, shagbark, and shellbark | .73 | 51.0 | 25.2 | 11,400 | 21,200 | 2,250,000 | 23,100 | 10,070 | 2,500 | 2,200 |
| Hard maple | Sugar | .60 | 42.5 | 13.7 | 10,700 | 16,100 | 1,830,000 | 19,500 | 8,780 | 1,670 | 2,500 |
| Mahogany | African | .46 | 33.0 | 11.1 | 9,200 | 12,200 | 1,540,000 | ...... | 6,200 | 1,170 | 1,520 |
| Do | True | .50 | 36.0 | 8.3 | 9,100 | 11,200 | 1,380,000 | ...... | 6,680 | 1,300 | 1,700 |
| Oak | Bur, cow, post, and white | .65 | 46.5 | 12.3 | 8,150 | 14,100 | 1,580,000 | 18,100 | 7,490 | 1,680 | 2,100 |
| Poplar | Cucumber, magnolia, and yellow | .45 | 32.5 | 11.1 | 7,450 | 11,800 | 1,580,000 | 15,200 | 6,550 | 810 | 1,310 |
| Walnut | Black | .52 | 39.0 | 9.7 | 10,000 | 14,300 | 1,680,000 | 15,400 | 7,810 | 1,350 | 1,170 |
| *Conifers* | | | | | | | | | | | |
| Cedar | Port Orford | .42 | 31.0 | 10.6 | 8,350 | 13,800 | 2,000,000 | 16,900 | 7,300 | 960 | 1,440 |
| Douglas fir | Coast | .45 | 35.0 | 7.9 | 8,200 | 12,000 | 2,060,000 | 12,100 | 8,440 | 940 | 1,080 |
| Fir | Amabilis, grand, noble, and white | .38 | 27.0 | 8.4 | 6,550 | 10,200 | 1,610,000 | 10,900 | 6,280 | 680 | 1,020 |
| Hemlock | Western | .40 | 29.0 | 6.0 | 5,800 | 9,600 | 1,460,000 | 10,600 | 6,540 | 600 | 950 |
| Pine | Eastern white | .36 | 27.0 | 6.4 | 6,950 | 9,500 | 1,420,000 | 9,300 | 6,320 | 760 | 1,070 |
| Do | Western white | .40 | 29.5 | 9.3 | 6,900 | 10,300 | 1,620,000 | 13,600 | 6,780 | 700 | 540 |
| Do | Norway | .46 | 33.0 | 10.5 | 10,600 | 13,600 | 1,830,000 | 16,600 | 8,110 | 950 | 1,380 |
| Spruce | Red, Sitka, and white | .36 | 27.0 | 9.0 | 6,750 | 10,300 | [2] 1,510,000 | 11,800 | [3] 5,970 | 720 | [4] 1,100 |

[1] Minimum specific gravity is based on oven dry weight and oven dry volume. [2] 1,600,000. [3] 5,500. [4] 750.

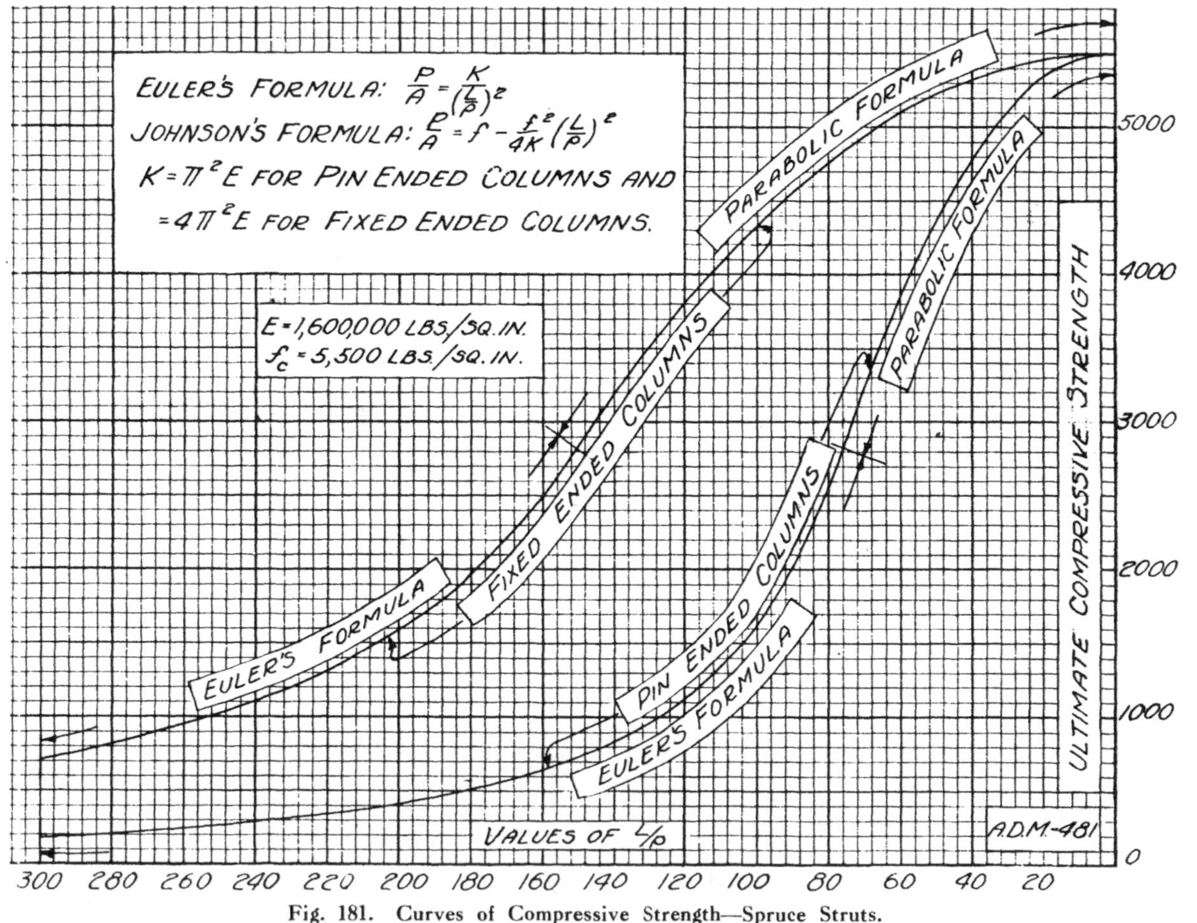


Fig. 181. Curves of Compressive Strength—Spruce Struts.

## THE EFFECT ON THE STRENGTH PROPERTIES OF WOOD OF SPIRAL AND DIAGONAL GRAIN

In determining the true slope of grain in any piece of wood two kinds of sloping grain must be considered, namely: (1) diagonal grain or the slope of the annual growth layers with respect to the axis of the piece and (2) spiral grain or the slope of the wood fibers within these annual layers. Diagonal grain is caused by the fact that the stock is not sawn parallel to the annual growth rings. Its maximum value occurs on a radial face. Spiral grain is a maximum on a face tangential to the growth rings. Frequently in inspection when a piece has both spiral and diagonal grain it is wrongly classified as having a grain with a slope equal to the slope of either the diagonal or spiral grain, depending on which has the greater slope. The point should be emphasized that when two slopes are present the resultant slope is greater than either single slope. Classification on the basis of absolute or resultant slope is the only safe and correct method. Suppose the slope of the diagonal grain is 1 in 15 and that of the spiral grain 1 in 20, the absolute slope = $\sqrt{\left(\frac{1}{20}\right)^2 + \left(\frac{1}{15}\right)^2} = \frac{1}{12}$ or a slope of 1 in 12. Classification by the method of greater slope may admit material 15 per cent lower in average strength properties.

Tests show that the damaging effect of combined spiral and diagonal grain is the same as that of spiral or diagonal grain of the same absolute slope, and that spiral grain is no more harmful than diagonal grain of the same degree. However, it is more likely to escape detection, and spiral grained material is more subject to deterioration from season checks than diagonal grained.

Table XXXI gives quite complete data on the effect of spiral grain on the strength properties of Sitka spruce and Douglas fir. Plots showing the decrease in the value of the different properties bring out that some properties are affected by slightly sloping (1 in 35) grain, that others are not influenced appreciably until the slope becomes 1 in 15 or greater. In general, the decrease is gradual until a slope of from 1 in 20 to 1 in 15 is reached. In this zone there is a sharp change in the slope of the curves and the rate of decrease becomes much more rapid. Sloping grain affects the modulus of elasticity least and the work and impact properties most. Hence, in material for slender struts sloping grain is not nearly as important as in members subject to severe shock and vibration. For most types of airplane construction it may be stated that when the absolute slope exceeds 1 in 30 the design must be changed to allow for the decrease in the ultimate stresses, and in no case should material be used in which the absolute slope is greater than 1 in 15. This ruling applies to highly stressed members.

It is suggested that spiral or diagonal grained stock can be effectively used in the webs of built-up spars, or as the center lamination of a strut built up with three laminations.

Few data are available on the effect of sloping grain on the strength properties of woods other than spruce or Douglas fir, and until such data are available, the percentage of reduction obtained by tests on spruce and Douglas fir should be used. Recent tests show that ash is affected by sloping grain to the same degree as spruce and Douglas fir.

## *THE RELATIVE MERITS OF WOODS*

In Table XXX the woods are arranged in their order of relative merit based on their properties per unit weight. In calculating this table certain properties were weighted as follows:

| | |
|---|---|
| Work to maximum load in static bending..... | 2.0 |
| Fiber stress at elastic limit in static bending.. | 1.5 |
| Modulus of rupture ...................... | 2.0 |
| Modulus of elasticity ...................... | 2.0 |
| Fiber stress at elastic limit in impact bending.. | 1.5 |
| Ultimate compressive strength parallel to grain | 1.0 |
| | 10.0 |

Work to maximum load represents the ability of the timber to absorb shock, after the elastic limit is passed, with a slight permanent or semi-permanent deformation and with some injury to the timber. This property is a measure of the combined strength and toughness of a material under bending stresses, and is, therefore, of great importance in airplane parts which are subjected to severe shock as in chassis struts, to large, suddenly applied loads, or to severe vibration. On small high-powered machines toughness is particularly essential. A wood may be high in its elastic properties but if it is low in toughness it will not be as suitable for most airplane construction as a somewhat weaker, but tougher wood. A wood lacking in toughness is said to be brashy. Brashness may be either natural or artificial. Wood may be said to be naturally brash, when one of its characteristic features is low work properties. However, any species may contain brash material due to obtaining lumber from very old mature trees, or to its being affected by fungus growth or decay. Wood may be said to be brash due to artificial causes when it has been subjected to high temperatures or very rapid drying.

The stresses that are developed in service should always be within the elastic limit of the material. Occasionally, however, under very severe conditions the elastic limit will be exceeded, and if the load continues for more than a short time gradual failure will ensue. Since working stresses are ordinarily based on ultimate stresses rather than on elastic limit values, a material with an elastic limit close to its ultimate strength is much superior to a material of the same ultimate strength but with a low elastic limit.

The weights given the different properties in determining the relative merit of the woods, as shown in Table XXX, were chosen with the purpose of selecting woods best fitted for general airplane work. In wing beams, all the properties which were considered in preparing

## TABLE XXX

## PROPERTIES OF WOODS IN TERMS OF PROPERTIES OF SPRUCE

| Wood.[1] | Weight per cubic foot. | Static bending. | | | | Impact bending. | Ultimate compression. | Sum of weighted values, divided by 10. | Col. 9, divided by Col. 2. |
|---|---|---|---|---|---|---|---|---|---|
| | | Work to maximum load. | Fiber stress at elastic limit. | Modulus of rupture. | Modulus of elasticity. | Fiber stress at elastic limit. | Strength parallel to grain. | | |
| 1 | 2 | 3 | 4 | 5 | 6 | 7 | 8 | 9 | [2]10 |
| Cedar (Port Orford) | 1.15 | 1.18 | 1.24 | 1.34 | 1.32 | 1.43 | 1.22 | 1.29 | 1.12 |
| Birch | 1.65 | 2.08 | 1.81 | 1.80 | 1.42 | 1.97 | 1.66 | 1.79 | 1.08 |
| Norway pine | 1.22 | 1.17 | 1.57 | 1.32 | 1.21 | 1.40 | 1.36 | 1.32 | 1.08 |
| Hickory | 1.89 | 2.80 | 1.69 | 2.06 | 1.49 | 1.96 | 1.69 | 1.99 | 1.05 |
| Black Ash | 1.28 | 1.71 | 1.26 | 1.37 | 1.12 | 1.05 | 1.17 | 1.30 | 1.01 |
| Spruce | 1.00 | 1.00 | 1.00 | 1.00 | 1.00 | 1.00 | 1.00 | 1.00 | 1.00 |
| Fir (Grand, etc.) | 1.00 | .93 | .97 | .99 | 1.07 | .92 | 1.05 | .986 | .986 |
| Black cherry | 1.29 | 1.23 | 1.55 | 1.30 | 1.01 | 1.21 | 1.34 | 1.26 | .977 |
| Western white pine | 1.09 | 1.03 | 1.02 | 1.00 | 1.07 | 1.15 | 1.14 | 1.06 | .973 |
| Poplar | 1.20 | 1.23 | 1.10 | 1.14 | 1.05 | 1.29 | 1.10 | 1.15 | .960 |
| African mahogany | 1.22 | 1.23 | 1.36 | 1.18 | 1.02 | .... | 1.04 | 1.17 | .959 |
| Basswood | .88 | .89 | .99 | .92 | .99 | .89 | .91 | .933 | .953 |
| Hard maple | 1.57 | 1.52 | 1.58 | 1.56 | 1.21 | 1.65 | 1.47 | 1.49 | .950 |
| Commercial white ash | 1.46 | 1.53 | 1.34 | 1.42 | 1.07 | 1.48 | 1.31 | 1.36 | .930 |
| Douglas fir | 1.30 | .88 | 1.21 | 1.16 | 1.36 | 1.03 | 1.41 | 1.16 | .893 |
| Eastern white pine | 1.00 | .71 | 1.03 | .92 | .94 | .79 | 1.06 | .893 | .893 |
| Black walnut | 1.44 | 1.08 | 1.48 | 1.39 | 1.11 | 1.31 | 1.31 | 1.26 | .877 |
| Elm | 1.68 | 2.08 | 1.26 | 1.53 | 1.05 | 1.47 | 1.33 | 1.47 | .872 |
| Hemlock, western | 1.07 | .67 | .86 | .93 | .97 | .90 | 1.10 | .888 | .830 |
| True mahogany | 1.33 | .92 | 1.35 | 1.09 | .91 | .... | 1.12 | 1.06 | .797 |
| Oak (Bur, cow, etc.) | 1.72 | 1.37 | 1.21 | 1.37 | 1.05 | 1.53 | 1.26 | 1.29 | .750 |

[1] The only species permitted are those given in Table XXIX, col. 2.

[2] In computing values in col. 10, the weight per cubic foot was raised only to the first power, which is correct for beams where the depth would not vary with the species of wood chosen, but is not correct for certain members, such as struts. In cases where a species is unusually low in any one strength property, as shown in cols. 3—8, inclusive, it may not be as suitable as another species whose index figure in col. 10 is lower.

## *TABLE XXXI*

### *RATIOS OF STRENGTH VALUES OF SPIRAL AND DIAGONAL TO STRAIGHT GRAINED MATERIAL*

| Slope of Grain | Static Bending | | | | | | | | Impact Bending | |
|---|---|---|---|---|---|---|---|---|---|---|
| | Fiber Stress at Elastic Limit* | | Modulus of Rupture* | | Modulus of Elasticity* | | Work to Maximum Load* | | Maximum Drop* | |
| **SITKA SPRUCE — Absolute Slope Classification** | | | | | | | | | | |
| **1:40 or less | 100.0 | 88.2 | 100.0 | 89.4 | 100.0 | 92.3 | 100.0 | 75.0 | 100.0 | 76.2 |
| 1:25 | 97.8 | 87.4 | 97.4 | 88.2 | 97.6 | 90.5 | 86.0 | 65.6 | 92.1 | 70.6 |
| 1:20 | 97.0 | 87.0 | 96.2 | 85.9 | 96.1 | 88.6 | 78.9 | 57.8 | 87.3 | 68.2 |
| 1.15 | 96.3 | 86.7 | 92.1 | 81.2 | 93.0 | 85.2 | 67.2 | 46.8 | 77.8 | 57.1 |
| 1:12.5 | 94.4 | 84.8 | 88.7 | 77.5 | 90.7 | 82.6 | 56.2 | 39.8 | 67.4 | 48.4 |
| 1:10 | 89.6 | 79.6 | 83.0 | 72.0 | 86.8 | 78.5 | 45.3 | 32.0 | 55.5 | 39.7 |
| 1:7.5 | 81.8 | 70.7 | 72.4 | 60.8 | 78.8 | 70.8 | 34.4 | 24.2 | 43.6 | 30.9 |
| 1:5 | 66.2 | 55.9 | 55.8 | 45.4 | 64.4 | 55.2 | 24.2 | 15.6 | 30.9 | 22.2 |
| **DOUGLAS FIR — Absolute Slope Classification** | | | | | | | | | | |
| 1:40 or less | 100.0 | 88.3 | 100.0 | 89.0 | 100.0 | 87.9 | 100.0 | 80.2 | 100.0 | 87.6 |
| 1:25 | 96.4 | 83.7 | 93.0 | 83.6 | 96.2 | 85.7 | 83.4 | 66.2 | 99.0 | 84.1 |
| 1:20 | 93.2 | 80.4 | 89.6 | 78.4 | 94.2 | 83.9 | 76.1 | 57.9 | 95.5 | 81.1 |
| 1:15 | 88.1 | 74.7 | 84.8 | 70.8 | 91.5 | 80.9 | 66.2 | 48.0 | 87.1 | 72.6 |
| 1:12.5 | 83.9 | 70.7 | 81.8 | 65.9 | 89.6 | 78.1 | 60.4 | 42.2 | 79.6 | 64.7 |
| 1:10 | 77.9 | 65.3 | 75.4 | 59.2 | 86.4 | 74.2 | 53.8 | 35.6 | 69.2 | 53.2 |
| 1:7.5 | 69.2 | 56.6 | 64.6 | 50.6 | 79.4 | 66.2 | 44.7 | 28.9 | 51.8 | 38.3 |
| 1:5 | 54.3 | 44.6 | 46.0 | 41.2 | 59.6 | 49.0 | 32.3 | 21.5 | 35.3 | 21.9 |

*The values in the first column under each property are the ratios of the average of the spiral or diagonal grained material to the average for straight grained material, expressed in per cent. The values in the second column under each property are the ratios of the "most probable value below mean" of the spiral or diagonal grained material to the average for straight grained material expressed in per cent. "The most probable value below the mean" corresponds closely with the arithmetical average of all the specimens having properties lower than the average properties of all the specimens. It is an indication of the reliability or variability of spiral and diagonal grained stock.

**All wood having a slope of grain of 1 to 40 or less was considered as straight grained.

Table XXX are of importance, and, therefore, the order of merit as indicated by this table is closely correct for this type of member. But in other portions of the structure only certain properties are of importance. For example, in selecting the best wood for a tail skid only the modulus of rupture, the work to maximum load and the fiber stress at the elastic limit in impact bending need be considered. In the case of a slender strut the modulus of elasticity is the most important property, though consideration should also be given the work to maximum load. For short struts, on the other hand, the ultimate compressive strength becomes more influential than the modulus of elasticity. It should be noted that in certain members, especially slender struts, where the strength varies as the moment of inertia, that in comparing woods of different specific gravity it is incorrect to use the first power of the ratio of the weights of the woods. The heavier species gives a smaller strut for the same strength, but, since the strength varies directly as the cube of the diameter, the material in the smaller strut is not so effectively placed, thus decreasing the advantage of the heavier, stronger material. In designing members subjected to different kinds of stresses an engineer should decide what properties are important and what their relative importance is. By the aid of the comparative values in Table XXX he can then quickly determine which wood is best adapted to any particular purpose, always bearing in mind that on large, heavy airplanes the shock absorbing properties are less essential than on fast, high-powered pursuit airplanes where vibration is severe. Column 10 of this table shows numerically the relative advantages of the different woods as far as their strength properties alone are concerned. Other factors besides strength properties must be considered also.

## *COMMENTS ON DIFFERENT SPECIES*

Characteristics other than strength must be considered in deciding on the value of a wood for a certain purpose. The comments which follow are based on a report by Mr. J. A. Newlin, in charge of timber mechanics at the Forest Products Laboratory.

A material should dry readily, stay to its place well, and not develop serious defects such as shakes, checks or rot. The way a wood works under the tool, the finish it takes, and all manufacturing conditions must be considered, as well as the size of the trees and the defects normally found in the species. Frequently high grade stock of an inferior species is better than low grade material of a superior species.

*Port Orford Cedar*—Recent data on this species is not quite as favorable as that given in Table XXIX.

*Douglas Fir*—This wood is considerably harder to dry than spruce and is more inclined to shakes, and to check during manufacture, and to develop these defects in service. It is inclined to break in long splinters and to shatter when hit (a serious defect in certain types of military airplanes). The use of Douglas fir in the manufacture of wing beams will require considerably more care than is necessary with spruce, but it should give very excellent results when substituted for spruce in the same sizes.

*Western White Pine*—It is more difficult to dry than eastern white pine, but could be very satisfactorily substituted for spruce in spruce sizes.

*Eastern White Pine*—Tests show this wood to be somewhat below spruce in hardness and rather low in shock resisting ability. It, however, runs quite uniform in its strength properties, is very easily kiln dried without damage, works well, stays to its place well and is recommended as a substitute for spruce in spruce sizes.

*Black and Commercial White Ash*—These woods are well adapted to steam bending.

*Basswood*—Basswood is one of the best species to receive nails without splitting and is used extensively for webs, veneer cores, etc.

*Birch*—Sweet and yellow birches are quite heavy, hard and stiff. They have a very uniform texture and take a fine finish. On account of their hardness and resistance to wear they can be used to face other woods to protect them against abrasion. Birch is extensively used in propeller and plywood construction.

*Black Cherry*—This is a very desirable propeller wood.

*Poplar*—Of the three species listed under this name, cucumber is considerably the best, while yellow poplar is the weakest. Cucumber is one of the few hardwoods which gives promise of being a good substitute for spruce. Magnolia is similar to cucumber. Yellow poplar, while rather low in shock resisting ability, has good working qualities, ability to retain its shape, and freedom from checks, shakes and such defects, characteristics which make it a fairly satisfactory substitute for spruce. It presents no manufacturing difficulties. Poplar is used to some extent in propeller construction.

*Elm*—This species is low in stiffness though very resistant to shock. It steam bends well, and if properly dried can be used for longerons as a substitute for ash which is somewhat lighter. Considerably more care is necessary in the drying of elm in order to have it remain in shape, as it twists and warps badly when not held quite firm. White and slippery elm when of the same density as rock elm may be substituted for the latter. A lighter grade of white elm could probably be used to excellent advantage in the bent work at the end of wings, rudders, elevators, etc.

*Sugar Maple*—Sugar maple is quite heavy, hard and stiff. It should be used along with birch in propeller manufacture. It has very uniform texture and takes a fine finish. Like birch it is often used to protect other woods against abrasion.

*Oak*—The oaks need not be considered as substitutes for spruce, but they play an important part in the manufacture of propellers. They are all quite heavy and hard, and are extremely variable in their strength properties. White oaks, as a rule, shrink and swell more slowly with changes in the weather than do red oaks. Because radial shrinkage in oaks is about half the tangential shrinkage, quarter-sawn oak is much superior to plain sawn oak for propeller construction. Southern grown oaks are much more difficult to dry than are the northern oaks. Northern white oaks when quarter sawn and carefully dried give very satisfactory propellers. Quarter sawn northern red oaks are fairly satisfactory for propeller construction, but they have the disadvantage of being more subject to defects in the living tree, decay more readily, and change more rapidly with change in weather conditions.

*Black Walnut*—This wood need not be considered at all as a substitute for spruce. But it probably makes the best propeller of any of the native species. Black walnut is somewhat difficult to dry, but has very excellent power of retaining its place and has good hardness to resist wear.

*Summary*—Data available indicate strongly that the following species can be substituted for spruce in highly stressed parts: Port Orford Cedar, Coast Type Douglas Fir, Amabilis, Grand, Noble and White Firs, Eastern and Western White Pine, Yellow Poplar, Cucumber and Magnolia. Certain other woods give good promise of furnishing spruce substitutes, though more work is necessary to overcome known difficulties before they can be definitely recommended; viz, Lodgepole Pine, Norway Pine, and Redwood.

175. *Properties of Duralumin*—The chemical composition of the duralumin made by the Aluminum Co. of America, is as follows:

| | |
|---|---|
| Aluminum ........................ | 94.5 per cent |
| Copper ........................... | 4.0 per cent |
| Magnesium ...................... | 0.5 per cent |
| Manganese ...................... | 0.5 per cent |
| Impurities ....................about | 0.5 per cent |

These impurities must contain no lead, tin or zinc.

The important physical properties are:

| | |
|---|---|
| Specific gravity ............... | 2.8 |
| Melting point ................. | 650° C. |
| Ultimate tensile strength ........ | 55,000 lb. per sq. in. |
| Yield point in compression ...... | 27,000 lb. per sq. in. |
| Ultimate shearing strength (rivets) | 35,000 lb. per sq. in. |
| Ultimate bearing strength (rivets) | 90,000 lb. per sq. in. |
| Modulus of elasticity .......... | 10,000,000 lb. per sq. in. |
| Elongation in 2 in. ............. | 18—22 per cent. |

Duralumin must not be bent around a pin with a radius less than four times the thickness of the duralumin.

To prevent cracking all edges should be rounded and, when possible, sections should be so designed that the edges will be at or near the neutral axis.

Bent plate fittings, with bent lugs which must resist vibration should be made from sheet steel instead of duralumin. For stressed parts, which while in flight are exposed to an increase in temperature of more than 100 deg. C., the use of duralumin is objectionable unless a correspondingly smaller strength value is used in computations. Cold has no harmful influence on duralumin. The joint between iron and steel and duralumin can be made without electrolytic action occurring. Pieces, which for better working must be heated, must in all cases be re-tempered after completion.

*TABLE XXXII*

*PROPERTIES OF CIRCULAR TUBING*

| Outside Diam. inches | Gauge 22 | 20 | 18 | Wall Thickness in Inches | | | | | | |
|---|---|---|---|---|---|---|---|---|---|---|
| | .028 | .035 | .049 | 1/16 | 3/32 | 1/8 | 5/32 | 3/16 | 7/32 | 1/4 |
| | | | | Moments of Inertia | | | | | | |
| 1/2 | .00116 | .00139 | .0018 | .0021 | .0026 | .00288 | | | | |
| 5/8 | .00234 | .00283 | .0037 | .0044 | .0057 | .00652 | | | | |
| 3/4 | .00414 | .00504 | .0067 | .0080 | .0106 | .01246 | .0137 | .0146 | | |
| 7/8 | .0067 | .00816 | .0109 | .0132 | .0178 | .02128 | .0239 | .0257 | .0270 | |
| 1 | .0101 | .0124 | .0166 | .0203 | .0277 | .03356 | .0381 | .0416 | .0442 | .0460 |
| 1–1/8 | .0145 | .0178 | .0240 | .0295 | .0407 | .04985 | .0572 | .0631 | .0677 | .0711 |
| 1–1/4 | .0201 | .0247 | .0334 | .0412 | .0573 | .07075 | .0819 | .0911 | .0984 | .1043 |
| 1–3/8 | .0269 | .0331 | .0450 | .0556 | .0778 | .09683 | .1129 | .1264 | .1373 | .1467 |
| 1–1/2 | .0352 | .0432 | .0588 | .0730 | .1008 | .1287 | .1509 | .1699 | .1859 | .1994 |
| 1–3/4 | .0562 | .0695 | .0950 | .1181 | .1678 | .2119 | .2508 | .2849 | .3147 | .3405 |
| 2 | .0845 | .1047 | .1432 | .1787 | .2556 | .3250 | .3873 | .4431 | .4928 | .5369 |
| 2–1/4 | .1210 | .1498 | .2058 | .2571 | .3698 | .4727 | .5663 | .6514 | .7283 | .7978 |
| 2–1/2 | .1665 | .2066 | .2840 | .3557 | .5137 | .6594 | .7935 | .9165 | 1.029 | 1.132 |
| 2–3/4 | .2220 | .2760 | .3794 | .4767 | .6909 | .8899 | 1.075 | 1.246 | 1.404 | 1.549 |
| 3 | .2885 | .3594 | .4950 | .6240 | .9047 | 1.169 | 1.415 | 1.645 | 1.860 | 2.059 |
| | | | | Radii of Gyration | | | | | | |
| 1/2 | .1672 | .1649 | .1604 | .1563 | .1474 | .1398 | | | | |
| 5/8 | .2113 | .2090 | .2044 | .2001 | .1907 | .1822 | | | | |
| 3/4 | .2555 | .2531 | .2484 | .2441 | .2344 | .2253 | .2171 | .2096 | | |
| 7/8 | .2996 | .2972 | .2925 | .2881 | .2782 | .2688 | .2601 | .2519 | .2446 | |
| 1 | .3438 | .3414 | .3367 | .3322 | .3221 | .3125 | .3034 | .2948 | .2868 | .2795 |
| 1–1/8 | .3880 | .3856 | .3808 | .3763 | .3661 | .3563 | .3469 | .3380 | .3296 | .3217 |

## *TABLE XXXII (Continued)*

| Outside Diam. inches | Gauge 22 | 20 | 18 | Wall Thickness in Inches | | | | | | |
|---|---|---|---|---|---|---|---|---|---|---|
| | .028 | .035 | .049 | 1/16 | 3/32 | 1/8 | 5/32 | 3/16 | 7/32 | 1/4 |
| **Radii of Gyration** | | | | | | | | | | |
| 1-1/4 | .4322 | .4297 | .4250 | .4204 | .4101 | .4002 | .3906 | .3815 | .3727 | .3644 |
| 1-3/8 | .4762 | .4739 | .4691 | .4646 | .4542 | .4441 | .4344 | .4250 | .4160 | .4075 |
| 1-1/2 | .521 | .5181 | .5133 | .5087 | .4983 | .4881 | .4783 | .4688 | .4595 | .4507 |
| 1-3/4 | .609 | .607 | .602 | .5970 | .5865 | .5762 | .5662 | .5564 | .5469 | .5376 |
| 2 | .697 | .695 | .690 | .6854 | .6748 | .6644 | .6542 | .6442 | .6343 | .6250 |
| 2-1/4 | .786 | .783 | .778 | .7737 | .7631 | .7526 | .7423 | .7322 | .7223 | .7126 |
| 2-1/2 | .874 | .871 | .867 | .8621 | .8513 | .8409 | .8305 | .8203 | .8102 | .8004 |
| 2-3/4 | .963 | .960 | .955 | .9504 | .9397 | .9291 | .9187 | .9084 | .8983 | .8883 |
| 3 | 1.051 | 1.049 | 1.043 | 1.0384 | 1.028 | 1.017 | 1.007 | .9966 | .9864 | .9763 |
| **Areas** | | | | | | | | | | |
| 1/2 | .0415 | .0511 | .0694 | .0859 | .1197 | .1473 | .1688 | .1841 | .1933 | .1964 |
| 5/8 | .0525 | .0649 | .0887 | .1104 | .1565 | .1964 | .2301 | .2577 | .2792 | .2945 |
| 3/4 | .0635 | .0786 | .1079 | .1349 | .1933 | .2454 | .2915 | .3313 | .3651 | .3927 |
| 7/8 | .0745 | .0924 | .1272 | .1595 | .2301 | .2945 | .3528 | .4050 | .4510 | .4909 |
| 1 | .0855 | .1061 | .1464 | .1841 | .2669 | .3436 | .4142 | .4786 | .5369 | .5891 |
| 1-1/8 | .0965 | .1199 | .1656 | .2086 | .3037 | .3927 | .4756 | .5522 | .6228 | .6872 |
| 1-1/4 | .1075 | .1336 | .1849 | .2332 | .3406 | .4418 | .5369 | .6259 | .7087 | .7854 |
| 1-3/8 | .1185 | .1473 | .2041 | .2577 | .3774 | .4909 | .5983 | .6995 | .7946 | .8836 |
| 1-1/2 | .1295 | .1611 | .2234 | .2823 | .4142 | .5400 | .6596 | .7731 | .8805 | .9818 |
| 1-3/4 | .1515 | .1886 | .2618 | .3313 | .4878 | .6381 | .7824 | .9204 | 1.0523 | 1.1781 |
| 2 | .1735 | .2161 | .3003 | .3804 | .5615 | .7363 | .9051 | 1.0677 | 1.2242 | 1.3745 |
| 2-1/4 | .1955 | .2436 | .3388 | .4295 | .6351 | .8345 | 1.0278 | 1.2149 | 1.3960 | 1.5708 |
| 2-1/2 | .2174 | .2710 | .3773 | .4786 | .7087 | .9327 | 1.1505 | 1.3622 | 1.5678 | 1.7672 |
| 2-3/4 | .2394 | .2985 | .4158 | .5277 | .7823 | 1.0308 | 1.2732 | 1.5094 | 1.7396 | 1.9635 |
| 3 | .2614 | .3260 | .4543 | .5768 | .8560 | 1.1290 | 1.3960 | 1.6567 | 1.9114 | 2.1599 |

176. *Miscellaneous Tables and Charts.*

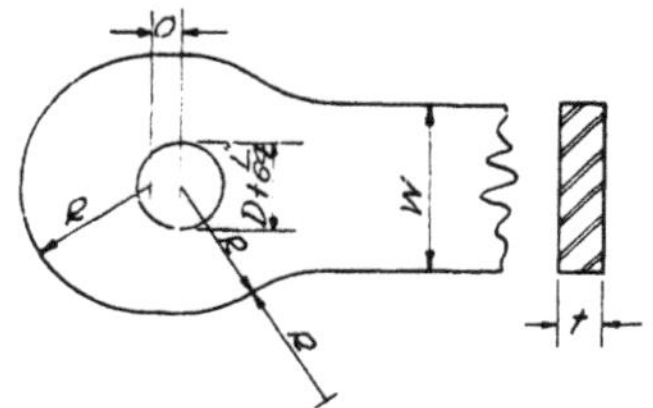


**Standard Lugs**

*TABLE XXXIII*
*PROPERTIES OF STANDARD LUGS*

| Tie Rod or Cable Rating | t | Shank | | R | Eye | | Pin Diameter** |
|---|---|---|---|---|---|---|---|
| | | W | Ultimate Strength* | | O | Ultimate Strength* | D |
| 1,000 | 3/32 | 5/16 | 1,610 | 7/32 | 3/64 | 1,205 | 3/16 |
| 1,900 | 1/8 | 11/32 | 2,360 | 9/32 | 1/16 | 2,470 | 3/16 |
| 2,600 | 1/8 | 15/32 | 3,220 | 3/8 | 5/64 | 3,330 | 1/4 |
| 3,400 | 3/16 | 13/32 | 4,180 | 11/32 | 1/16 | 4,350 | 1/4 |
| 4,600 | 3/16 | 9/16 | 5,800 | 7/16 | 3/32 | 5,630 | 5/16 |
| 6,100 | 3/16 | 23/32 | 7,400 | 9/16 | 7/64 | 7,560 | 3/8 |
| 8,500 | 1/4 | 3/4 | 10,310 | 9/16 | 7/64 | 10,100 | 3/8 |
| 12,500 | 5/16 | 7/8 | 15,040 | 21/32 | 9/64 | 14,750 | 7/16 |
| 17,500 | 3/8 | 1-1/32 | 21,800 | 3/4 | 5/32 | 20,300 | 1/2 |

*Based on quarter hard carbon steel of an ultimate tensile strength of 55,000 lbs. per sq. in.
**Diameter of hole in eye is pin diameter + 1/64 in.

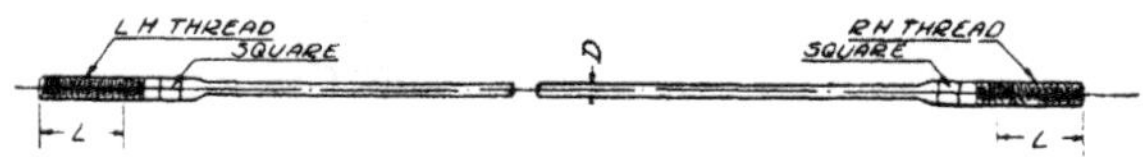


**Swaged Tie Rods**

## *TABLE XXXIV*

## *PROPERTIES OF SWAGED TIE RODS*

| Size and Threads* per 1 in. | D | L Length of Usable Thread | Area of Cross Section in sq. in. (Swaged) Portion) | Ultimate Strength in lbs. (minimum) Threaded Portion | Swaged Portion |
|---|---|---|---|---|---|
| 6—40 | .101 | .715 | .0080 | 1,150 | 1,000 |
| 10—32 | .134 | 1.000 | .0141 | 2,200 | 1,900 |
| 12—28 | .155 | 1.063 | .0189 | 3,000 | 2,600 |
| 1/4 —28 | .180 | 1.219 | .0255 | 4,000 | 3,400 |
| 5/16—24 | .223 | 1.250 | .0391 | 6,550 | 5,700 |
| 3/8 —24 | .274 | 1.375 | .0590 | 9,800 | 8,500 |
| 7/16—20 | .326 | 1.500 | .0835 | 13,200 | 11,500 |
| 1/2 —20 | .377 | 1.625 | .1117 | 17,800 | 15,500 |

*Standard U. S. Threads.

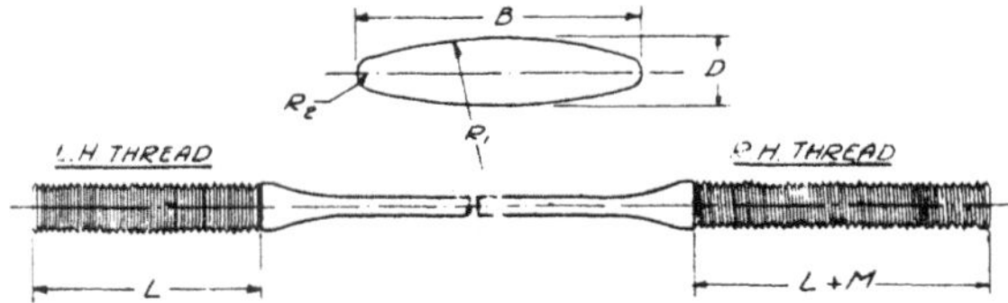


Streamline Wires

## *TABLE XXXV*
## *PROPERTIES OF STANDARD STREAMLINE WIRES*

| Size and Threads* per 1 in. | B | D | L Length of Usable Thread | Area of Cross Section in sq. in. (Swaged Portion) | Ultimate Strength in lbs. (minimum) Threaded Portion | Streamline Portion |
|---|---|---|---|---|---|---|
| 6—40 | .192 | .048 | 1.0 | .0071 | 1,350 | 1,000 |
| 10—32 | .256 | .064 | 1.1 | .0125 | 2,700 | 1,900 |
| 1/4 —28 | .348 | .087 | 1.3 | .0234 | 5,000 | 3,400 |
| 5/16—24 | .440 | .110 | 1.5 | .0376 | 8,050 | 5,700 |
| 3/8 —24 | .540 | .135 | 1.7 | .0563 | 12,400 | 8,500 |
| 7/16—20 | .636 | .159 | 1.9 | .0781 | 16,700 | 11,500 |
| 1/2 —20 | .732 | .183 | 2.0 | .1026 | 22,800 | 15,500 |

*Standard U. S. Threads.
M equals 1/2 in. for lengths less than 8 ft. For greater lengths, M must be increased 1/4 in. for each 1 ft. increase in length.

## *TABLE XXXVI*
## *STRENGTH AND WEIGHT OF STEEL WIRE CABLE*

| Diameter in inches | | 19 Strands Non-Flexible Cable: Approx. Weight in lbs. per 100 ft. | 19 Strands Non-Flexible Cable: Breaking Strength in lbs. (minimum) | 7x19 Extra-Flexible Cable: Approx. Weight in lbs. per 100 ft. | 7x19 Extra-Flexible Cable: Breaking Strength in lbs. (minimum) |
|---|---|---|---|---|---|
| .375 | 3/8 | | | 26.45 | 14,400 |
| .344 | 11/32 | | | 22.53 | 12,500 |
| .312 | 5/16 | 20.65 | 12,500 | 17.71 | 9,800 |
| .281 | 9/32 | | | 14.56 | 8,000 |
| .250 | 1/4 | 13.50 | 8,000 | 12.00 | 7,000 |
| .218 | 7/32 | 10.00 | 6,100 | 9.50 | 5,600 |
| .187 | 3/16 | 7.70 | 4,600 | 6.47 | 4,200 |
| .156 | 5/32 | 5.50 | 3,200 | 4.44 | 2,800 |
| .125 | 1/8 | 3.50 | 2,100 | 2.88 | 2,000 |
| .109 | 7/64 | 2.60 | 1,600 | | |
| .094 | 3/32 | 1.75 | 1,100 | | |

TABLE 37.

DATA ON ENGINE WEIGHTS AND HORSEPOWERS

| No. | Engine | No. of cyl. | Model | Normal H.P. | R.P.M. at normal H.P. | Maximum H.P. | R.P.M. at maximum H.P. | Total dry weight | Weight of engine dry/normal H.P. | Weight of engine Water | Fuel consumption at normal R.P.M. lbs./HP hr. | Oil consumption at normal R.P.M. lbs./HP hr. |
|---|---|---|---|---|---|---|---|---|---|---|---|---|
| 1 | Liberty | 6 | 6 | 231.5 | 1700 | 240 | 1850 | 567.5 | 2.46 | 21 | .526 | .0256 |
| 2 | " | 8 | 8 | 290 | 1700 | 317 | 2000 | 638.0 | 2.20 | 28 | .490 | .0211 |
| 3 | " | 12 | 12 | 421 | 1700 | 449 | 1940 | 843.6 | 2.01 | 45 | .496 | .0317 |
| 4 | Hispano-Suiza | 8 | A or I | 154 | 1450 | 204 | 2070 | 467.0 | 3.03 | 44 | .515 | .023 |
| 5 | " " | 8 | E | 190 | 1800 | 221 | 2230 | 476.0 | 2.51 | 44 | .493 | .0193 |
| 6 | " " | 8 | H | 325 | 1800 | 376 | 2240 | 632.0 | 1.95 | 58 | .521 | .055 |
| 7 | Packard | 8 | 1A-744 | 179.5 | 1600 | 199 | 2000 | 541.8 | 3.02 | 25¾ | .484 | .056 |
| 8 | " | 12 | 1A-1116 | 282 | 1600 | 354 | 2240 | 733.2 | 2.60 | 41 | .460 | .0347 |
| 9 | " | 12 | 1A-2025 | 540 | 1800 | 584 | 2000 | 1126.5 | 2.09 | 58 | .503 | .0342 |
| 10 | Hall Scott | 6 | L6-a | 214 | 1700 | 214 | 1700 | 548.9 | 2.56 | 25 | .590 | |
| 11 | Lawrance | 3 | | 56.5 | 1600 | 65 | 1950 | 147.4 | 2.61 | 0 | .512 | .0404 |
| 12 | Curtiss Kirkham | 12 | K-12 | 396.5 | 2250 | 406 | 2420 | 678.5 | 1.71 | 38¼ | .499 | .0532 |
| 13 | Napier Lion | 12 | | 472.5 | 1925 | 496 | 2050 | 858.0 | 1.82 | | .504 | .0335 |
| 14 | Benz | 6 | | 221.5 | 1400 | 238 | 1600 | 638.2 | 2.88 | 19 | .491 | .0468 |
| 15 | Mercedes | 6 | | 184 | 1400 | 192.5 | 1700 | 659.0 | 3.58 | 27¾ | .449 | .026 |
| 16 | Fiat | 6 | A-12Bis | 301 | 1600 | 339 | 1900 | 924.0 | 3.07 | 33½ | .579 | .0626 |
| 17 | Mercedes | 6 | | 252 | 1400 | 268 | 1600 | 936.0 | 3.72 | 54 | .543 | .0366 |
| 18 | Bayern | 6 | 4 | 234 | 1400 | 254 | 1600 | 643.5 | 2.75 | | .421 | .0326 |
| 19 | Rolls Royce Eagle | 12 | 8 | 359 | 1800 | 375 | 2000 | 933.0 | 2.60 | 26 | .50 | .028 |
| 20 | Fiat | 12 | A-14 | 650 | 1550 | 725 | 1730 | 1739.3 | 2.68 | 60 | .528 | .055 |
| 21 | Isotta Fraschini | 6 | V-6 | 260 | 1650 | 285 | 1850 | 572.0 | 2.20 | | .436 | .021 |
| 22 | Curtiss | 8 | OX-5 | 90 | 1400 | 95 | 1450 | 385.0 | 4.27 | 17 | .60 | .030 |
| 23 | Le Rhone | 9 | | 80 | | 85 | 1300 | 260.0 | 3.25 | 0 | | |
| 24 | " " | 9 | "110" | 130 | 1250 | 131 | 1300 | 330.0 | 2.54 | 0 | .567 | .09 |
| 25 | Gnome | 9 | B-2 | 100 | 1200 | 106 | 1230 | 310.0 | 3.10 | 0 | .785 | .17 |
| 26 | ABC Dragonfly | 9 | | 340 | 1650 | 350 | 1750 | 550.0 | 1.62 | 0 | .585 | .028 |

The data for engines 1 to 16 inclusive were obtained from Standard Engine Reports issued by McCook Field and are reliable and comparable. The remaining data were taken from the best sources available, such as technical magazines and special reports. The horsepowers and fuel consumptions for the first 16 engines were obtained from dynamometer tests under the best conditions. In flight, the power developed will be somewhat lower and the fuel consumption higher.

The weights of the engines dry include, carburetor, ignition, oil and water pumps, and propeller hub, complete, but not battery or switch and voltage regulator.

## *TABLE XXXVIII*

## *MISCELLANEOUS POWER PLANT WEIGHTS*

| | Curtiss OX5 LeRhone "110" | Hispano-Suiza E | Hispano-Suiza H | Liberty 12 |
|---|---|---|---|---|
| Propeller (light wood) ..... | 20 | 30 | 35 | 45 |
| Propeller (heavy wood).... | | | 55 | 60 |
| Radiator (nose type) empty. | | 55 | 75 | 115 |
| Radiator water ........... | | 35 | 45 | 65 |
| Water in piping .......... | 5 | 5 | 5–10 | 5–10 |
| Radiator Connections and Shutter ................ | 15 | 15 | 15 | 25 |
| 2 Short Manifolds ........ | | | 15 | 22 |
| 2 Long Manifolds ......... | | 23 | | |
| I Set of Stacks........... | | | 8 | 15 |

Specific gravity of best test gasoline.......... .71
Specific gravity of good test gasoline......... .72
Weight of good test gasoline............... { 6.0 lbs./U. S. gal. / 7.2 lbs./Imperial gal.
Weight of engine oil ...................... { 7.5 lbs./U. S. gal. / 9.0 lbs./Imperial gal.
Weight of water ........................ 8.33 lbs./U. S. gal.

Weight of plain sheet steel gasoline tanks, .60 to .80 lbs./U. S. gal.

Weight of leakproof gasoline tanks, 2.0 to 2.5 lbs./gal. for small gravity tanks, 1.5 to 1.8 lbs./gal. for 80 to 30 gal. tanks, and .8 to 1.2 lbs./gal. for 500 to 100 gal. tanks.

Weight of standard storage battery for ignition system, 10.5 lbs.

Weight of standard switch and voltage regulator for ignition system, 2.4 lbs.

Propeller, radiator, and radiator water weights do not vary appreciably with different types of engines, but with their horsepower.

Contrary to last foot note to Table XXXVII the weights of standard storage battery and switch and voltage regulator are included in the dry weights for the three Packard engines, for the Hall-Scott L6-A, and for the Lawrance radial.

# TABLE XXXIX
## PROPERTIES OF STEELS
### U. S. Army Specifications

| Spec. No. | Description | Ultimate Tensile Strength | Yield Pt. | Elongation in 2 in., % | Reduction in area, % |
|---|---|---|---|---|---|
| STEEL BARS AND BILLETS - Cold Drawn or Cold Rolled | | | | | |
| 10028-A | Carbon | 65,000 | 50,000 | 15 | 35 |
| 10029 | Alloy | 100,000 | 80,000 | 17 | 40 |
| 100[illegible] | " | 125,000 | 100,000 | 16 | 50 |
| 10034 | Dead Soft Carbon | 70,000 | --- | -- | -- |
| STEEL BARS AND BILLETS - Hot Rolled, Annealed, Heat Treated, or Cold Drawn | | | | | |
| 10030-B | Low Carbon | 70,000 | 45,000 | 20 | -- |
| 10036-B | Medium " | 80,000 | 60,000 | [illegible] | 45 |
| 10037-B | Half Hard Carbon | 95,000 | 70,000 | 18 | 45 |
| 10045-B | Alloy | 100,000 | 80,000 | 20 | 50 |
| 10046-B | " | 125,000 | 95,000 | 17 | 50 |
| 10047-C | " | 150,000 | 115,000 | 14 | 45 |
| 10048-B | " | 175,000 | 150,000 | 12 | 40 |
| 10049-B | " | 200,000 | 160,000 | 10 | 35 |
| 10050-B | " | 225,000 | 190,000 | 9 | 30 |
| For Case Hardening | | | | | |
| 10051-A | Alloy | 130,000 | 120,000 | 18 | -- |
| 10052-A | " | 160,000 | 150,000 | 14 | -- |
| 10053-A | " | 180,000 | 170,000 | 1[illegible] | -- |
| STEEL SHEETS AND STRIPS - Cold Rolled and Annealed | | | | | |
| 10200-B | Extra Soft Carbon | 42,000 | 24,000 | [illegible] | -- |
| 10201-B | Quarter Hard " | 55,000 | 36,000 | [illegible] | -- |
| 102[illegible] | Medium " | 55,000 | 36,000 | [illegible] | -- |
| 10203 | Spring Steel | | | | |
| 10206-B | Alloy Steel | 100,000 | 75,000 | 15 | -- |
| 10211 | Mild Car. Van. | 60,000 | 40,000 | [illegible] | -- |
| 102[illegible] | Soft Carbon | 48,000 | 30,000 | 25 | -- |
| STEEL FORGINGS - Refining Anneal Before Final Heat Treatment | | | | | |
| 10250 | Annealed Carbon | 65,000 | 36,000 | 25 | 56 |
| 10251-A | Heat Treat. " | 90,000 | [illegible] | 2[illegible] | 45 |
| 10252 | Alloy | 1[illegible] | 80,000 | 20 | 50 |
| 10254-A | " | 130,000 | 115,000 | 17 | 50 |
| 10255 | Heat Treat. Carbon | 95,000 | 70,000 | 15 | 45 |
| STEEL TUBES - Cold Drawn | | | | | |
| 10225-A | Soft Carbon | 5[illegible],000 | 36,000 | -- | -- |
| 10226-A | Medium Carbon | [illegible],000 | 60,000 | | |
| 10227 | Alloy Grade #1 | 110,000 | 90,000 | 15 in 2 in.<br>5 in 8 in. | |
| | #2 | 85,000 | 60,000 | 25 in 2 in.<br>10 in 8 in. | |
| 10[illegible] | Alloy Axle | 200,000 | | 5 in 2 in. | |
| 10230-A | " | " | | 3 in 4 in. | |

COMPOSITION & PROPERTIES OF STEELS

Air Service, Engineering Division.

Note. 1. Physical properties are after suitable heat treatment, or cold-rolling or cold-drawing.

2. Elongations for steel sheets and strips are taken in 4 in. instead of 2 in.

3. In specifications 10,200-B and 10,202-A the strength required in sheets is less than in strips.

| | Ultimate Strength Strips | Ultimate Strength Sheets | Yield Point Strips | Yield Point Sheets |
|---|---|---|---|---|
| 10,200-B | 42,000 | 38,000 | 24,000 | 20,000 |
| 10,202-A | 55,000 | 50,000 | 36,000 | 30,000 |

4. The steel numbers marked • are permitted under the specifications; those marked ● are the steel numbers recommended and in common use. In the alloy steels the analysis to be used depends on the service to which the part is to be subjected.

5. In specifications 10,028-A and 10,029 the properties given are for steel in billet form. For steel in the form of small bars, the following properties are required:

| Specification 10,028-A | Ultimate | Yield Point | Elong. | Reduction |
|---|---|---|---|---|
| Bars not over .375 in. in diam. or width. | 70,000 | 50,000 | 10 | 40 |
| Bars .375 to .75 in. in diam. or width. | 69,000 | 50,000 | 13 | 38 |
| Bars .75 to 1.50 in. in diam. or width. | 68,000 | 50,000 | 15 | 36 |
| Bars over 1.50 in. in diam. or width. | 65,000 | 50,000 | 15 | 36 |

| Steel Number | Carbon | Manganese | Phosphorus | Sulphur | Nickel | Chromium |
|---|---|---|---|---|---|---|
| 1010 | 0.05-0.15 | 0.30-0.60 | 0.045 | 0.050 | | |
| 1015 | .10-0.20 | .30-0.60 | .045 | .050 | | |
| 1020 | .15-0.25 | .30-0.60 | .045 | .050 | | |
| X1020 | .15-0.25 | .60-0.80 | .045 | .08-.09 | | |
| 1025 | .20-0.30 | .50-0.80 | .045 | .050 | | |
| 1030 | .25-0.35 | .50-0.80 | .045 | .050 | | |
| 1035 | .30-0.40 | .50-0.80 | .045 | .050 | | |
| 1045 | .40-0.50 | .50-0.80 | .045 | .050 | | |
| X1045 | .40-0.50 | .70-1.00 | .025 | .045 | | |
| 1050 | .45-0.55 | .50-0.80 | .045 | .050 | | |
| 1055 | .50-0.60 | .50-0.80 | .040 | .045 | | |
| 1065 | .60-0.70 | .50-0.70 | .040 | .045 | | |
| 1070 | .65-0.75 | .50-0.70 | .040 | .045 | | |
| 1080 | .75-0.90 | .25-0.50 | .040 | .045 | | |
| 1095 | .90-1.05 | .25-0.50 | .040 | .045 | | |
| 2315 | .10-0.20 | .30-0.6[illegible] | .040 | .045 | 3.25-3.[illegible] | |
| 2320 | .15-0.25 | .30-0.60 | .040 | .045 | 3.25-3.[illegible] | |
| 2325 | .20-0.30 | .50-0.80 | .040 | .045 | 3.25-3.75 | |
| 2330 | .25-0.35 | .50-0.80 | .040 | .045 | 3.25-3.75 | |
| 2335 | .30-0.40 | .50-0.80 | .040 | .045 | 3.25-3.75 | |
| 2340 | .35-0.45 | .50-0.80 | .040 | .045 | 3.25-3.75 | |
| 2515 | .10-0.20 | .30-0.60 | .040 | .045 | 4.50-5.50 | |
| 3120 | 0.15-0.25 | 0.30-0.60 | 0.040 | 0.045 | 1.00-1.50 | 0.45-0.75 |
| 3130 | .25-0.35 | 0.50-0.80 | 0.040 | 0.045 | 1.00-1.50 | .45-0.75 |
| 3135 | .30-0.40 | .50-0.80 | .040 | .045 | 1.00-1.50 | .45-0.75 |
| 3140 | .35-0.45 | .50-0.80 | .040 | .045 | 1.00-1.50 | .45-0.75 |
| 3145 | .40-0.50 | .50-0.80 | .040 | .045 | 1.00-1.50 | .45-0.75 |
| 3215 | .10-0.20 | .30-0.60 | .040 | .045 | 1.75-2.25 | .90-1.25 |
| 3230 | .25-0.35 | .30-0.60 | .040 | .045 | 1.75-2.25 | .90-1.25 |
| 3240 | .35-0.45 | .30-0.60 | .040 | .045 | 1.75-2.25 | .90-1.25 |
| 3245 | .40-0.50 | .30-0.60 | .040 | .045 | 1.75-2.25 | .90-1.25 |
| X3315 | .10-0.20 | .30-0.60 | .040 | .045 | 2.75-3.25 | .70-1.00 |
| X3330 | .25-0.35 | .45-0.75 | .040 | .045 | 2.75-3.25 | .70-1.00 |
| X3340 | .35-0.45 | .45-0.75 | .040 | .045 | 2.75-3.25 | .70-1.00 |
| 3315 | .10-0.20 | .30-0.60 | .040 | .045 | 3.25-3.75 | 1.25-1.75 |
| 3330 | .25-0.35 | .30-0.60 | .040 | .045 | 3.25-3.75 | 1.25-1.75 |
| 3340 | .35-0.45 | .30-0.60 | .040 | .045 | 3.25-3.75 | 1.25-1.75 |
| 3440 | .35-0.45 | .30-0.60 | .040 | .045 | 4.00-5.00 | 1.00-1.50 |
| | | | | | Chromium | Vanadium |
| 6120 | .15-0.25 | .30-0.60 | .040 | .045 | .60-0.90 | 0.15 |
| 6125 | .20-0.30 | .50-0.80 | .040 | .045 | .80-1.10 | .15 |
| 6130 | .25-0.35 | .50-0.80 | .040 | .045 | .80-1.10 | .15 |
| 6135 | .30-0.40 | .50-0.80 | .040 | .045 | .80-1.10 | .15 |
| 6140 | .35-0.45 | .50-0.80 | .040 | .045 | .80-1.10 | .15 |
| 6145 | .40-0.50 | .50-0.80 | .040 | .045 | .80-1.10 | .15 |
| 6150 | .45-0.55 | .50-0.80 | .040 | .045 | .80-1.10 | .15 |

## TABLE XL—PROPERTIES OF NON-FERROUS METALS

COMPOSITION & PROPERTIES OF NON-FERROUS METALS

AIR SERVICE, ENGINEERING DIVISION.

| Spec. No. | Description | | Ultimate Tensile Strength | Yield Point | Elong. in 2 in., % | Reduction in Area, % | Copper | Tin | Iron | Lead | Zinc | Phosphorus | Manganese | Aluminum | Hardening Constituents | Other Impurities | Nickel | Special 99.5% Aluminum | Standard No. 1 - 99.0% Aluminum, (Min) | Standard No. 2 - 98.0% Aluminum, (Min) | Must be either lake or electrolytic copper and have a purity of at least 99.88% |
|---|---|---|---|---|---|---|---|---|---|---|---|---|---|---|---|---|---|---|---|---|---|
| CASTINGS | | | | | | | | | | | | | | | | | | | | | |
| 11020 | Gun Metal | | 30,000 | 15,000 | 15 | | 87-89 | 9-11 | Max.0.2 | Max.0.2 | 1-3 | | | | | | | | | | |
| 11021 | Manganese Bronze | | 70,000 | 30,000 | 15 | | 56-60 | | | Max.0.15 | 37-41 | | | | Max.3.0 | | | | | | |
| 11022-A | Phosphor Bronze | | 20,000 | 10,000 | 8 | 10 | 79-81 | 9-11 | | 9-11 | | .1-0.3 | | | | 0.25 | | | | | |
| 11023 | Aluminum Alloy, Alloy No. 1 | | 18,000 | | 1.5 | | 7-8.5 | | | | | | | Rem. | | Max.1.7 | | | | | |
| 11024 | Aluminum Alloy, Alloy No. 2 | | 18,000 | | | | 9½-10½ | | | | | | | Rem. | | Max.1.7 | | | | | |
| 11026-A | Red Brass | | | | | | 84-86 | 4-6 | | 4-6 | 4-6 | | | | | 0.25 | | | | | |
| 11028 | Bronze | | 35,000 | 19,000 | 9 | | Rem. | 10-11.5 | Max.0.2 | Max.0.2 | 2-3 | | | | | | | | | | |
| 11029 | Yellow Brass | | | | | | 6[illegible]-65 | | | 2-4 | 31-36 | | | | | Max.0.50 | | | | | |
| BARS | | | | | | | | | | | | | | | | | | | | | |
| 11030-B | Naval Brass or Equiv. Alloy | | 62,000 | 37,000 | 25 | 37 | 59-63 | 0.5-1.5 | Max.0.10 | Max.0.30 | Rem. | | | | | | | | | | |
| 11033 | Aluminum Alloy | | 45,000 | 20,000 | 15 | | | | | | | | | | | | | | X | | |
| 11034 | Aluminum | | 18,000 | | 6 | | | | | | | | | | | | | | X | X | |
| 11035 | Phosphor Bronze | | 55,000 | 30,000 | 25 | | 94 | 3.5 | Max.0.10 | Max.0.20 | Max.0.30 | 0.05-0.50 | | | | | | | | | |
| 11036 | Manganese Bronze | | 70,000 | 35,000 | 30 | | 57-60 | Min.0.50 | Max.1.00 | Max.0.20 | 37-40 | | Max.0.30 | | | Max.0.10 | | | | | |
| 11037 | Free Cutting or Screw Mach. Br. | | 45,000 | | | | [illegible]-63 | | Max.0.10 | 2.25-3.25 | Rem. | | | | | Max.0.25 | | | | | |
| SHEETS AND STRIPS | | | | | | | | | | | | | | | | | | | | | |
| 11040 | Copper Sheet | Soft | 30,000 | | 25 | | | | | | | | | | | | | | | | X |
| | | Hard | 35,000 | | 18 | | | | | | | | | | | | | | | | |
| 11042 | Naval Brass or Equiv. Alloy | | 56,000 | 27,000 | 35 | | 59-63 | 0.5-1.5 | Max.0.10 | Max.0.3 | Rem. | | | | | | | | | | |
| 11043-A | Soft Brass Sheet | | 40,000 | | | | 79-85 | | Max.0.08 | Max.0.15 | Rem. | | | | | | | | | | |
| 11044-B | Medium Hard Rolled Alloy | | 18,000 | | 5 to 7 | | | | | | | | | | | | | | X | X | |
| 11045 | Monel Metal Sheet or Strip | | 60,000 | 25,000 | 25 | | Rem. | | Max.3.5 | | | | Max.3.5 | | | Max.0.8 | 60 Min. | | | | |
| 11047-A | Phosphor Bronze Strip | | 85-115,000 | 65,000 to 95,000 | 5 to 20 | | 91-93 | Rem. | Max.0.10 | | Max.0.20 | Max.0.15 | | | | | | | | | |
| 11051-B | Hard Rolled Sheet Aluminum | | 22,000 | | 2 | | | | | | | | | | | | | | X | X | |
| 11054 | Aluminum Alloy Sheet | Temp. 1 | 55,000 | 25,000 | 15 | | | | | | | | | | | | | | X | | |
| | | Temp. 2 | 50,000 | 25,000 | 20 | | | | | | | | | | | | | | | | |
| 11058 | Soft Annealed Aluminum Sheet | | 12,000 | | 10-20-30 | | | | | | | | | | | | | | X | X | |
| TUBING | | | | | | | | | | | | | | | | | | | | | |
| 11046-A | Seamless Brass | | | | | | 79-82 | | Max.0.10 | Max.0.[illegible]0 | Rem. | | | | | | | | | | X |
| 11048-A | Seamless Copper | | | | | | | | | | | | | | | | | | | | X |
| 11053-A | Aluminum | | 22,000 | | 30 | | | | | | | | | | | | | | X | X | |
| 11055 | Aluminum Alloy | Temp. 1 | 55,000 | 3[illegible],000 | 12 | | | | | | | | | | | | | | X | | |
| | | Temp. 2 | 50,000 | 3[illegible],000 | 20 | | | | | | | | | | | | | | | | |
| WIRE | | | | | | | | | | | | | | | | | | | | | |
| 11015 | Brass, for Brazing | | | | | | 78-82 | | Max.0.1 | Max.0.[illegible] | Rem. | | | | | Max.1.25 | | | | | |
| 11052 | Brass Spring | | [illegible] | 75,000 | 3 | 57 | [illegible]5-68 | | Max.0.[illegible] | [illegible] | Rem. | | | | | | | | | | |
| 11056 | Soft Copper | | | | | | | | | | | | | | | | | | | | X |
| 11057 | Phosphor Bronze Spring | | 1[illegible],000 | | 4 | | 94 Min. | Min.4.5 | Max.0.10 | Max.0.10 | Max.0.20 | 0.05-.50 | | | | | | | | | |

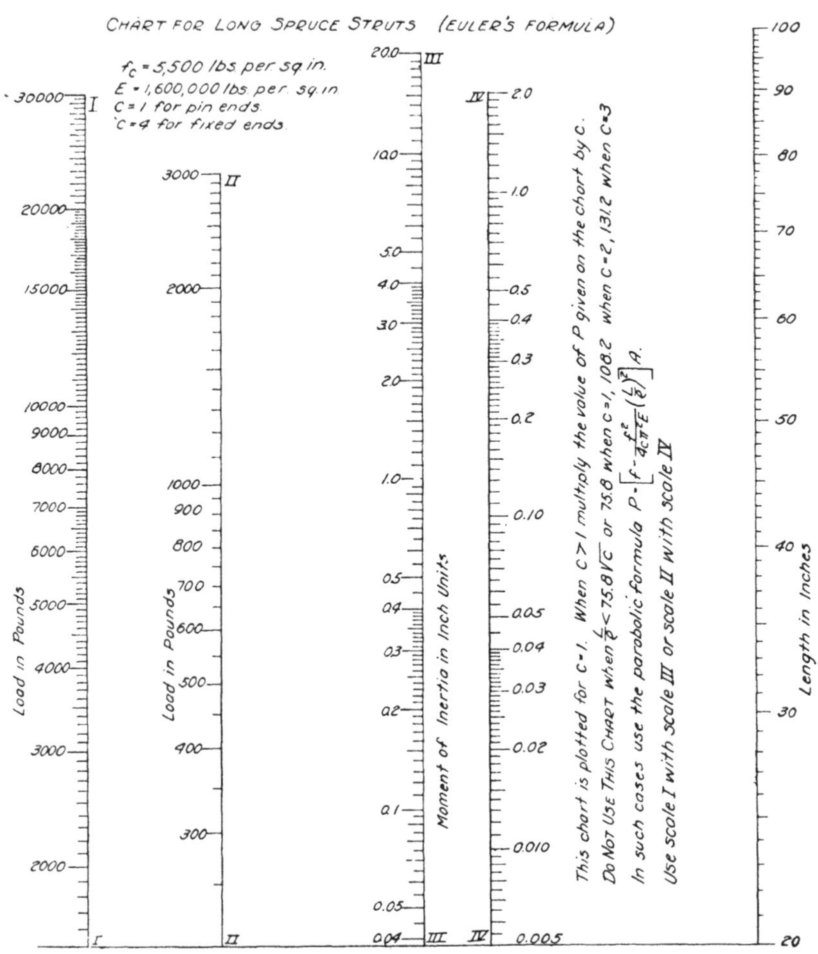


**Fig. 182. Chart for Long Spruce Struts (Euler's Formula)**

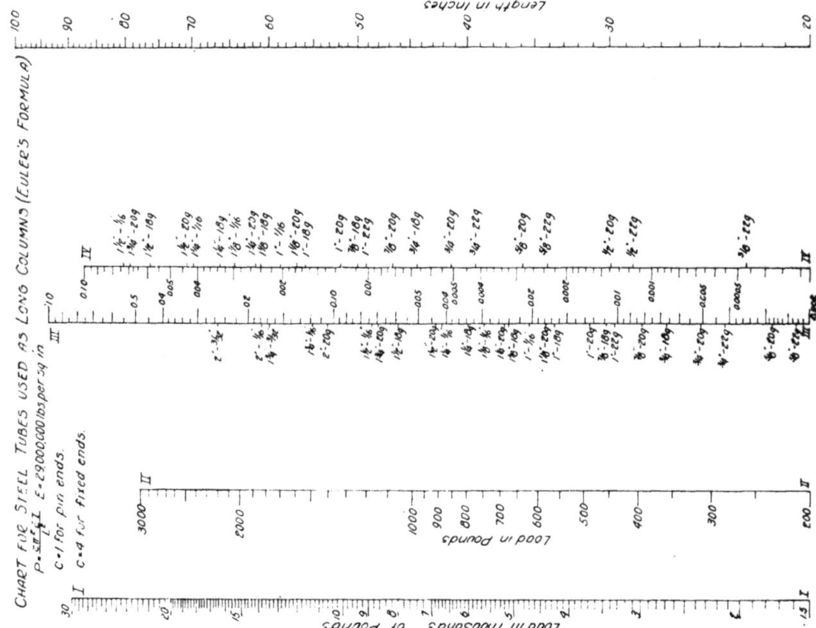


This chart is plotted for C=1. When C>1 multiply the value of P given on the chart by C.

DO NOT USE THIS CHART when L is LESS than the value given in the tables. In such cases use the charts for the parabolic formula $P = \left[f - \frac{f^2}{4cn^2E}\left(\frac{L}{\rho}\right)^2\right]A$

The sizes given in the tables are in order of weight.

Use scale I with scale III or scale II with scale IV.

Scales III and IV are graduated on one side for Moments of Inertia, I.

This chart may be used for any steel strut whose $\frac{L}{\rho} > 126\sqrt{C} = 178$ when C=2 and 218 when C=3

| O.D. | Gage | C=1 | C=2 | C=3 |
|---|---|---|---|---|
| Specification 10225 | | | | |
| 1/2 | 22 | 20.8 | 29.4 | 36.0 |
| 1/2 | 20 | 20.4 | 28.9 | 35.4 |
| 5/8 | 22 | 26.2 | 37.0 | 45.5 |
| 3/4 | 22 | 31.7 | 44.8 | 55.0 |
| 5/8 | 20 | 26.0 | 36.7 | 45.0 |
| 3/4 | 20 | 31.4 | 44.4 | 54.4 |
| 7/8 | 20 | 36.8 | 52.1 | 64.0 |
| 1 | 20 | 42.3 | 60.0 | 73.5 |
| 3/4 | 18 | 30.8 | 43.6 | 53.5 |
| 1 1/8 | 20 | 47.8 | 67.5 | 83.0 |
| 7/8 | 18 | 36.3 | 51.4 | 63.0 |
| 1 | 18 | 41.8 | 59.0 | 72.5 |
| 1 1/8 | 18 | 47.2 | 66.9 | 82.0 |
| 1 | 1/16 | 41.2 | 59.4 | 71.5 |
| 1 1/4 | 18 | 53.7 | 74.5 | 91.4 |
| 1 1/8 | 1/16 | 46.7 | 66.0 | 81.0 |
| 1 1/2 | 18 | 63.6 | 90.0 | 110.2 |
| 1 1/4 | 1/16 | 52.2 | 74.0 | 90.5 |
| 1 1/2 | 1/16 | 63.0 | 89.1 | 109.2 |
| 1 3/4 | 1/16 | 74.0 | 104.8 | 128.2 |
| 2 | 1/16 | 84.9 | 120.3 | 147.3 |
| 1 3/4 | 3/32 | 72.7 | 103.0 | 126.0 |
| 2 | 3/32 | 83.5 | 118.2 | 145.0 |
| Specification 10227 | | | | |
| 1 | 22 | 27.9 | 39.4 | 48.3 |
| 1 | 20 | 27.7 | 39.2 | 48.0 |
| 1 1/8 | 20 | 31.3 | 44.2 | 54.1 |
| 1 1/4 | 20 | 34.8 | 49.2 | 60.3 |
| 1 | 18 | 27.3 | 38.6 | 47.3 |
| 1 1/2 | 20 | 42.0 | 59.4 | 72.8 |
| 1 1/8 | 18 | 30.9 | 43.7 | 53.5 |
| 1 | 1/16 | 27.9 | 38.1 | 46.7 |
| 1 1/4 | 18 | 34.5 | 48.7 | 59.6 |
| 1 3/4 | 20 | 49.2 | 69.5 | 85.3 |
| 1 1/8 | 1/16 | 30.5 | 43.2 | 53.0 |
| 2 | 20 | 56.4 | 79.8 | 97.8 |
| 1 1/4 | 1/16 | 34.1 | 48.2 | 59.0 |
| 1 1/2 | 1/16 | 41.3 | 58.3 | 71.5 |
| 1 3/4 | 1/16 | 48.4 | 68.5 | 83.9 |
| 2 | 1/16 | 55.6 | 78.6 | 96.4 |
| 1 3/4 | 3/32 | 47.6 | 67.3 | 82.5 |
| 2 | 3/32 | 54.7 | 77.4 | 94.8 |

This chart can be used for long Duralumin struts by dividing the values of P from the chart by 3 if $\frac{L}{\rho} > 88.5\sqrt{C} = 125$ when C=2, and 153 when C=3.

Fig. 183. Chart for Steel Tubes Used as Long Columns (Eulers Formula)

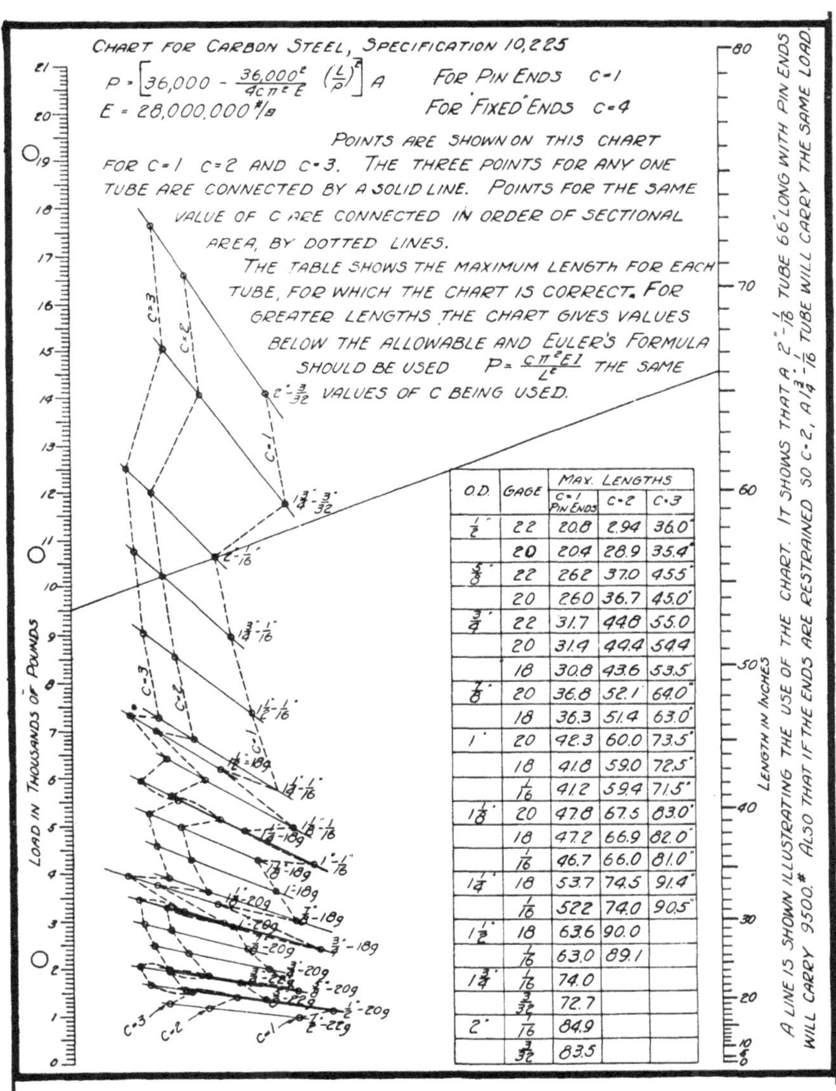


| O.D. | GAGE | MAX. LENGTHS | | |
|---|---|---|---|---|
| | | C=1 PIN ENDS | C=2 | C=3 |
| 1/2" | 22 | 20.8 | 2.94 | 36.0" |
| | 20 | 20.4 | 28.9 | 35.4" |
| 5/8" | 22 | 26.2 | 37.0 | 45.5" |
| | 20 | 26.0 | 36.7 | 45.0" |
| 3/4" | 22 | 31.7 | 44.8 | 55.0 |
| | 20 | 31.4 | 44.4 | 54.4 |
| | 18 | 30.8 | 43.6 | 53.5" |
| 7/8" | 20 | 36.8 | 52.1 | 64.0" |
| | 18 | 36.3 | 51.4 | 63.0" |
| 1" | 20 | 42.3 | 60.0 | 73.5" |
| | 18 | 41.8 | 59.0 | 72.5" |
| | 1/16 | 41.2 | 59.4 | 71.5" |
| 1 1/8" | 20 | 47.8 | 67.5 | 83.0" |
| | 18 | 47.2 | 66.9 | 82.0" |
| | 1/16 | 46.7 | 66.0 | 81.0" |
| 1 1/4" | 18 | 53.7 | 74.5 | 91.4" |
| | 1/16 | 52.2 | 74.0 | 90.5" |
| 1 1/2" | 18 | 63.6 | 90.0 | |
| | 1/16 | 63.0 | 89.1 | |
| 1 3/4" | 1/16 | 74.0 | | |
| | 3/32 | 72.7 | | |
| 2" | 1/16 | 84.9 | | |
| | 3/32 | 83.5 | | |

Fig. 184. Chart for Carbon Steel, Specification 10,225

$P = \left[90{,}000 - \frac{90{,}000^2}{4c\pi^2 E}\left(\frac{L}{\rho}\right)^2\right]A$. For Pin Ends $c = 1$

$E = 30{,}000{,}000$ lbs. per sq. in. For Fixed Ends $c = 4$

Points are shown on this chart for $c = 1$, $c = 2$, and $c = 3$. The three points for any one tube are connected by a solid line. Points for the same value of c are connected, in order of sectional area, by dotted lines.

The table shows the maximum length for each tube, for which the chart is correct. For greater lengths the chart gives values below the allowable and Euler's Formula should be used. $P = \frac{c\pi^2 E I}{L^2}$ the same values of c being used.

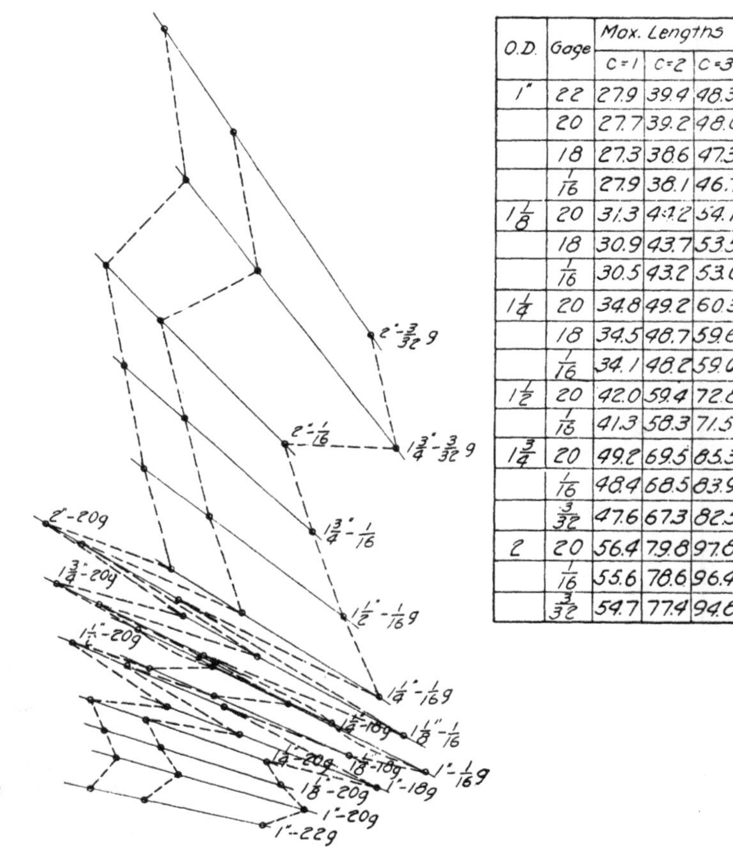


| O.D. | Gage | Max. Lengths | | |
|---|---|---|---|---|
| | | c = 1 | c = 2 | c = 3 |
| 1" | 22 | 27.9 | 39.4 | 48.3 |
| | 20 | 27.7 | 39.2 | 48.0 |
| | 18 | 27.3 | 38.6 | 47.3 |
| | 1/16 | 27.9 | 38.1 | 46.7 |
| 1 1/8 | 20 | 31.3 | 44.2 | 54.1 |
| | 18 | 30.9 | 43.7 | 53.5 |
| | 1/16 | 30.5 | 43.2 | 53.0 |
| 1 1/4 | 20 | 34.8 | 49.2 | 60.3 |
| | 18 | 34.5 | 48.7 | 59.6 |
| | 1/16 | 34.1 | 48.2 | 59.0 |
| 1 1/2 | 20 | 42.0 | 59.4 | 72.8 |
| | 1/16 | 41.3 | 58.3 | 71.5 |
| 1 3/4 | 20 | 49.2 | 69.5 | 85.3 |
| | 1/16 | 48.4 | 68.5 | 83.9 |
| | 3/32 | 47.6 | 67.3 | 82.5 |
| 2 | 20 | 56.4 | 79.8 | 97.8 |
| | 1/16 | 55.6 | 78.6 | 96.4 |
| | 3/32 | 54.7 | 77.4 | 94.8 |

Fig. 185. Chart for Alloy Steel Tubes, Specification 10,227

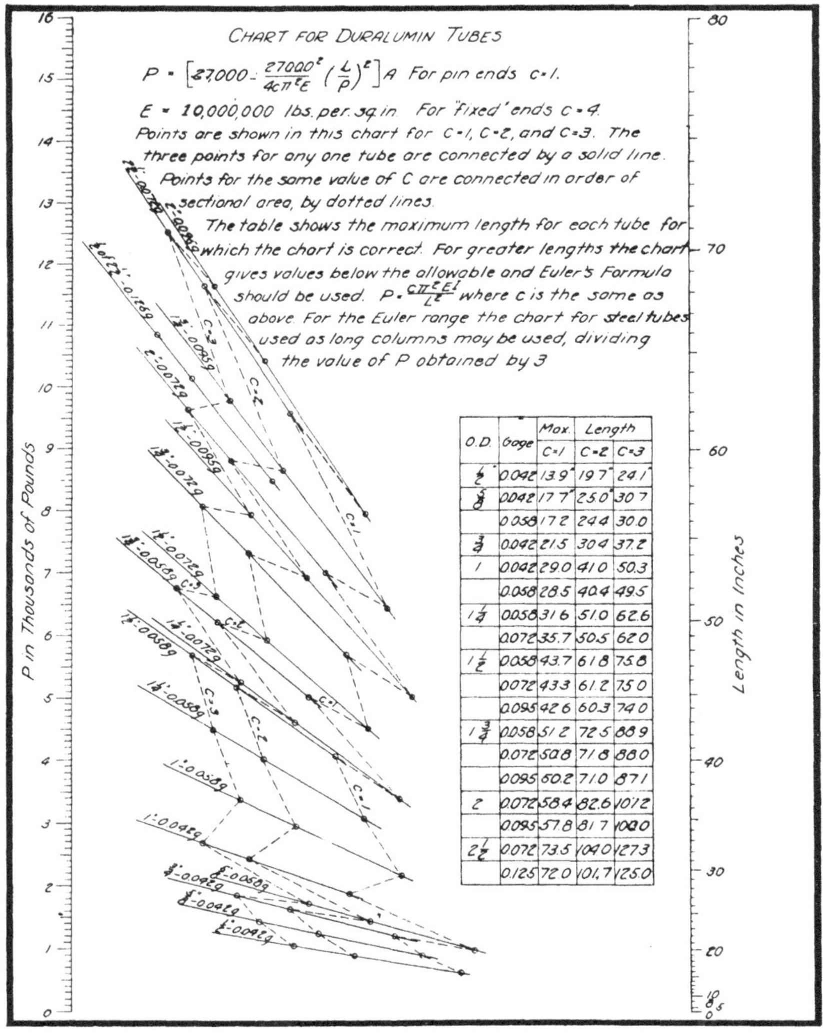


| O.D. | Gage | Max. Length C = 1 | C = 2 | C = 3 |
|---|---|---|---|---|
| 1/2" | 0.042 | 13.9" | 19.7" | 24.1" |
| 5/8 | 0.042 | 17.7" | 25.0" | 30.7 |
| | 0.058 | 17.2 | 24.4 | 30.0 |
| 3/4 | 0.042 | 21.5 | 30.4 | 37.2 |
| 1 | 0.042 | 29.0 | 41.0 | 50.3 |
| | 0.058 | 28.5 | 40.4 | 49.5 |
| 1 1/4 | 0.058 | 31.6 | 51.0 | 62.6 |
| | 0.072 | 35.7 | 50.5 | 62.0 |
| 1 1/2 | 0.058 | 43.7 | 61.8 | 75.8 |
| | 0.072 | 43.3 | 61.2 | 75.0 |
| | 0.095 | 42.6 | 60.3 | 74.0 |
| 1 3/4 | 0.058 | 51.2 | 72.5 | 88.9 |
| | 0.072 | 50.8 | 71.8 | 88.0 |
| | 0.095 | 50.2 | 71.0 | 87.1 |
| 2 | 0.072 | 58.4 | 82.6 | 101.2 |
| | 0.095 | 57.8 | 81.7 | 100.0 |
| 2 1/2 | 0.072 | 73.5 | 104.0 | 127.3 |
| | 0.125 | 72.0 | 101.7 | 125.0 |

Fig. 186. Chart for Duralumin Tubes

# INDEX

www.ingramcontent.com/pod-product-compliance
Lightning Source LLC
Chambersburg PA
CBHW021429180326
41458CB00001B/188

* 9 7 8 1 9 2 9 1 4 8 4 6 2 *